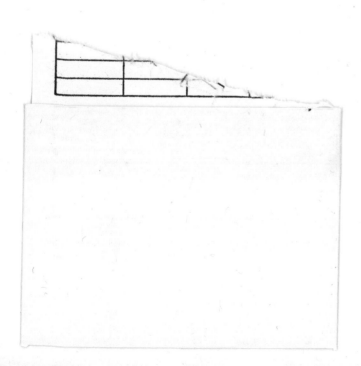

THE CELL SURFACE
IN DEVELOPMENT

THE CELL SURFACE
IN DEVELOPMENT

Edited by

A. A. MOSCONA

University of Chicago

A Wiley Biomedical-Health Publication

JOHN WILEY & SONS, New York · London · Sydney · Toronto

Library of Congress Cataloging in Publication Data:

Main entry under title:
The Cell surface in development.

  (A Wiley biomedical-health publication)
  Papers of a symposium organized by a scientific committee of the International Society of Developmental Biologists and held at the University of Montreal in Aug. 1973 in conjunction with the quadrennial congress of the Society.
    1. Cell membranes—Congresses. 2. Cell differentiation—Congresses. I. Moscona, Aron Arthur, 1922-   ed.   II. International Society of Developmental Biologists. [DNLM: 1. Cell membrane. QH601 M896c 1974]

QH601.C4   599'.08'75   74-7308

ISBN 0-471-61855-1

Printed in the United States of America

10 9 8 7 6 5 4 3 2 1

AUTHORS

Norman G. Anderson
Molecular Anatomy (MAN) Program
Oak Ridge National Laboratory
Oak Ridge, Tennessee

Gary Bennett
Department of Anatomy
McGill University
Montreal, Canada

Mark S. Bretscher
Medical Research Council Laboratory
of Molecular Biology
Cambridge, England

Jean-Pierre Changeux
Département de Biologie Moléculaire
Institut Pasteur
Paris, France

Joseph H. Coggin
Department of Microbiology
University of Tennessee
Knoxville, Tennessee

Albert Dorfman
Departments of Pediatrics and
Biochemistry, The Committee on
Developmental Biology, and
Joseph P. Kennedy, Jr. Mental
Retardation Research Center
Pritzker School of Medicine
University of Chicago
Chicago, Illinois

Gerald M. Edelman
The Rockefeller University
New York, New York

Irving Goldschneider
University of Connecticut
Health Center
School of Medicine
Farmington, Connecticut

Pei-Lee Ho
Department of Pediatrics and
Joseph P. Kennedy, Jr. Mental
Retardation Research Center
Pritzker School of Medicine
University of Chicago
Chicago, Illinois

C. Huang
Department of Biochemistry
University of Virginia School
of Medicine
Charlottesville, Virginia

Carol Jones
Institute of Cancer Research
University of Colorado
Medical Center
Denver, Colorado

C. P. Leblond
Department of Anatomy
McGill University
Montreal, Canada

Daniel Levitt
Department of Pediatrics and
Joseph P. Kennedy, Jr. Mental
Retardation Research Center
Pritzker School of Medicine
University of Chicago
Chicago, Illinois

B. J. Litman
Department of Biochemistry
University of Virginia School
of Medicine
Charlottesville, Virginia

A. A. Moscona
Departments of Biology and
Pathology, and the Committee on
Developmental Biology
The University of Chicago
Chicago, Illinois

Theodore T. Puck
Institute of Cancer Research
Department of Biophysics
University of Colorado
Medical Center
Denver, Colorado

Jean-Paul Revel
Division of Biology
California Institute of Technology
Pasadena, California

Nils R. Ringertz
Institute for Medical Cell Research
and Genetics
Medical Nobel Institute
Karolinska Institutet
Stockholm, Sweden

Saul Roseman
McCollum-Pratt Institute
The Johns Hopkins University
Baltimore, Maryland

Urs Rutishauser
The Rockefeller University
New York, New York

Leo Sachs
Department of Genetics
Weizmann Institute of Science
Rehovot, Israel

Judson D. Sheridan
Department of Zoology
University of Minnesota
Minneapolis, Minnesota

Richard L. Sidman
Department of Neuropathology
Harvard Medical School
Boston, Massachusetts

Patricia G. Spear
The Rockefeller University
New York, New York
Now at Department of Microbiology
The University of Chicago
Chicago, Illinois

T. E. Thompson
Department of Biochemistry
University of Virginia
School of Medicine
Charlottesville, Virginia

Paul Wuthier
University of Colorado
Medical Center
Denver, Colorado

Ichiro Yahara
The Rockefeller University
New York, New York

FOREWORD

The organization of an International Conference under the sponsorship of a Society whose Board Members are scattered all over the world is in itself a formidable task. The International Society of Developmental Biologists, however, has been fortunate to be able to rely on a group of dedicated people who have invested the best of their energies and of their competence in the organization of this Congress.

The original planning of the Congress was largely due to the activity and enthusiasm of the late Secretary of the Society, Edgar Zwilling, whose untimely death has been a great loss both to Developmental Biology and to our Society. It is in recognition of his efforts and of his work for the advances of our Society that the Board has decided to dedicate this Congress to his memory.

The responsibilities of Ed Zwilling were taken over by Robert DeHaan, in collaboration with the local Organizers, Professor B. Messier and Professor R. Gagnon of the University of Montreal; it is thanks to their coordinate efforts that the Congress shaped up so well. The scientific organization was laid primarily on the shoulders of Aron A. Moscona who has also accepted the additional responsibility of editing this volume. To all these colleagues goes the gratitude of the International Society of Developmental Biologists. Thanks are also due to all the speakers who have promptly complied with the request of handing in on time the manuscripts of their presentations. Indeed, it is because of their cooperation that the Society has been able to avail itself of the promptness and the excellent work performed by the publishers.

ALBERTO MONROY
President, I. S. D. B.

PREFACE

The periplast ... which has hitherto passed under the names of cell-wall, contents, and intercellular substance, is the subject of all the most important metamorphic processes, whether morphological or chemical, in the animal or the plant. By its differentiation every variety of tissue is produced.

T. H. HUXLEY (1853)

In recent years there has been something of a revolution in the science of cell surfaces. Its impact increasingly affects cell biology as a whole, and especially developmental biology and immunology. This book reviews progress made so far and aims to further foster interest in this area. It brings together some of the most important concepts and advances in cell surface biology as they relate to differentiation, and it reflects the sense of excitement shared by those working in this field.

Like most revolutions, this one too has its roots in history. The classical embryologists and immunologists, whose concepts often converged, thought of the cell surface in terms of a dynamic structure involved in interactions with the environment, in cell guidance, and in intercellular communication. Later, as embryology, immunology, and cell physiology drifted apart, this foresighted viewpoint gave way to narrower interests. Models of the erythrocyte "ghost" became the dominating prototype of the animal cell surface which was thus pictured as a static, uniform "pellicle" essentially similar in all types of cells, and serving primarily as a permeability barrier. The advent of modern molecular and developmental biology and immunology revived a broader interest in animal cell surfaces and prompted reexamination of the older concepts in the light of biological realities. The resulting tide of information—physical, biochemical, cytological, and immunological—has led to renewed awareness of the cell surface's significance and complexity. It is now abundantly clear that the cell surface is a dynamic system, which is heterogeneous in structure and composition and diverse in different kinds of cells. It undergoes changes in embryonic differentiation and in metaplasia, and is subject to genomic regulation and environmental modulation. Its versatile functions play decisive roles in the control of cell growth, cell movement, and cell recognition, and therefore in morphogenesis, differentiation, and immune response.

This book has a twofold purpose: to present a broad spectrum of up-to-date information on concepts and outlooks concerning physical, biochemical, and biological characteristics of cell surfaces; and to focus interest specifically on the relationship of the cell surface to mechanisms of differentiation and morphogenesis. These aims were met thanks to the enthusiastic cooperation of the participants. The resulting volume echoes T. H. Huxley's surprisingly foresighted intuition; its content strongly suggests that exploration of the cell surface may provide keys to the understanding of fundamental cellular processes, both normal and abnormal, particularly those involving morphogenesis, cell communication, and differential gene expression. Various technical and conceptual advances described in these chapters clearly indicate that this important frontier of cell science is now amenable to sophisticated exploration by physical, chemical, and genetic methods.

The symposium presented in this volume was organized by a Scientific Committee of the International Society of Developmental Biologists, which consisted of Alberto Monroy, James D. Ebert, Robert DeHaan, and myself. It was held at the University of Montreal in August 1973 in conjunction with the quadrennial Congress of the Society. I thank the symposium speakers for contributing papers of unusual interest and significance, which jointly set a landmark in the progress of cell surface science; I also acknowledge the Publisher's cooperation in assuring rapid publication.

A. A. MOSCONA

Chicago, Illinois
April 1974

CONTENTS

THE CELL SURFACE
IN DEVELOPMENT

Bilayers and Biomembranes: Compositional Asymmetries Induced by Surface Curvature

T. E. THOMPSON, C. HUANG, and B. J. LITMAN

Today the most widely held concept of biological membrane structure is a refinement of the Danielli hypothesis advanced in the early 1930s (Danielli and Davson, 1935). The most recent and complete statement of this basic idea is that of Singer and Nicolson (1972). The concept is composed of two essential elements.

1. The lipid component is a bilayer in structure and as such contributes to the membrane a set of barrier properties. The simplest expression of these properties is a general impermeability of the membrane to small water-soluble molecules. The lipid molecules comprising the bilayer are free to move diffusionally both within the plane of the bilayer and with a "flip-flop" motion from one bilayer face to the other (Kornberg and McConnell, 1971; Scandella et al., 1972).

2. The protein components are immersed to varying degrees in the bilayer and may be free to move within it. These protein components confer on the membrane its basic physiological functions. In their simplest form they are expressed as specific transport properties.

The evidence for this view of biological membrane structure is manifold, and new support is accumulating daily. In general, support has come from two complementary lines of investigation: one involving biological membranes, the other based on studies of experimental model systems. The study of models has been particularly useful in establishing the molecular structure and barrier properties of the organized lipid component, the bilayer (Thompson and Henn, 1970). This work has relied, for the most part, on the study of two different but related types of lipid bilayer models: the planar bilayer first described by Mueller et al. (1962) and the liquid-crystal or liposome dispersion first utilized by Bangham et al. (1965). The structures and properties of these systems and their relevance to biological membranes have been reviewed in detail (e.g., Mueller and Rudin, 1969; Thompson and Henn, 1970; Bangham, 1972).

Although the liposome dispersions introduced by Bangham permit an experimental approach to many interesting problems, several characteristics of the system limit its usefulness. In the first place, the liposomes vary greatly in size from objects which are visible in a light microscope down to vesicles only several hundred angstroms in diameter. Second, within each of the liposomes the lipid is organized into concentric bimolecular lamellae, each separated from its neighbor by an interspersed water lamella. This multicomponent character, together with the marked size heterogeneity, makes it difficult to relate biologically relevant properties to the surface area of the lipid lamellae.

These limitations can be avoided by the use of phospholipid dispersions in which the liposomes are spherical vesicles, homogeneous in size, each composed of a single continuous lipid bilayer enclosing a volume of aqueous solution. The preparation of vesicles of this type was first described by Huang (1969). These vesicles can, in effect, be regarded as phospholipid "lamellar macromolecules." Consequently they can be employed in experiments utilizing the variety of physical techniques which have been developed to examine the physical properties of macromolecular systems.

The purpose of this paper is to describe the results of recent experiments carried out on this type of vesicle system, which indicate that both the molecular packing and the intrabilayer composition are functions of the radius of curvature of the bilayer. These observations made on the model system suggest, on the one hand, that the changes in the curvature of biological membranes produced mechanically by physiological mechanisms may, in the regions of maximum curvature, modify the structure, composition, and functional properties of the lipid bilayer component of the membrane. On the other hand, these observations suggest that transbilayer compositional asymmetries generated biosynthetically in biological membranes may be expected to induce bilayer curvature. Thus asymmetries in composition may be related to the relatively small radii of curvature seen in the morphology of many membrane systems.

The discussion in the following sections will be concerned first with the preparation and physical properties of vesicles formed from a single phospholipid component. Attention will then be directed toward a consideration of the special properties of these vesicles which derive from their small radii of curvature. The discussion will center finally on the transbilayer concentration asymmetries in vesicles of this type that are composed of two lipid components.

VESICLE PREPARATION, STRUCTURE, AND PROPERTIES

The preparation of aqueous dispersions of homogeneous single-walled lipid vesicles involves two essential steps: sonication and molecular sieve chromatography (Huang, 1969; Huang and Thompson, 1974). The appropriate mixture of lipids, which have been colyophilized from an organic solvent such as chloroform, is first suspended in a dilute aqueous salt solution. Sonication of this suspension for several hours is carried out at 4°C under argon or nitrogen to prevent oxidation. After centrifugation to remove debris, the dispersion is clear but appears slightly blue because of Tyndall scattering. This dispersion is then subjected to molecular sieve chromatography on a Sepharose 4B column. A typical elution pattern of a phosphatidylcholine dispersion is shown in Fig. 1A. It consists of two distinct fractions, labeled I and II. The liposomes dispersed in Fraction I are heterogeneous in size and so large that they do not penetrate into the sieve and are thus eluted with the void volume. The liposomes in Fraction II are distributed within the internal volume of the sieve and elute as a broad symmetrical peak. It is apparent in Fig. 1A that the Fraction I peak is more pronounced if the turbidity measured at 300 nm is plotted against the tube number. This is due to the fact that the liposomes in this fraction are much larger than the vesicles in Fraction II and hence scatter more light per unit concentration of lipid phosphorus.

Figure 1B is a plot of the turbidity at 300 nm versus lipid phosphorus, taken from the data shown in Fig. 1A. Adjacent to each data point in the tube number. Subfractions of II contained in tubes 17–20 show a linear relation between turbidity and lipid phosphorus. A regression line based on the data points for these subfractions passes through the origin of the plot. It is the lipid vesicles in subfractions 17–20 which have been shown to be single walled and homogeneous in size by a number of criteria (Huang, 1969; Huang et al., 1970; Huang and Charlton, 1971; Huang and Charlton, 1972a, b; Redwood et al., 1972; Huang and Lee, 1973).

Some of the important physical parameters which characterize such vesicles prepared from egg phosphatidylcholine are summarized in Table 1 (Huang and Thompson, 1974). If the average molecular weight of the phosphatidylcholine of egg lecithin is taken to be 770, the average number of lipid molecules per vesicle can be calculated to be about 2600. The thickness of the bilayer is about 40 Å, based on the electron density profile determined by X-ray diffraction (Wilkins et al., 1971). If the external radius of the vesicles is taken to be 105 Å, as estimated from the diffusion coefficient, the ratio of the areas of the external to internal faces of the bilayer is then 2.6. If the packing density of the phospholipid in each face of the bilayer is assumed to be the same, the numbers of molecules in the external and internal faces are, respectively, 1875 and 725. It is also easy to show that the volume of the bilayer shell of the vesicle is almost exactly 4 times the volume of the interior aqueous compartment.

TABLE 1. PHYSICAL PROPERTIES OF HOMOGENEOUS FRACTION II VESICLES[a]

Sedimentation coefficient	$s^0_{20, w}$	$(2.63 \pm 0.10) \times 10^{-13}$ sec
Diffusion coefficient	$D^0_{20, w}$	$(2.03 \pm 0.4) \times 10^{-7}$ cm^2 sec^{-1}
Partial specific volume of vesicle wall	$\bar{v}$	0.9848 ± 0.0007 ml g^{-1}
Intrinsic viscosity	$[n]$	0.041 dl g^{-1}
Vesicle shell weight		2.0×10^6 daltons
Vesicle diameter		210 ± 8 Å

[a] Aqueous phase: 0.1 M KCl +0.01 M Tris, pH 8.0, 20°C.

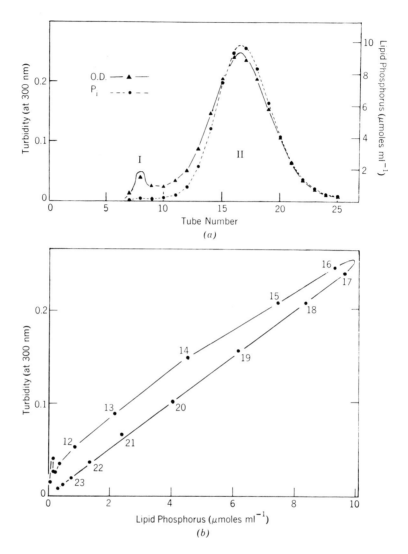

Figure 1. A typical molecular sieve chromatogram of a sonicated egg phosphatidylcholine vesicle dispersion run on a Sepharose 4B column (2.5 × 50 cm). Aqueous phase: 0.1 *M* KCl + 0.01 *M* Tris, pH 8.5 (*a*) The turbidity of eluate subfractions at 300 nm is indicated by triangles; the lipid phosphorus concentration (P_i) of each tube is given by the solid circles. (*b*) Turbidity at 300 nm versus P_i content of each tube in Fig. 1*a*. The tube numbers are adjacent to each data point. [From C. Huang and T. E. Thompson, in *Methods in Enzymology,* Vol. 32 (S. Fleischer, L. Packer, and R. Estabrook, eds:), Academic Press, New York.]

    Figure 2, a diagrammatic cross section through a vesicle, illustrates the disparity in the radii of curvature of the inner and outer faces of the vesicle wall and the extreme smallness of both radii. A second, less obvious characteristic of the vesicle wall is the difference in the signs of the radii of curvature for the two faces. Thus, relative to the bilayer wall, the outer surface is defined by a positive radius; the inner surface, by a negative radius. From the standpoint of molecular packing this means that the volume of bilayer subtended by unit area on the outer surface

STRUCTURE OF FRACTION II VESICLES

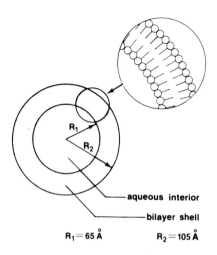

aqueous interior

bilayer shell

$R_1 = 65 \text{ Å}$ $R_2 = 105 \text{ Å}$

Figure 2. Diagrammatic cross section through a Fraction II vesicle. The insert illustrates the bilayer arrangement of phosphatidylcholine molecules in the vesicle wall.

decreases in cross section toward the center of the vesicle wall, whereas the volume subtended by unit area on the inner surface increases in cross section as the distance toward the center of the bilayer decreases. This situation implies, in simple terms, that the polar head groups may be more restrictively packed on the inner bilayer surface than on the outer, while in contrast the acyl chains of molecules on the inner surface may be less restrictively packed than the corresponding moieties of molecules on the outer surface.

The important characteristics of these dispersions are thus the homogeneity of vesicle size, the division by the vesicle wall of the aqueous phase into a continuous external compartment and an internal microscopically divided compartment, and the small radii of curvature defining the vesicle wall. It is important to note that the mean radius of curvature of these vesicles is quite similar to the minimum radius of curvature revealed in many biological membranes by electron microscopy (see, e.g., Fawcett, 1966).

Successful fractionation of the sonicated dispersions rests on the fact that vesicles with radii of 105 Å appears to be of minimum size. Sonication at higher energies, for longer times, or under other conditions does not reduce this minimum size. Variations in these parameters does, however, affect the proportion of small vesicles in the dispersion. For example, a shorter time of sonication results in an increase in Fraction I and a concomitant decrease in Fraction II. Prolonged sonication does not entirely eliminate Fraction I (Dawidowicz, 1973).

It is thus apparent that the minimum-size vesicles represent a unique configuration of the bilayer. As might be expected, several of the physical properties of these vesicles are distinctly different from the corresponding properties of liposomes of larger radius. For example, the proton magnetic resonance spectra obtained from aqueous dispersions of large liposomes and large multilamellar structures of the type studied by Bangham exhibit resonance lines which are so broad that they can be resolved only by special techniques (Seiter and Chan, 1973). However, upon sonication, the resonance lines sharpen dramatically and are easily resolved by low-field-strength equipment. In fact, the vesicles having a minimum radius of curvature are characterized by lines of maximum sharpness (Chapman et al., 1968; Plenkett et al., 1968; Sheard, 1969). Although the origin of this dependence of line width on liposome size has been the subject of considerable controversy, recent work strongly supports the view that the marked decrease of proton NMR line widths in small-radius vesicles is due to increased molecular motion in these structures (Sheetz and Chan, 1973; Seiter and Chan, 1973). A similar conclusion has been reached on the basis of spin-label studies (March et al., 1972).

This interpretation is directly supported by the fact that the partial specific volume of the bilayer in the small vesicles can be shown experimentally to be larger than the partial specific volume of the lipid in bilayers of larger radius of curvature (Sheetz and Chan, 1973). Table 2 presents values of the partial specific volume and external radius for both large and Fraction II vesicles prepared from egg phosphotidylcholine (Sears and Thompson, 1974). These results were obtained using a Parr cavity resonance density meter.

Studies of the phase transitions that phospholipids in bilayer array undergo have also provided evidence that the phospholipids in Fraction II vesicles have greater intramolecular motional freedom. These phase transitions have been shown by direct calorimetry and dilatometry measurements to be much broader in vesicles of minimum radius of curvature (Hinz and Sturtevant, 1972b; Sheetz and Chan 1973).

TRANSBILAYER COMPOSITIONAL ASYMMETRIES

It is clear from the preceding discussion that minimum-radius-of-curvature vesicles formed from a single phospholipid component have different molecular packing constraints than the bilayers of larger radii of curvature. The data presented in this section will show that, in a similar type of vesicle composed of two lipid components, the packing constraints lead to different relative concentrations of the two components in the inner and outer faces of the bilayer.

Small Vesicles Formed from Phosphatidylcholine and Phosphatidylethanolamine

Minimum-radius vesicles composed of a mixture of phosphatidlyethanolamine and phosphatidylcholine derived from egg yolk can be prepared by the procedure described above (Litman, 1973). In order to achieve complete mixing of the lipids in a single vesicle, it is necessary to colyophilize the desired mixture of lipids from a suitable organic solvent before suspension in the buffer solution for sonication.

TABLE 2. COMPARISON OF THE EXTERNAL RADII AND PARTIAL SPECIFIC VOLUMES OF LARGE AND FRACTION II VESICLES[a]

	Radius (Å)	$\bar{v}$ (ml g^{-1})
Fraction II vesicles	108	0.9827 ± 0.0004
Large vesicles	500	0.9496 ± 0.0004

[a] Aqueous phase: pure water.

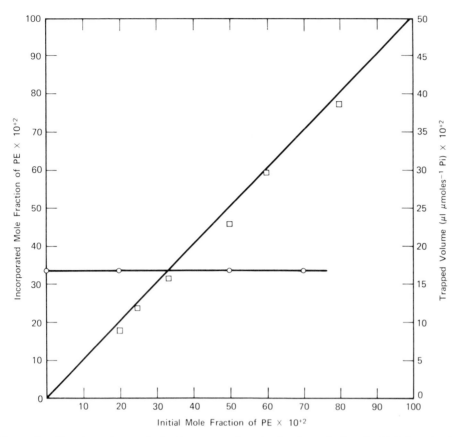

Figure 3. The dependence of the incorporated mole fraction of phosphatidylethanolamine (□) and the trapped volume (○) on the initial mole fraction of this lipid. Phosphatidylethanolamine is abbreviated as PE.

Figure 3 shows that after sonication and fractionation on a Sepharose 4B column, the composition of the Fraction II vesicles, over a wide range of phosphatidylethanolamine mole fraction, is essentially identical to the composition of the colyophilized mixture.

Although extensive studies to determine the size, shape, and shell weight of these vesicles have not yet been carried out, the volume of the internal aqueous compartment has been found to be independent of the mole fraction of phosphatidylethanolamine and identical to that found for pure phosphatidylcholine vesicles. These results are shown by the open circles defining the horizontal line in Fig. 3. These so-called trapped volume measurements are based on chemical determination of the mass of glucose trapped in the internal aqueous space per mole of lipid phosphorus during sonication in the presence of a known glucose concentration. By use of this determination, the trapped volume can be calculated, assuming the concentration of glucose within the internal compartment to be the same as the known glucose concentration during sonication. The trapped glucose is operationally defined as the glucose which cochromatographs with the Fraction II vesicles on a Sephadex G-50 column. Since the permeability of the vesicle wall to glucose is very small, free glucose in the external aqueous phase is easily separated from trapped glucose by passage of the vesicle dispersion over a molecular sieve column. The constancy of the trapped volume, as well as the fact that the turbidity per micromole of lipid phosphorus is constant, strongly supports the view that in this system vesicle size is independent of composition over the range from 0 to 0.7 mole fraction phosphatidylethanolamine.

In any two-component system of this type, it might be expected that the chemical composition of the system would be represented by a distribution about an average value. The narrower this distribution, the closer would be the composition of each vesicle to the composition of the total dispersion plotted in Fig. 3. Although chemical analysis of the subfractions of Fraction II has established that all subfractions have the same average composition, ultracentrifugation has shown that the width of the composition distribution about the average value broadens as the mole fraction of phosphatidylethanolamine increases (Litman, 1973).

Determination of the relative concentrations of phosphatidylethanolamine on the inner and outer faces of the vesicle wall can be made using the chemical reactivity of the primary amino group with 2,4,6-trinitrobenzenosulfonic acid (TNBS). Under mild conditions which do not disrupt the phospholipid bilayer, TNBS reacts with the ethanolamine moiety of phosphatidylethanolamine to yield the trinitrophenyl derivative of phosphatidylethanolamine. The absorption spectrum of this derivative permits quantitative determination of the number of amino groups in the system that are available for reaction. Since the bilayer can be shown to be impermeable to TNBS under a variety of conditions, if this reagent is added to the external aqueous compartment it will react only with phosphatidylethanolamine in the outer face of the bilayer. The total amount of phosphatidylethanolamine in both faces of the vesicle wall can be determined with this same reagent after disruption of the vesicle by propanol. In this manner the ratio of phosphatidylethanolamine in the outer bilayer face to the total amount of this phospholipid in the vesicle wall can be established with good precision.

Data of this type, determined as a function of the mole fraction of phosphatidylethanolamine in phosphatidylcholine vesicles, are plotted as the squares in Fig. 4. If the relative concentrations of phosphatidylethanolamine were the same in both inner and outer bilayer faces, the value of the ratio would be given by the ratio of the area of the outer face to the total bilayer surface as calculated from the geometric parameters of the vesicles (Fig. 2). This expected ratio is 0.72, as indicated by the dashed line in Fig. 4. It is quite apparent that the relative concentrations of phosphatidylethanolamine in the outer and inner bilayer faces can differ markedly from the expected values, depending on the composition of the total system.

A similar result can be obtained by direct determination of the amount of phosphatidylethanolamine on the inner surface of the vesicle wall. This determination is based on the fact that protonated primary amino groups do not react with TNBS. Thus, if the pH is maintained at 5 or lower during preparative sonication of the vesicles, unreactive TNBS present in the aqueous phase will be trapped in the interior aqueous compartment of the vesicle. If TNBS in the external aqueous phase is removed by molecular sieve chromatography, and the pH is the system then adjusted to 8.5, the TNBS trapped within the vesicle will react with the phosphatidylethanolamine on the inner surface of the vesicle wall. The results of these experiments are shown in Fig. 4 as the open circles.

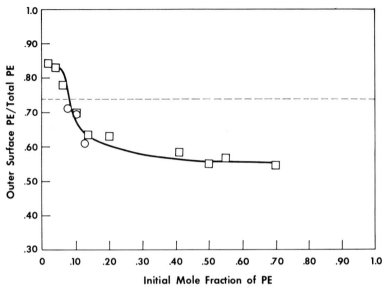

Figure 4. The ratio of the content of phosphatidylethanolamine (PE) in the outer surface to the total PE content of the vesicle as determined by selective labeling of the outer surface (□) and the inner surface (○) phosphatidylethanolamine.

Vesicles Formed from Phosphatidylcholine and Cholesterol

The relative amounts of phosphatidylcholine in the outer and inner faces of the walls of small vesicles composed of mixtures of phosphatidylcholine and cholesterol have recently been determined using proton NMR and paramagnetic shift agents (Huang et al., 1974). These studies clearly show that at mole fractions of cholesterol in the range 0.3–0.5 there is a marked increase in the concentration of phosphatidylcholine in the outer bilayer face relative to the inner face. By inference, a reciprocal disproportionation of the concentration of cholesterol exists in the two faces of the vesicle wall. At cholesterol concentrations below mole fraction 0.3, however, the relative concentrations of the two lipid components are the same in both faces.

The differential identification of phosphatidylcholine on the inner and outer bilayer faces is based on the following observations. Paramagnetic ions such as Mn^{2+} or lanthanides in the $3+$ oxidation state interact with the phosphate group of phosphatidylcholine and cause a shift or broadening of the NMR signal due to the N-methyl protons. However, addition of paramagnetic ions at suitable concentrations to an aqueous dispersion of phosphatidylcholine vesicles produces a shift or broadening of only a part of the N-methyl proton signal. Since interaction of the paramagnetic ion is necessary for perturbation of proton resonance, and since it is easy to show that the bilayer is impermeable to the paramagnetic ion, the perturbed signal must originate from the protons on the outer surface of the vesicle wall, whereas the unperturbed signal must arise from N-methyl protons on the inner surface (Bystrov et al., 1971; Kostelnik and Castellano, 1972; Levine et al., 1973; Huang et al., 1974). Quantitative determination of the magnitudes of the perturbed and unperturbed signals can thus be used to estimate the relative amounts of phosphatidylcholine on the two faces of the bilayer. If, in addition, the geometry of the vesicle is known, this information can be used to calculate the relative concentrations of this lipid component in the bilayer faces in the manner described in the preceding section.

Spectrum A, Fig. 5, is the NMR signal from the N-methyl protons of egg phosphatidylcholine in a small-vesicle preparation. Spectrum B, Fig. 5, shows the splitting of the signal caused by the addition of praseodymium (Pr^{3+}) to the external aqueous phase of this preparation. The larger peak on the left arises from the N-methyl protons on the outer face of the vesicle wall, while the smaller unperturbed peak on the right is due to phosphatidylcholine molecules on the inner wall in contact with the praseodymium-free interior aqueous compartment of the vesicle. As a paramagnetic shift reagent, Pr^{3+} is particularly useful because of the large shift produced at low concentrations of this ion. The actual magnitude of the shift depends on the praseodymium/phospholipid ratio (Huang et al., 1974).

The ratios of the areas occupied by egg phosphatidylcholine on the outer and inner surfaces of small vesicles formed from various mixtures of this lipid with cholesterol are given in Table 3 (Huang et al., 1974). The vesicles used in this study were formed by prolonged sonication, in dilute salt solution, of mixtures of these lipids colyophilized from benzene. In order to achieve the relatively high concentration of vesicles required for the NMR study, the sonicated dispersions were not sharply fractionated by molecular sieve chromatography. Although vesicles formed in this manner are single walled, the preparations are somewhat less homogeneous than Fraction II vesicles. Their average external radius of curvature is about 125 Å, slightly larger than the R_2 value given in Fig. 2 and Table 1 for Fraction II vesicles. In other respects the vesicles are very similar. Their compositions are essentially the same as the gross compositions of the colyophilized lipid

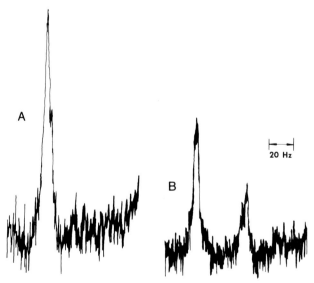

Figure 5.   Splitting of N-methyl proton NMR signal by praseodymium in aqueous dispersion of egg phosphatidylcholine vesicles. Spectrum A, without Pr^{+3}; spectrum B, [Pr^{+3}]/[phospholipid phosphorus] = 0.152. These data were obtained with a Perkin-Elmer R-20a NMR Spectrometer operating at 60kHz, 33°C.

TABLE 3. RATIO OF AMOUNTS OF PHOSPHATIDYLCHOLINE ON OUTER AND INNER SURFACES OF PHOSPHATIDYLCHOLINE-CHOLESTEROL VESICLES[a]

Mole fraction Cholesterol	Amount Outer Surface[b] / Amount Inner Surface
0.0	2.2
0.22	2.1
0.25	2.3
0.30	2.9
0.50	2.9

[a] Aqueous phase: 0.1 M KCl + M Tris, pH 8.0.

[b] Ratio of areas of NMR peaks due to N-methyl proton resonances from phosphatidylcholines in the outer and inner surfaces of the vesicle wall.

mixtures from which they are prepared (Huang et al., 1974; Huang and Newman, 1974).

It is apparent from the data in Table 3 that, at cholesterol concentrations less than mole fraction 0.3, the ratio of the concentration of phosphatidylcholine on the outer face to that on the inner face is close to the expected value of 2.2. This value is based on an external radius of 125Å and a bilayer thickness of 40 Å. However, at 0.3 mole fraction cholesterol and above, the ratio increases to 2.9, indicating a marked increase in the concentration of phosphatidylcholine on the outer surface of the vesicle relative to that on the inner surface.

Vesicles Formed from Phosphatidylcholine and Phosphatidylglycerol

Recently, Michaelson et al. (1973) have used both proton and phosphorus NMR, together with paramagnetic shift reagents, to investigate the structure of small vesicles in aqueous dispersion formed from an equimolar mixture of egg phosphatidylcholine and phosphatidylglycerol derived from it by the action of phospholipase-D. These authors report a striking transbilayer disproportionation of concentration in this system. The vesicle dispersion examined is similar to that discussed in the preceding section. Although no detailed physical data are available, it is very likely that the average outer radius of the vesicles is close to 125 Å.

The phosphorus NMR signals from this element in phosphatidylcholine and phosphatidylglycerol in the mixed-lipid vesicle proved to be distinguishable from each other. The situation was utilized by the authors to determine independently the amounts of each phospholipid in both the outer and inner faces of the vesicle wall. Their results are given in Table 4, row 1. It is apparent from the data that a substantial fraction of the total phosphatidylglycerol is in the outer surface.

TABLE 4. MOLAR COMPOSITION OF SMALL VESICLES[a]

Total Vesicle	Outer Surface	Inner Surface
1. PG/PC = 1	PG/PC = 2.0	PG/PC = 0.33
2. PE/PC = 1	PE/PC = 0.65	PE/PC = 3.0
3. Chol/PC = 1	Chol/PC = 0.89	Chol/PC = 1.3

[a] Abbreviations: PG, phosphatidylglycerol; PC, phosphatidylcholine; PE, phosphatidylethanolamine; Chol, cholesterol.

For comparison, calculated component molar ratios are given in Table 4 of lipids in the outer and inner vesicle wall surfaces for the phosphatidylethanolamine-phosphatidylcholine (row 2) and the phosphatidylcholine-cholesterol (row 3) systems discussed in the preceding sections. The values presented are only for equimolar total concentrations of each pair of lipids. In these two systems the amounts in the vesicle surfaces of only one of the two lipids of the pair are known. The calculated values in rows 2 and 3 are thus based on the assumption that the molecular packing density of the pair of lipids is the same in both faces of the bilayer.

DISCUSSION

It is interesting to examine the possible bases for the transbilayer compositional asymmetries found in these three binary lipid systems of small-radius vesicles. The presence of phosphatidylcholine in all three systems cannot in itself account for the compositional asymmetry, since in two systems the outer surface is enriched in this lipid while in the third it is markedly depleted. It is apparent that the basic explanation must rest in the dissimilarities in the interactions between the two lipids of each pair and in the smallness of size and the disparity in size of the radii of curvature of the two bilayer faces.

The relative enrichment of phosphatidylethanolamine in the inner face of the vesicle wall above mole fraction 0.1 is most probably due to the propensity of this lipid at high concentrations to form hexagonal rather than bilayer phases. Pure phosphatidylethanolamine does not form stable vesicles, nor do mixtures of this lipid with phosphatidylcholine above mole fraction 0.7 phosphatidylethanolamine (Litman, 1973). At room temperature phosphatidylethanolamine alone in water has been shown by several studies to exist as a mixture of lamellar and type II hexagonal (H_{II}) phases (Husson, 1967; Junger and Reinaurer, 1969; Tinker and Pinteric, 1971). Phosphatidylcholine, on the other hand, exists only as a lamellar phase under these conditions (Luzzati, 1968). The characteristic feature of the H_{II} phase is a hexagonal array of water-filled cylinders, formed with the phosphatidylethanolamine polar head groups, lining the inner surface of each cylinder. The radius of curvature of this inner surface can be as small as 40 Å. The transbilayer compositional asymmetry seen in mixed phosphatidylcholine-phosphatidylethanolamine spherical vesicle systems above mole fraction 0.1 of the latter component may occur because the smaller radius of curvature of the inner surface of the vesicle wall presents conditions more comparable to the packing geometry attained with the very small radii of curvature found in the H_{II} phase. Since phosphatidylcholine cannot exist in a hexagonal phase, it seems reasonable that its packing requirement will be well satisfied in a lamellar outer surface which is relatively free of phosphatidylethanolamine. This interpretation is supported by the low-angle X-ray diffraction studies of Luzzati and Husson (1962) on mixed-lipid systems. These workers observed that in systems consisting of one phase the various lipid components are homogeneoulsy mixed. However, under conditions giving rise to the coexistence of both lamellar and hexagonal phases, separation of different molecular species into the different phases is observed. The relation between these observations of Luzzati and Husson and the transbilayer composition of asymmetries in small vesicles is apparent if each monolayer of bilayer comprising the vesicle wall is regarded as a microscopic phase.

For the phosphatidylglycerol-phosphatidylcholine small-vesicle system, Michaelson et al. (1973) suggest that the transbilayer asymmetry of composition that they observed is due to electrostatic interaction between the negatively charged phosphatidylglycerol molecules. It seems quite reasonable that the repulsive forces between negative charges are less on the outer surface than on the more sharply curved inner surface of the vesicle wall, at constant charge density. Recently Israelachvili (1973) provided a quantitative formulation of this idea which closely predicts the results obtained by Michaelson and his coworkers. This theoretical treatment further predicts that the transbilayer asymmetry of concentration should increase with decreasing radius of curvature and with decreasing ionic strength of the aqueous phases. Thus far, the dependence of asymmetry on these parameters has not been investigated experimentally.

Much evidence suggests that the molecular organization of cholesterol-containing phosphatidylcholine bilayers is complex. For example, studies utilizing NMR (Darke et al., 1972), calorimetry (Hinz and Sturtevant, 1972a), and X-ray diffraction (Engleman and Rothman, 1972) have provided strong evidence that below about 0.3 mole fraction cholesterol two phases coexist in the bilayer. One of these phases consists of pure phosphatidylcholine; the other, of a 1–2 cholesterol-phosphatidylcholine mixture. At 0.3 mole fraction cholesterol and above, only a single mixed phase exists. In addition, studies by Johnson (1973) and recent work by Sears and Thompson (1974) indicate that the apparent partial specific volume of the mixed lipid exhibits a maximum at about 0.3 mole fraction cholesterol. The sedimentation coefficient of small vesicles composed of mixtures of these lipids also shows a sharp increase at about 0.3 mole fraction cholesterol (Johnson, 1973; Huang and Newman, 1974).

This information, however, provides little basis for understanding the compositional asymmetry seen at 0.3 mole fraction cholesterol and above. It is interesting that the appearance of this asymmetry occurs at roughly the same concentration of cholesterol at which other physical parameters of the vesicle also change. Huang et al. (1974) have argued that the asymmetry may reflect interactions with those component phosphatidylcholines of egg lecithin which have a high degree of unsaturation. Experiments designed to test this idea are now in progress.

Whatever the molecular explanation for these transmembrane compositional asymmetries may be, it is clear that the asymmetries must be the direct result of the small radii of curvature of the bilayer of the vesicle wall. Thus in these small-vesicle systems a relation exists between bilayer composition and morphology. The implications of this situation for biological membranes rest on the widely accepted idea that the bilayer is the structural form of most of the lipid in biological membranes. These implications take on greater significance in view of the fact that the functional properties at any locus in a biological membrane must be directly related to the organization and properties of the molecules composing the locus. Thus, in regions of biological membranes exhibiting very small radii of curvature, a relation may exist between membrane barrier properties, as they are a reflection of the properties of a lipid bilayer, and the curvature of the membrane. If the lipid composition asymmetry is generalized to include protein components, a relation may exist between many other membrane properties and curvature.

It seems reasonable to suppose that the cause-effect nature of this relation could operate in either of two ways. On the one hand, changes in curvature driven by physiological mechanisms such as microtubule or microfilament formation could cause changes in composition and hence in function. On the other hand, changes in the relative compositions of components in opposite bilayer faces, accomplished in the cell by biosynthetic mechanisms, could result in alterations of membrane curvature, which in turn could have functional significance.

Whether such relations actually exist in biological membranes is not known at the present time. It is certain, however, that minimum-radii-of-curvature structures of the order of 100 Å are exhibited by many biological membranes of diverse origin (Fawcett, 1966). Considerable evidence suggests that at least in the membrane of the erythrocyte there is a marked compositional asymmetry, not only in the lipid components, but also in the protein components (Bretscher, 1973; Verkleij et al., 1973). Recent studies on bovine retinal rod outer-segment disc membranes indicate a marked asymmetry in the distribution of phosphatidylserine and phosphatidylethanolamine (Smith and Litman, 1974). The asymmetry observed in the above cases, however, is associated with radii of curvature much larger than 100 Å. A search for transbilayer compositional asymmetries in regions of biological membrane exhibiting very small radii of curvature is currently under way in our laboratory.

SUMMARY

Plasma membranes and the membranes of subcellular organelles define surfaces characterized by radii of curvature which range from infinity down to a minimum of about 100 Å. Most current concepts of biological membrane structure are based on the premise that a major portion of the lipid component is present in bilayer form and as such constitutes a barrier matrix for the organization of protein and carbohydrate components. Recent work utilizing aqueous dispersions of spherical vesicles, each composed of a single continuous lipid bilayer surrounding an internal aqueous compartment, indicates that both molecular packing and intrabilayer composition are functions of the radius of curvature of the bilayer. Nuclear magnetic resonance and partial specific volume studies show that, above the liquid-crystalline transition temperature, maximum fluidity of the phospholipid acyl side chains exists in bilayers of vesicles having a minimum radius of curvature. Studies of proton NMR line broadening by paramagnetic ions in vesicles with a minimum radius of curvature formed from mixtures of phosphatidylcholine and cholesterol indicate a higher cholesterol concentration in the inner surface of the vesicle wall than in the outer surface. Chemical modification studies show that in these same-sized vesicles containing phosphatidylcholine and phosphatidylethanolamine there is a disproportionation of phosphatidylethanolamine toward the inner surface of the vesicle wall. These observations on model systems suggest that changes in curvature of biological membranes produced by physiological mechanisms (e.g., microfilaments, microtubules) may, in the regions of maximum curvature, modify the structure, composition, and functional properties of the lipid bilayer component of the membrane. Conversely, compositional asymmetries generated biosynthetically may induce regions of curvature in natural membranes.

ACKNOWLEDGMENTS

The work of the authors discussed in this article was supported in part by National Institutes of Health Grants GM 14628, GM 17452, and EY 00548.

REFERENCES

Bangham, A. D. (1972). Lipid bilayers and biomembranes. *Ann. Rev. Biochem.* **41**. 753–776.

Bangham, A. D., M. M. Standish, and J. C. Watkins (1965). Diffusion of univalent ions across the lamellae of swollen phospholipids. *J. Mol Biol.* **13**, 238–252.

Bretscher, M. S. (1973). Membrane structure: Some general principles. *Science* **181**, 622–629.

Bystrov, V. F., N. I. Dubrovina, L. I. Barsukov, and L. D. Bergelson (1971). NMR differentiation of the internal and external phospholipid membrane surfaces using paramagnetic Mn^{+2} and Eu^{+3} ions. *Chem. Phys. Lipids* **6**, 343–350.

Chapman, D., D. J. Fluck, S. A. Plenkett, and G. G. Shipley (1968). Physical studies on phospholipids. X. The effect of sonication on aqueous dispersions of egg yolk lecithin. *Biochim. Biophys. Acta* **163**, 255–261.

Danielli, J. F., and H. Davson (1935). A contribution to the theory of the permeability of thin films. *J. Cell Comp. Physiol.* **5**, 495–508.

Darke, A., E. G. Finer, A. G. Flook, and M. C. Phillips (1972). Nuclear magnetic resonance study of lecithin-cholesterol interactions. *J. Mol. Biol.* **63**, 265–279.

Dawidowicz, E. (1973). Unpublished observation.

Engleman, D. M., and J. E. Rothman (1972). The planar organization of lecithin-cholesterol bilayers. *J. Biol. Chem.* **247**, 3694–3697.

Fawcett, D. (1966). *An Atlas of Fine Structure: The Cell.* W. B. Saunders Co., Philadelphia.

Hinz, H., and J. M. Sturtevant (1972a). Calorimetric investigation of the influence of cholesterol on the transition properties of bilayers formed from synthetic L-α-lecithins in aqueous suspension. *J. Biol. Chem.* **247**, 3697–3700.

Hinz, H., and J. M. Sturtevant (1972b). Calorimetric studies of dilute aqueous suspensions of bilayers formed from synthetic L-α-lecithins. *J. Biol. Chem.* **247**, 6071–6075.

Huang, C. (1969). Studies on phosphatidylcholine vesicles: Formation and physical characteristics. *Biochemistry* **8**, 344–352.

Huang, C., and J. P. Charlton (1971). Studies on phosphatidylcholine vesicles: Determination of partial specific volume by sedimentation velocity method. *J. Biol. Chem.* **246**, 2555–2560.

Huang C., and J. P. Charlton (1972a). Studies on the state of phosphatidylcholine molecules before and after ultrasonic and gel-filtration treatments. *Biochem. Biophys. Res. Commun.* **46**, 1660–1666.

Huang, C., and J. P. Charlton (1972b). Interaction of phosphatidylcholine vesicles with 2-*p*-toluidinylnaphthalene-6-sulfonate. *Biochemistry* **11**, 735–740.

Huang, C., and L. P. Lee (1973). Diffusion studies on phosphatidylcholine vesicles. *J. Am. Chem. Soc.* **95**, 234–239.

Huang, C., and G. Newman (1974). Unpublished observations.

Huang, C., and T. E. Thompson (1974). Model membranes: Preparation of homogeneous, single-walled phosphatidylcholine vesicles. In *Methods in Enzymology*, Vol. 32 (S. Fleischer, L. Packer, and R. W. Estabrook, eds.). Academic Press, New York (in press).

Huang, C., J. P. Charlton, C. I. Shyr, and T. E. Thompson (1970). Studies on phosphatidylcholine vesicles with thiocholesterol and a thiocholesterol-linked spin label incorporated in the vesicle wall. *Biochemistry* **9**, 3422–3426.

Huang, C., J. Sipe, S. T. Chow, and R. B. Martin (1974). Differential interaction of cholesterol with phosphatidylcholine on the inner and outer surfaces of lipid bilayer vesicles. *Proc. Nat. Acad. Sci. U.S.A.* **71**, 359–362.

Husson, F. (1967). Structure des phases liquide-cristallines de différents phospholipides, monoglycérides, sphingolipides, anhydres ou en présence d'eau. *J. Mol. Biol.* **25**, 363–382.

Israelachvili, J. N. (1973). Theoretical considerations on the asymmetric distribution of charged phospholipid molecules on the inner and outer layers of curved bilayer membranes. *Biochim. Biophys. Acta* **323**, 659–663.

Johnson, S. M. (1973). The effect of charge and cholesterol on the size and thickness of sonicated phospholipid vesicles. *Biochim. Biophys. Acta* **307**, 27–41.

Junger, E., and H. Reinauer (1969). Liquid-crystalline phases of hydrated phosphatidylethanolamine. *Biochim. Biophys. Acta* **183**, 304–308.

Kornberg, R. D., and H. M. McConnell (1971). Inside-outside transitions of phospholipids in vesicle membranes. *Biochemistry* **10**, 1111–1120.

Kostelnik, R. J., and S. M. Castellano (1972). 250 MHz Proton NMR study of sonicated egg yolk lecithin in D_2O. II. The doublet character of the resonance signal of the choline methyl groups. *J. Magn. Resonance* **7**, 219–223.

Levine, Y. K., A. G. Lee, N. J. M. Birdsall, J. C. Metcalf, and J. D. Robinson (1973). The interaction of paramagnetic ions and spin labels with lecithin bilayers. *Biochim. Biophys. Acta* **291**, 592–607.

Litman, B. J. (1973). Lipid model membranes. Characterization of mixed phospholipid vesicles. *Biochemistry* **12**, 2545–2554.

Luzzati, V. (1968). X-ray diffraction studies on lipid-water systems. In *Biological Membranes* (D. Chapman, ed.). Academic Press, New York, pp. 71–123.

Luzzati, V., and F. Husson (1962). The structure of the liquid-crystalline phase of lipid-water systems. *J. Cell Biol.* **12**, 207–219.

March, D., A. N. Phillips, A. Watts, and P. F. Knowles (1972). A spin-label study of fractionated egg phosphatidylcholine vesicles. *Biochem. Biophys. Res. Commun.*, **49**, 641–648.

Michaelson, D. M., A. F. Horwitz, and M. P. Klein (1973). Transbilayer asymmetry and surface homogeneity of mixed phospholipids in cosonicated vesicles. *Biochemistry* **12**, 2637–2644.

Mueller, P., and D. O. Rudin (1969). Translocators in bimolecular lipid membranes: Their role in dissipative and conservative bioenergy transductions. In *Currents in Bioenergetics*, Vol. III (D. R. Sanadi, ed.). Academic Press, New York, pp. 157–242.

Mueller, P., D. O. Rudin, H. T. Tien, and W. C. Wescott (1962). Reconstitution of excitable cell membrane structure *in vitro*. *Circulation* **26**, 1167–1171.

Plenkett, S. A., A. G. Flook, and D. Chapman (1968). Physical studies on phospholipids. XI. Nuclear resonance studies of lipid-water systems. *Chem. Phys. Lipids* **2**, 273–290.

Redwood, W. R., S. Takashima, H. P. Schwan, and T. E. Thompson (1972). Dielectric studies on homogeneous phosphatidylcholine vesicles. *Biochim. Biophys. Acta* **255**, 557–566.

Scandella, C. J., P. Devaux, and H. M. McConnell (1972). Rapid lateral diffusion of phospholipids in rabbit sarcoplasmic reticulum. *Proc. Nat. Acad. Sci. U.S.A.* **69**, 2056–2060.

Sears, B., and T. E. Thompson (1974). Unpublished observations.

Seiter, C. H. H., and S. Chan (1973). Molecular motion in lipid bilayers: A nuclear magnetic resonance line width study. *J. Am. Chem. Soc.* **95**, 7541–7558.

Sheard, D. (1969). Internal mobility in phospholipids. *Nature* **223**, 1057–1059.

Sheetz, M. P., and S. Chan (1973). Effect of sonication on the structure of lecithin bilayers. *Biochemistry* **11**, 4573–4581.

Singer, S. J., and G. L. Nicolson (1972). The fluid mosaic model of the structure of cell membranes. *Science* **175**, 720–731.

Smith, H. G., and B. J. Litman (1974). Unpublished observations.

Thompson, T. E., and F. A. Henn (1970). Experimental phospholipid model membranes. In *Membranes of Mitochondria and Chloroplasts* (E. Racker, ed.). Van Nostrand-Reinhold Co., New York, pp. 1–52.

Tinker, D. O., and L. Pinteric (1971). On the identification of lamellar and hexagonal phases in negatively stained phospholipid-water systems. *Biochemistry* **10**, 860–865.

Verkleij, A. J., R. F. A. Zwaal, B. Roelofsen, P. Comfurius, D. Kastelijn, and L. L. M. van Deenen (1973). The asymmetric distribution of phospholipids in the human red cell membrane: A combined study using phospholipases and freeze-etch electron microscopy. *Biochim. Biophys. Acta* **323**, 178–193.

Wilkins, M. H. F., A. E. Blaurock, and D. M. Engleman (1971). Bilayer structure in membranes. *Nature New Biol.* **230**, 72–76.

CHAPTER TWO

Some General Principles of Membrane Structure

MARK S. BRETSCHER

I should like to summarize some of the more recent experiments which may shed light on the structure of biological membranes. Naturally, any cellular membrane is designed to fulfill a particular biological function, so that no two membranes are likely to have the same chemical components; in a similar way, the cytoplasms of different cells are different, reflecting the specialization of different cells. The aim, therefore, is to try to abstract a picture of a membrane which is general and applies to all cells. That this can be done is shown by the fact that almost all membranes are built from a basic structure—the lipid bilayer.

The lipid bilayer, recognized as early as 1925 by Gorter and Grendel, is the dominating influence on the physical and chemical properties of biological membranes. The specific properties of a membrane are provided by the proteins associated with it—and protein usually accounts for almost two thirds of the mass of a membrane. It is obvious that many molecules, such as amino acids, sugars, and inorganic ions, which do not spontaneously cross a lipid bilayer, must be transported across a membrane, presumably by a protein molecule. In view of the fact that most soluble proteins have one dimension that is at least about 40 Å long, and that a bilayer has approximately the same thickness, it would seem reasonable to suppose that many proteins in the membrane should extend right across the bilayer, from one side to the other. Surprisingly, there has been much reluctance to accept this view, in spite of the suggestive evidence in favor of it. This attitude is probably due to the early suggestion of Danielli and Davson (1935) that protein

---

Dedicated to the memory of the late Dr. R. J. Winzler.

is associated only with either surface of the bilayer, spread out in a thin layer over each surface; this view seems to have dominated the field until very recently.

PROTEINS

There are two kinds of experimental evidence that proteins actually extend across a bilayer. In the first, a bilayer is cracked down its middle plane by freeze-fracture, and a replica of its interior surfaces made and examined in the electron microscope. This technique first allowed Branton (1966) to see in beautiful detail many small particles in the interior of the erythrocyte membrane; these particles are presumably protein. The experiment does not actually show that these particles extend across the bilayer, but it is suggestive.

In a quite different approach, I have tried to tackle the problem of how proteins are organized in a membrane by chemical methods. The intention here was to look at a typical cell, the human erythrocyte, by labeling the outer surface with a membrane-impermeable reagent which was highly radioactive—to tag the proteins on the outer surface. The reagent that I used for this purpose was ^{35}S-formylmethionylsulfone methyl phosphate (FMMP), a powerful acylating reagent (Bretscher, 1971a). Alongside such an experiment one can look at which proteins are exposed on either surface by labeling fragmented erythrocyte membranes; in this case proteins on both surfaces of the membrane should be labeled. When this approach is applied to the human erythrocyte, we find the following: only two main components become labeled from outside the cell (Fig. 1, gel a). There must, of course, be many other proteins on the outside, but presumably they either are present in very much smaller amounts or else are quite unreactive. Hence, on the outside, this evidence indicates just two main components. One is a protein with a molecular weight of some 105,000 daltons, which I call component a. The other is the major glycoprotein so extensively studied by the late Dr. R. J. Winzler and his colleagues and, more recently, by Marchesi and his coworkers. On these SDS-gels it migrates just ahead of component a.

By contrast, when ghosts are labeled, every protein visible in an SDS-polyacrylamide gel becomes labeled—all proteins are exposed on at last one surface (Fig. 1, gel c). No protein seems to be hidden away inside the bilayer (not that the bilayer would be big enough to accommodate it anyway). This experiment shows that much more protein is associated with the inner surface of the erythrocyte membrane than with the external surface—a result which I think is important in understanding membranes.

Perhaps the most interesting question to ask at this stage is whether either of the two proteins labeled from the external surface extends across the membrane. In other words, is more of either protein labeled from both sides of the membrane in ghosts than is labeled from the outer surface in the intact cell? I have tried to examine this possibility in the cases of both component α and the glycoprotein. Because of the relative simplicity of the glycoprotein experiments, I shall restrict myself here to that molecule (Bretscher, 1971c).

The molecular orientation of the glycoprotein molecule in the membrane has, in part, been elucidated by Winzler. He showed that all the carbohydrate is attached at about ten different sites along the backbone of the polypeptide chain; the poly-

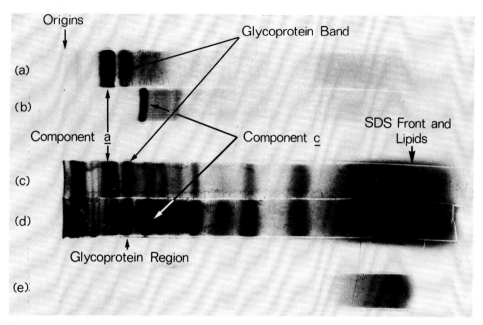

Figure 1. Autoradiogram of SDS-polyacrylamide gels of erythrocyte membranes (gels a–d) and synthetic phosphatidylethanolamine (e) labeled with FMMP. The erythrocytes were labeled as follows: (a) labeled intact cells; (b) cells treated with pronase and then labeled; (c) labeled ghosts; (d) labeled ghosts derived from cells which had been digested with pronase. Cells (A, Rh+) were washed in 0.15 M NaCl. Pronase digestion of intact cells was in 0.15 M NaCl, 0.1 M NH$_4$HCO$_3$ and 0.5 mg-ml^{-1} pronase at 37°C for 1 hour. Cells or ghosts (0.2 ml were labeled in 0.05 M NaHCO$_3$, 0.05 M Na$_2$CO$_3$, 0.05 M NaCl buffer with about 10^8 cpm FMMP at 0°C for 10 minutes. They were then washed in 0.15 M NaCl. Cells were lysed by dilution into 0.02 M Tris-Cl, pH 7.4, containing 3×10^{-5} M CaCl$_2$. Ghosts were dissolved in SDS and fractionated on 7.5% acrylamide gels in phosphate buffer under standard conditions. After running, they were sliced and a center section was autoradiographed. The gels were *not* washed or stained before autoradiography.

peptide is small, being around 90 residues in length. All the carbohydrate is found external to the bilayer, attached to the N-terminal segment of this polypeptide. This N-terminal region can be cleaved by treating erythrocytes with proteolytic enzymes (Winzler, 1969; Bender et al., 1971). Winzler also found that this small protein, when isolated in pure form and digested with proteolytic enzymes, produced a very insoluble core, presumably the part of the molecule associated with the hydrophobic interior of the membrane. But where is the rest of the glycoprotein—is it all in the membrane, or does it traverse it?

As already shown (Fig. 1, gel a), the glycoprotein is labeled when the membrane-impermeable reagent, ^{35}S-FMMP, is added to intact erythrocytes. A fingerprint of the labeled peptides obtained by the digestion of this molecule with trypsin and chymotrypsin contains no discrete, radioactive peptides. Apart from a large smear of radioactivity around the origin (similar to that seen in Fig. 2a), the fingerprint is a blank.

When the region of an SDS-gel which contains the glycoprotein, labeled in cell ghosts, is eluted and fingerprinted, many labeled peptides are found (Fig. 2a).

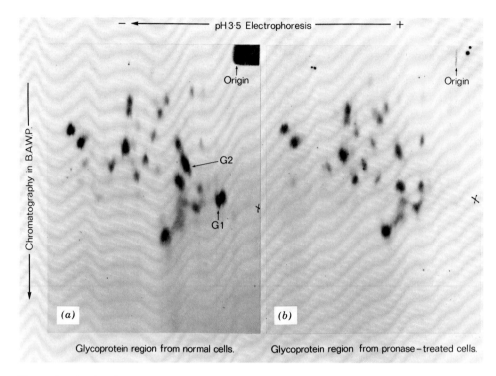

Glycoprotein region from normal cells.   Glycoprotein region from pronase–treated cells.

Figure 2. Crude glycoprotein fingerprint: autoradiogram of FMMP-labeled peptides. The glycoprotein regions of gels similar to those in fig. 1 (gels c and d) were cut out of a gel, and the labeled material was eluted, freed of SDS, and digested with trypsin and chymotrypsin. The digest was fractionated by paper electrophoresis at pH 3.5 and by chromatography in butanol-acetic acid-water-pyridine, 30:6:24:20. Autoradiogram of material from the band off one gel took about 2 weeks.

Most of these peptides have nothing to do with the glycoprotein; the glycoprotein comigrates with another major polypeptide (located on the inner surface of the membrane), which contributes most of the labeled peptides. The peptides belonging to the glycoprotein can be deduced by fingerprinting the material from the same region of the gel which has labeled ghosts fractionated on it, but from which the glycoprotein is removed. The latter condition is easily achieved by prior treatment of the intact cells with pronase, followed by cell lysis and labeling of the ghosts with FMMP. This shows that, of all the strongly labeled peptides seen in Fig. 2a, only two are derived from the glycoprotein molecule, peptides G1 and G2. That both these peptides do arise from the glycoprotein is shown by isolating the glycoprotein from labeled ghosts and fingerprinting it. The resulting map, containing just two discrete labeled peptides, G1 and G2, is shown in Fig. 3.

The properties of peptides G1 and G2 which we can thus deduce are the following:

1. They do not become labeled when intact cells are labeled, but only when ghosts are labeled. This shows that they come from the inner surface of the erythrocyte membrane.

2. When the external surface of the erythrocyte is proteolysed, so as to digest the

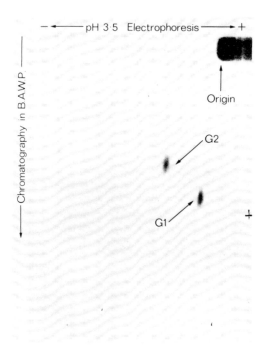

Figure 3. Pure glycoprotein fingerprint: autoradiogram of FMMP-labeled peptides. The labeled glycoprotein, isolated from a 5% SDS-acrylamide gel, was rerun on a 7.5% gel, taking advantage of the anomalous mobility of glycoproteins on SDS gels (Bretscher, 1971c). The pure glycoprotein was fingerprinted as described in Fig. 2.

external region of the glycoprotein, the protein carrying the two peptides, G1 and G2, disappears from its normal position on an SDS-polyacrylamide gel. This shows that peptides G1 and G2 are derived from a molecule which is partly exposed on the outer surface of the cell.

3. These two peptides, G1 and G2, are part of the glycoprotein molecule.

These data, therefore, reveal that the glycoprotein molecule extends across the erythrocyte membrane (Bretscher, 1971c). The carbohydrate is located near the N-terminus, as shown by Winzler, and this is on the external surface of the cell. Because of the small size of the protein (some 90 amino acids long: Kathan and Winzler, 1963), it is difficult to arrange it in any way other than having the C-terminal end at the cytoplasmic surface. This would require a hydrophobic segment of some 25 residues to extend across the bilayer, an arrangement for which good evidence now exists (Winzler, 1969; Segrest et al., 1972). Peptides G1 and G2 therefore must arise from the C-terminal end of the glycoprotein, which is in the red cell cytoplasm.

In similar experiments (Bretscher, 1971b) I have shown that the other major erythrocyte protein, component *a*, also extends across the bilayer. A nice confirmation of this has recently been described by Steck (1972). All the remaining proteins of the erythrocyte membrane which we can see on an SDS-gel are presumably located at the inner surface.

LIPIDS

The phospholipids are arranged in a bilayer, but is any greater order superimposed on this liquid-crystalline state? There are several lines of evidence (see Bretscher, 1972a, 1973) that there is a greater degree of organization. There are four common kinds of phospholipids in erythrocytes; these are shown in Fig. 4. Two of them, phosphatidylcholine and sphingomyelin, have phosphorylcholine as their head groups; two, phosphatidylethanolamine and phosphatidylserine, have aminoalkyl-phosphoryl head groups. The possibility of an asymmetry in the phospholipid composition of the two halves of a bilayer arose when I found that very little phospholipid could be acylated by FMMP in intact cells; much more was reactive in cell ghosts (Bretscher, 1971d, 1972b). Since only phosphatidylserine and phosphatidylethanolamine have free amino groups which could, in principle, react with FMMP, this suggested that both these phospholipids are located primarily at the cytoplasmic side of the bilayer [see Fig. 1, in which labeled phospholipid, which migrates at the SDS front, is labeled more heavily in ghosts (gels c and d) than in intact cells (gels a and b)]. This would leave both sphingomyelin and phosphatidylcholine (rather similar lipids) at the external surface. There are many other chemical data which support this conclusion (Maddy, 1964; Knauf and Rothstein, 1971; Bangham et al., 1958), as well as enzymatic evidence based on the susceptibility of erythrocytes from different animals to lysis in the presence of phospholipases. For example, Turner (1957) found that sheep erythrocytes, unlike most other erythrocytes, have no phosphatidylcholine but contain a much larger-than-usual proportion of sphingomyelin. Again, unlike other erythrocytes, these sheep cells are resistant to the action of crude cobra-venon phospholipase A. It is tempting to put these two observations in parallel, as did Turner, and suggest that the replacements by sphingomyelin of phosphatidylcholine is responsible for the difference in susceptibility to lysis by the venon. Obviously, the chemical basis for

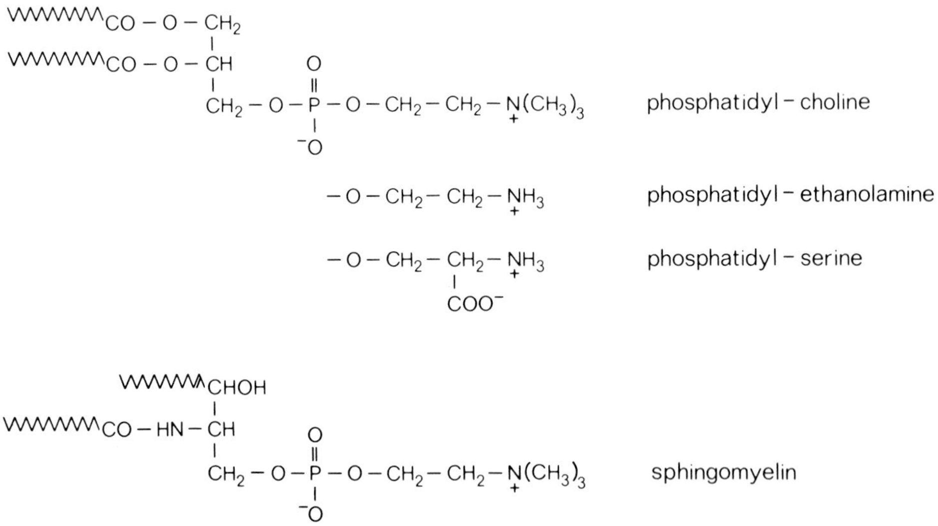

Figure 4. The main phospholipids found in higher organisms (not drawn to scale).

this would be that sphingomyelin, unlike phosphatidylcholine, has no 2-acyl linkage to a glycerol moiety and so cannot be hydrolyzed. Furthermore, it suggests that sphingomyelin (in sheep erythrocytes) and therefore also the replaced phosphatidyl-choline (in other erythrocytes) make up a large portion of the accessible lipid surface on the exterior of the erythrocyte.

The existence of an asymmetric lipid bilayer implies that phospholipid molecules do not flip from one side of the bilayer to the other very readily. The actual rate of migration (or flip-flop) of a spin-labeled phospholipid in a synthetic phospho-lipid vesicle has been determined by Kornberg and McConnell (1971). They discovered that the absolute rate is very slow—a half-life of many hours at 30°C. Two points should be emphasized. First, the presence of either cholesterol or proteins which extend across the bilayer would be expected to slow the rate of flip-flop. In other words, the flip-flop rate in an erythrocyte membrane where both these forces are acting may be very much slower than the above rate. Second, it is possible that natural membranes have enzymes which catalyze the transmem-brane motion of phospholipids—flippases. The reason for this emerges if we consider a hypothetical scheme as to how asymmetrical bilayers *may* be synthe-sized—and here, of course, I am assuming that membranes of all higher cells follow what I believe is true of the erythrocyte. Since it is likely that all phospho-lipid synthesis occurs at the cytoplasmic surface of a membrane, there must be a flippase to catalyze the transfer of some phospholipids across the membrane to form the outer half of the bilayer (see Fig. 5). If this flippase had a specificity for,

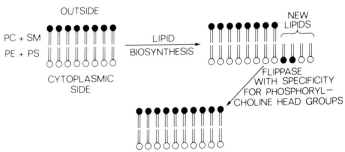

Figure 5. Schematic representation of a hypothetical mechanism for the biosynthesis of an asymmetric bilayer. A flippase embedded in the membrane equalizes the pressure in each half of the bilayer by catalyzing the transfer of phosphorylcholine phospholipids from one side to the other: ⫯ , choline phospholipids; ⫯ , aminophospholipids.

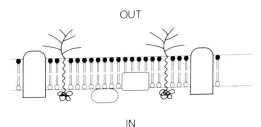

Figure 6. Schematic representation of the manner in which proteins may be accommodated in an asymmetric lipid bilayer in the erythrocyte membrane (not drawn to any particular scale).

say, phosphorylcholine head groups, the suggested asymmetric bilayer would automatically follow: choline lipids outside–amino lipids inside. This means that great caution must be exercised in interpreting flip-flop measurements in real membranes. We can expect some of the rates to be faster than those in synthetic bilayers and therefore presumably enzymatically catalyzed.

GENERAL CONCLUSIONS

Our picture of the erythrocyte membrane is based, then, on a lipid bilayer having compositional asymmetry in its phospholipid (and possibly also glycolipid) components. The glycoprotein and another major protein (component *a*) are located in a fixed orientation across the membrane, and many more proteins are associated with the inner surface of the bilayer. We do not know whether (and if so, to what extent) some of these latter proteins replace, or indeed penetrate, the inner half of the bilayer, although the difficulty experienced in removing most of them by washing the ghosts with strong salt solutions implies that they are probably inserted into the bilayer. If this is so, there are two points to notice.

First, the base composition of erythrocyte phospholipids shows that there is usually more external (choline) than internal (amino) phospholipids. The volume deficit in the inner half of the bilayer caused by this disparity may well be filled in by protein; that is, a substantial portion (maybe 30%) of the cytoplasmic half of the bilayer may be protein. Alternatively, the suggested asymmetry of phospholipids may be incomplete; some choline phospholipids may be located within the cytoplasmic half of the bilayer.

Second, the erythrocyte membrane also has compositional asymmetry within its hydrophobic phase. Phosphatidylcholine, sphingomyelin, and glycolipid each contain very few polyunsaturated fatty acid residues; by contrast, phosphatidylethanolamine and phosphatidylserine are rich in these components, particularly arachidonic acid (see Rouser et al., 1968). These polyunsaturated residues in the inner half of the bilayer may provide a less ordered phase, which may in turn be a better solvent for accommodating proteins than the outer half of the bilayer.

The compositional asymmetry of a bilayer means that membranes always have a polarity, even in the absence of any associated proteins. Cellular membranes can usually be given a gross polarity: one side is adjacent to the cytoplasm, the other side is remote from it. The usual model for membrane fusion requires that this topological polarity be maintained. Thus, when secretory vesicles or membrane viruses fuse with the plasma membrane, or when two cells fuse, the cytoplasmic side of each membrane always remains on the cytoplasmic side of the fused membrane, whereas the external side is always kept external. The asymmetry described here simply provides a molecular basis for defining each side of the bilayer.

The insertion into the bilayer of proteins which extend across the membrane provides another, more easily measured, polarity. How do these proteins come to be in the membrane? Two general classes of protein seem to be synthesized by a cell: cytoplasmic and secreted proteins. Cytoplasmic ribosomes synthesize cytoplasmic proteins such as RNA polymerase, hemoglobin, or β-galactosidase. Membrane-bound ribosomes synthesize proteins for secretion and eventual export from the cell (Siekevitz and Palade, 1960). The decision as to whether or not a protein is to be secreted—that is, on which class of ribosome it will be synthesized—is presumably genetically defined. A messenger RNA must therefore have coded in it information which determines the class of ribosome to which it attaches. Selection at this level determines the fate of the protein product. If the messenger RNA attaches to membrane-bound ribosomes, it seems probable that the secreted protein is extruded through the membrane as it is synthesized. This general scheme is consistent with recent studies of the specificity of protein synthesis by membrane-bound ribosomes.

It is widely assumed that membrane proteins are synthesized by membrane-bound ribosomes—that they are a special class of secreted protein. This view, I think, needs to be revised. There are several reasons why I believe that, rather, they are synthesized as cytoplasmic products which then diffuse into the membrane and become lodged there. At a simple-minded level, consider a protein which only partly penetrates the inner surface of a bilayer. Were it synthesized by a membrane-bound ribosome and (let us suppose) "fed" into the membrane as it was assembled, how would it find its way to the cytoplasmic, rather than the external, side of the bilayer? The presence of an asymmetric lipid bilayer could help; but since phospholipids do not flip-flop and all the available evidence suggests that the same is true for proteins, this seems unlikely. It is more plausible to imagine that such a protein arrived there by solution in the membrane from the cytoplasm. Now consider a protein which extends across the bilayer. Two problems arise. First, how does it find its correct orientation in the bilayer if it is synthesized by membrane-bound ribosomes? This is a question similar to the one above for a partially penetrating protein. Second, at a different level, I find it difficult to see how the protein could *fold up* properly in an environment which at once is very hydrophobic and is bounded by very hydrophilic (aqueous) regions. The main forces which are usually cited as holding proteins together are hydrophobic bonds and ionic interactions. If a protein tried to fold up *in* a membrane, I believe that the forces which would determine its folding would not be the usual peptide-peptide interactions but that the whole process would be dominated by the very discontinuous environment provided by the lipid bilayer. If one bears in mind the fact that a transmembrane protein, like component *a*, is probably larger in each dimension (say, $50 \times 50 \times 100$ Å) than a lipid bilayer (say, 40 Å), the implausibility of its folding up inside the membrane is obvious. I therefore believe that, for membrane proteins, synthesis is by cytoplasmic ribosomes, followed by solution in the membrane. Whether a cytoplasmically synthesized protein remains as a soluble component, or whether it partially dissolves in the inner surface of the bilayer, or whether it dissolves in the membrane so that it traverses the bilayer, is determined by the nature of the protein. For a transmembrane protein, the nature of the protein will ensure its correct polarity of insertion into the membrane; once insertion occurs, it seems likely that glycosylation from the external side of the membrane will ensure a permanent abode for that protein across the membrane. In other words, the carbohydrate may act as an irreversible lock on the protein to hold it fast in the membrane.

There is one piece of evidence which argues strongly in favor of this hypothesis. The lactose operon of *Escherichia coli* has three genes, which code for β-galactosidase (z), lactose permease (y), and thiogalactoside transacetylase (a). These three genes are transcribed and translated as a single messenger RNA (Kepes,

1969). It is then very likely (especially in view of the polar nature of some z-gene mutants on the y-gene) that each of these genes is translated by the same class of ribosome. Since both the first and the last genes on the messenger RNA (those for β-galactosidase and the transacetylase) are cytoplasmic proteins, the middle gene, coding for lactose permease (which probably spans the bacterial plasma membrane), must also be synthesized as a cytoplasmic protein.

This general conclusion—that a protein which traverses the membrane has arrived in this position by diffusion from the cytoplasm—could explain why the major proteins exposed on the outer surface of the erythrocyte seem so chemically unreactive: both the glycoprotein and component a appear to have few amino groups on their external surfaces. In the case of the glycoprotein, the part of it on the cell exterior is built from rather neutral amino acids (almost one half is accounted for by serine plus threonine: Kathan and Winzler, 1963). This N-terminal segment of the protein (without its carbohydrate) may spontaneously diffuse across the membrane and later be fixed there by the addition of carbohydrate from the external side of the membrane.

SUMMARY

The arrangement of lipids and the major proteins of the erythrocyte membrane has been discussed. The conclusions from this discussion I list here as a set of general guidelines for the structure of membranes of higher organisms: some of these rules may be wrong. However, it seems useful now to be as precise as possible at a molecular level and thereby focus attention on various specific points.

1. The basic structure of a membrane is a lipid bilayer with (*a*) choline phospholipids and glycolipids in the external half and (*b*) amino (and possibly some choline) phospholipids in the cytoplasmic half. There is effectively no lipid exchange across the bilayer (unless enzymatically catalyzed).

2. Some proteins extend across the bilayer. In such cases, they will in general have carbohydrate on their surface remote from the cytoplasm. This carbohydrate may prevent the protein from diffusing out of the membrane into the cytoplasm: it acts as a lock on the protein.

3. Just as lipids do not flip-flop, proteins do not rotate across the membrane. Lateral motion or rotation of lipids and proteins in the plane of the bilayer may be expected.

4. Most membrane protein is associated with the inner, cytoplasmic surface of the membrane. Proteins are not usually associated exclusively with the outer half of the lipid bilayer.

5. Membrane proteins are a special class of cytoplasmic proteins, not of secreted proteins.

REFERENCES

Bangham, A. D., B. A. Pethica, and G. V. F. Seaman (1958). *Biochem. J.* **69**, 12.

Bender, W. W., H. Garan, and H. C. Berg (1971). *J. Mol. Biol.* **58**, 783.

Branton, D. (1966). *Proc. Nat. Acad. Sci. U.S.A.* **55**, 1048.

Bretscher, M. S. (1971a). *J. Mol. Biol.* **58**, 775.

Bretscher, M. S. (1971b). *J. Mol. Biol.* **59**, 351.

Bretscher, M. S. (1971c). *J. Mol. Biol.* **231**, 229.

Bretscher, M. S. (1971d). *Biochem. J.* **122**, 40.

Bretscher, M. S. (1972a). *Nature New Biol.* **236**, 11.

Bretscher, M. S. (1972b). *J. Mol. Biol.* **71**, 523.

Bretscher, M. S. (1973). *Science* **181**, 622.

Danielli, J. F., and H. Davson (1936). *J. Cell Physiol.* **5**, 495.

Gorter, E., and F. Grendel (1957). *J. Exp. Med.* **41**, 439.

Kathan, R. H., and R. J. Winzler (1963). *J. Biol. Chem.* **238**, 21.

Kepes, A. (1969). *Prog. Biophys. Mol. Biol.* **19**, 201.

Knauf, P. A. and A. Rothstein (1971). *J. Gen. Physiol.* **58**, 190.

Kornberg, R. D., and H. M. McConnell (1971). *Biochemistry* **10**, 1111.

Maddy, A. H. (1964). *Biochem. Biophys. Acta.* **88**, 390.

Rouser, G., G. J. Nelson, S. Fleischer, and G. Simon (1968). In *Biological Membranes* (D. Chapman, ed.). Academic Press, New York, p. 5.

Segrest, J. P., R. L. Jackson, V. T. Marchesi, R. B. Guyer, and W. Terry (1972). *Biochem. Biophys. Res. Commun.* **49**, 964.

Siekevitz, P., and G. E. Palade (1960). *J. Biophys. Biochem. Cytol.* **7**, 619.

Steck, T. L. (1972). In *Membrane Research.* Academic Press, New York, p. 71.

Turner, J. C. (1957). *J. Exp. Med.* **105**, 189.

Winzler, R. J. (1969). In *Red Cell Membrane* (G. A. Jamieson and T. J. Greenwalt, eds.). J. B. Lippincott Co., Philadelphia, p. 157.

CHAPTER THREE

Elaboration and Turnover of Cell Coat Glycoproteins

C. P. LEBLOND AND GARY BENNETT

Even though our views of the plasma membrane of animal cells are evolving, the old concept of a bimolecular lipid layer has retained its importance. It is now known, however, that, in addition to proteins residing on the surfaces of the lipid layer, there are others which extend into or through this layer. Many of these membrane proteins are actually glycoproteins whose carbohydrate-rich portions extend outward from the outer surface of the membrane (Singer and Nicolson, 1972).

In the electron microscope, however, the outer surface of most cells appears naked in routinely stained preparations (Fig. 1). Only exceptionally are fine, irregular processes (referred to as fuzzy coat) seen at the outer surface, as for instance, at the apex of the epithelial cells of the intestine or in some unicellular organisms (Fig. 2). Thus, until recently, it was generally believed that the plasma membranes of most animal cells contained no material external to its outer leaflet.

More than 30 years ago, however, Chambers (1940) claimed that the dissecting microscope revealed the existence at the cell surface of a layer of distinct consistency, which, he assumed, served as a cement substance between cells. Chambers' observations were largely ignored, and it remained for histochemical and chemical findings to provide further evidence for the presence of external material. Some of this evidence was rather indirect; for example, when embryonic or tissue culture cells were trypsinized, they lost their adhesiveness and dissociated. This result was attributed to removal by the enzyme of an extraneous coat which had cemented the cells together (Moscona, 1952).

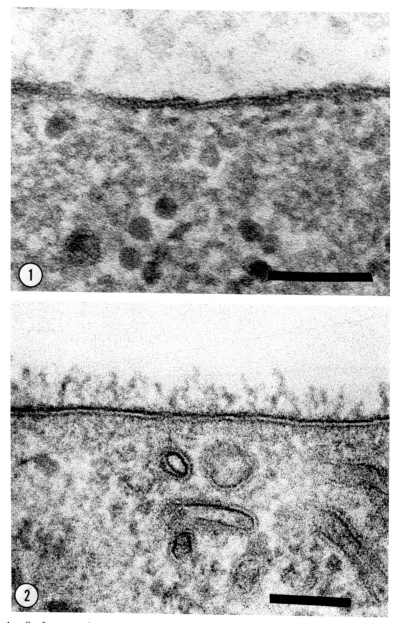

Figure 1. Surface membrane of a mast cell. Glutaraldehyde fixation. Stained with uranyl and lead. 280,000X. When cut perpendicularly, the surface membrane is seen as a three-layered structure composed of an inner and an outer electron-dense leaflet with an electron-lucent layer in between. In this cell the outer leaflet appears bare. (Courtesy of Mr. Julius Batky.)

The marker corresponds to 0.1 μ in Figs. 1 and 2 but to 1 μ in all other electron micrographs.

Figure 2. Surface membrane of a unicellular flagellate inhabiting the lumen of a mouse colon. Glutaraldehyde fixation. Stained with uranyl and lead. 200,000X. The trilayered nature of the surface membrane is clearly seen. Attached to and radiating out from the outer electron-dense leaflet are the filaments of the fuzzy coat. (Courtesy of Dr. M. Weinstock.)

30

HISTOCHEMICAL EXAMINATION OF THE CELL SURFACE

The surface of the cells was examined in sections stained by various reagents used to locate carbohydrate groups. The first method employed was the periodic acid-Schiff (PA-Schiff) reaction, which detects the vicinyl hydroxyls, also called 1,2-glycol groups, of carbohydrates. It was known that significant reactions are produced by the glycogen and glycoproteins present in sections; but since glycogen is readily eliminated by amylase treatment, the PA-Schiff technique may be considered to be specific for glycoproteins (Leblond et al., 1957). This technique was found to stain the extraneous coat of oocytes (Leblond, 1950), as well as cell surfaces which exhibit a fuzzy coat, that is, the apical surfaces of epithelial cells in the intestine, epididymis, and kidney (Leblond, 1950; Burgos, 1964), as well as the surface of amoebae (Bairatti and Lehmann, 1954). But what about the apparently naked plasma membranes of most cells? Here, PA-Schiff staining was reported in only two locations, that is, the surfaces of stratified squamous epithelial cells (Wislocki et al., 1951) and the lateral surfaces of intestinal epithelial cells (Puchtler and Leblond, 1958). In the 1960s our group undertook a systematic study of the staining of cell surfaces by the PA-Schiff technique and found that in over fifty cell types of the rat, that is, in all examined cells, a definite staining could be detected along the whole cell surface (Rambourg et al., 1966). For instance, in the Meibomian gland, a type of sebaceous gland, PA-Schiff staining delineated the periphery of each cell (Fig. 3). In the epithelium of small intestine (Fig. 4), the columnar cells exhibited not only heavily stained apical surfaces, but also moderately stained lateral surfaces. The staining of the lateral cell membranes was interrupted, however, at the "tight junction" of the terminal bars—a fact suggesting that the fusion of the outer leaflets of adjacent cell membranes prevents the occurrence of a layer of PA-Schiff-stained material. In other words, this layer would be found only when the surface of the outer leaflet is free.

Colloidal iron, a stain which specifically detects acidic carbohydrate groups (Mowry, 1963), was also found to stain all surfaces of all cell types studied (Rambourg et al., 1966). Thus, in nervous tissue (Fig. 5), the periphery of neurons was heavily stained, while in seminiferous epithelium (Fig. 6), the interface of the epithelial cells was indicated by a heavy line of stain.

Histochemical stains specific for carbohydrates were devised for use in the electron microscope, particularly periodic acid combined with aldehyde reagents, as in the periodic acid-chromic acid-silver methenamine technique, which has a specificity very similar to that of the PA-Schiff technique in the light microscope. When this technique was applied to the kidney, staining was noted along the various membranes of the epithelial cells, as may be seen in Fig. 9, which represents several cells in a proximal convoluted tubule. A particularly heavy staining was noted on the outer surfaces of the long microvilli making up the apical brush border, as well as along the inner surfaces of the invaginations found in this region. Lysosomes were variously stained, often very intensely (Fig. 9). Staining of lateral membranes was partly or completely interrupted at tight junctions. Figure 10 shows a cross section of small intestinal columnar epithelial cells, in which heavy staining is localized at the narrow space between the lateral cell membranes of adjacent cells. In addition, stacks of Golgi saccules are also stained, as well as the inner surfaces of cytoplasmic vesicles (Rambourg et al., 1969). Staining was,

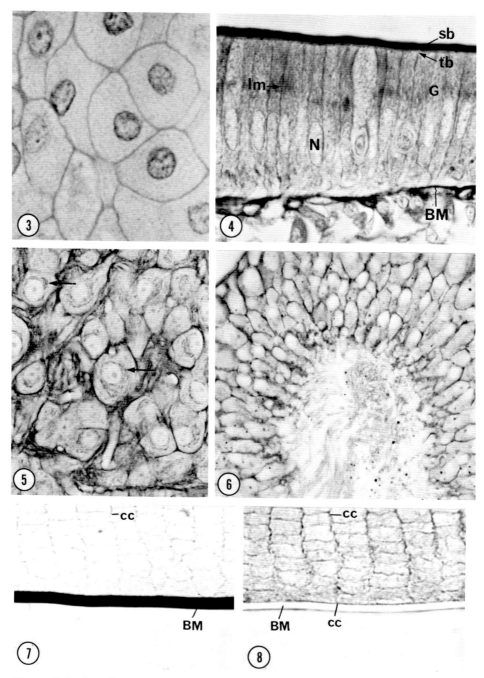

Figures 3–8. Paraffin sections of rat tissues stained with the periodic acid-Schiff or colloidal iron technique. [From A. Rambourg, M. Neutra, and C. P. Leblond, *Anat. Rec.* **154**, 41–71. (1966).]

Figure 3. Meibomian gland from the upper eyelid of the rat. Bouin fixation. PA-Schiff toluidine blue stain. Photography with green filter. 1200X. The cells of the Meibomian gland, a type of sebaceous gland, are outlined by a PA-Schiff-stained strip composed of the fused cell coats of adjacent cells. The nuclei are stained by toluidine blue.

in fact, observed at the surfaces of all examined cells (Rambourg and Leblond, 1967). Later, staining of cell surfaces was observed in the electron microscope with phosphotungstic acid at low pH (Rambourg, 1969), a technique which produces a pattern of reaction identical to that obtained with the methods combining periodic acid and aldehyde reagents (Rambourg, 1971).

Finally, staining of the cell surface was observed in the electron microscope with the cationic dyes colloidal iron (Benedetti and Emmelot, 1967) and thorium (Rambourg and Leblond, 1967), as well as with ruthenium red (Luft, 1971). All three of these indicate the presence of acid groups.

The staining of the cell surface by the PA-Schiff technique in the light microscope and by techniques with similar specificity in the electron microscope clearly revealed the presence of 1,2-glycol-containing carbohydrates at the cell surface, whereas cationic stains demonstrated the existence of acid groups. These might be present in acid mucopolysaccharides, but, as pointed out by Rambourg et al. (1966), they could equally well consist of sialic acid. Since glycoproteins contain side chains with 1,2-glycol-containing hexoses and sialic acid, they were probably responsible for the observed reactions. Glycolipid molecules may also be present, although these should be largely washed out of the sections during processing. Direct chemical analysis of surface membranes from animal cells reveals a carbohydrate content of 2–10%; a small fraction of this is accounted for by glycolipids, but most consists of glycoproteins (Winzler, 1970). Mucopolysaccharides have not been shown to be present at the surface of cells *in vivo*, although heparin sulfate has been demonstrated at the surface of cultured mammalian cells (Kraemer and Tobey, 1972). The absence of uronic acid residues in the isolated microvilli of intestinal cells (which are presumed to contain a high concentration of cell surface carbohydrates) clearly rules out the presence of mucopolysaccharides at this site (Forstner, 1969). The overall conclusion is that a layer composed of the carbo-

---

Figure 4.   Epithelium of small intestine of rat. Bouin fixation. PA-Schiff toluidine blue stain. 1200X. A row of columnar cells is seen in which the nucleus (N) is lightly stained with toluidine blue. The striated border (sb) composed of the apical microvilli is intensely stained by PA-Schiff. Above the nucleus, the Golgi zone (G) is lightly stained, while the fused cell coats of lateral cell membranes (lm) are moderately stained by the same technique and appear as dark lines, except for an unstained portion just below the striated border. This unstained portion is presumed to be the tight junction of the terminal bar (tb). The basement membrane (BM) is the intensely stained band below the cells.

Figure 5.   Supraoptic nucleus of central nervous system. Bouin fixation. Colloidal iron stain. 540X. The surface of the neurons is outlined intensely. Within the cells, the Golgi region is stained (arrows).

Figure 6.   Seminiferous epithelium. Carnoy fixation. Colloidal iron stain. 470X. The network in the epithelium consists of colloidal iron-stained lines at the interface between the cells.

Figure 7.   Posterior portion of the lens capsule. Bouin fixation. PA-Schiff stain. 1200X. The dark band in the lower part of the picture is the strongly stained lens capsule, which is in fact the basement membrane of the lens (BM). Above this, the lens fibers are separated from each other by faintly stained lines corresponding to their cell coats (cc).

Figure 8.   Posterior portion of the lens capsule. Bouin fixation. Colloidal iron stain. 1200X. The lens capsule now appears as an unstained light band (BM). The thin stained line above it represents the basal cell coat of the lower row of lens fibers (cc). The stained line below the lens capsule may represent absorbed mucopolysaccharides from the vitreous humor.

Comparison of Figs. 7 and 8 suggests that the basement membrane and the cell coat of adjacent cells may be distinguished in colloidal iron but not PA-Schiff preparations.

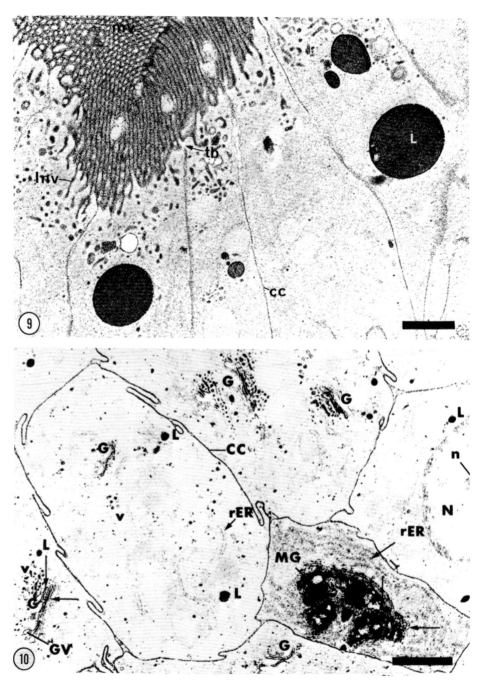

Figure 9. Kidney proximal tubule cells. Glutaraldehyde fixation. Stained by the periodic acid-chromic acid-silver methenamine technique. 1350X. Specific staining occurs along the lateral membranes (cc) of the cells and the surfaces of the microvilli of the apical brush border (mv). Stain is also taken up by invaginations of the apical cell surface (Inv) and by lysosomes (L); tb: unstained tight junction of terminal bar. (Courtesy of Dr. W. Hernandez.)

Figure 10. Cross section through the Golgi region of rat duodenal villus epithelium cells. Glutaraldehyde fixation. Stained by the periodic acid-chromic acid-silver methenamine tech-

hydrate side chains of acid glycoproteins is localized at the outer surface of the plasma membrane of all cells.

This layer is referred to as cell coat, a term first used by Gasic and Berwick (1962) when they observed staining of ascites cell surfaces by colloidal iron. Bretscher (Chapter 2 of this volume) reported that surface glycoproteins prevent the access of pronase to the other proteins of the plasma membrane. Hence they constitute not only a morphological but also a physiological barrier at the cell surface. The name cell coat fits these properties. Stanley Bennett (1963) proposed to apply one term, glycocalyx, to both cell coat and basement membrane. However, the basement membrane (also called the basal lamina), which the PA-Schiff technique shows as a heavily stained line separating epithelia from connecting tissue (Fig. 4) is a structure with different properties. Thus, unlike the cell coat, it does not stain with colloidal iron, as may be seen by comparing the basement membrane of the lens (also called the lens capsule) after PA-Schiff (Fig. 7) and colloidal iron staining (Fig. 8). Because the basement membrane does not stain with colloidal iron, the cell coat which fills the space between this membrane and the cell surface stands out (Fig. 8), whereas it is not distinguishable from the basement membrane after PA-Schiff staining (Fig. 7). For this reason and others, the cell coat is a structure essentially different from the basement membrane, and each should retain its own name.

PASSAGE OF LABELED SUBSTANCES, PARTICULARLY ³H-FUCOSE, TO THE SURFACE OF THE COLUMNAR CELLS IN THE INTESTINAL EPITHELIUM

Biochemical studies carried out on fractions of the columnar cells of the small intestine (Forstner, 1971; Alpers, 1972) have shown the presence of several glycoproteins in the microvillar apical border, including the digestive enzymes maltase, sucrase, β-naphthylamidase, and alkaline phosphatase (Forstner, 1971). These enzymes could be removed from the still-intact membranes of the centrifuged microvilli by a short treatment with papain, a result indicating that they are located at the surface, where they presumably constitute some of the cell coat glycoproteins.

Glycoproteins are composed of a main chain, which is polypeptidic, and of

---

nique. 1600X. Several columnar cells and a heavily stained goblet cell (lower right) are separated by a sharp line (cc), wavy in places. This line is attributed to the fusion of the stained "cell coat" found on the outer surfaces of adjacent cells. The columnar cell at right shows a weak, nonspecific staining of nucleus (N) and nucleolus (n). Within some of the columnar cells one or several stained Golgi stacks (G) may be seen. In some stacks, it is possible to identify a forming face with saccules that are little or not stained (horizontal arrow in the cell at lower left) and a mature face with saccules that are strongly stained and are associated with vesicles (vertical arrow). At the edge of the saccules, the Golgi vacuoles are outlined by a stained line (GV). Finally, the columnar cells contain a few intensely stained dense bodies, presumed to be lysosomal in nature (L), as well as stained small vesicles (v). In contrast, the cisternae of rough endoplasmic reticulum (rER) show only a slight, nonspecific staining of the ribosomes; their content is unreactive.

In the goblet cell at lower right, the group of mucous globules is intensely stained (MG). On one side, the stained Golgi saccules are distinguishable, with the forming face at the horizontal arrow and the intensely stained mature face at the vertical arrow. A special feature of this cell type is the presence of a weakly stained content in the cisternae of rER. [From A. Rambourg, W. Hernandez, and C. P. Leblond, *J. Cell Biol.* **40**, 395–414 (1969).]

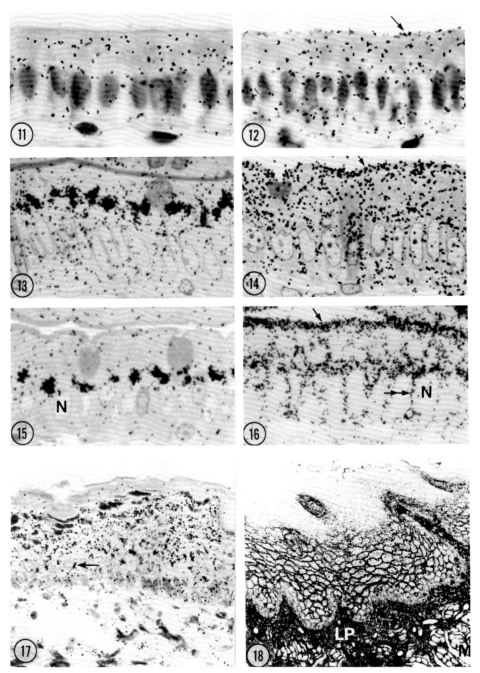

Figures 11 and 12. Radioautographs of rat duodenal epithelial cells at the base of a villus after ^{3}H-proline injection. Paraffin sections stained with hematoxylin and eosin; 2-day exposure. 1050X.

Figure 11. Ten minutes after ^{3}H-proline injection. A diffuse reaction occurs over the cytoplasm of the columnar cells, suggesting uptake of the label by the rough endoplasmic reticulum. No silver grains appear over the apical surface of the columnar cells.

Figure 12. Four hours after ^{3}H-proline injection. In addition to the diffuse reaction over the

carbohydrate side chains. The amino sugars N-acetylglucosamine and N-acetylgalac-tosamine, combined with mannose, usually form the inner residues of the side chains; galactose residues usually occur next to the free ends, whereas fucose or sialic acid occupies the end position (Spiro, 1970). This is believed to be the case in cell coat glycoproteins (Winzler, 1970).

The elaboration of cell coat glycoproteins was examined by radioautography after injection of either labeled amino acids as precursors of the polypeptide chain or labeled sugars as precursors of the residues in the carbohydrate side chains. The first radioautographic studies of this type were reported by Ito (1965), using intestinal epithelial cells incubated *in vitro* with labeled amino acids or glucose. He observed that, with either substance, the label appeared first in the cytoplasm and later in the microvillar apical border of intestinal cells. This observation indi-cated that each cell was the source of its own cell coat material. After injection of the amino acid *³H-proline* into rats (unpublished study), we found that, at 10 minutes, a diffuse radioautographic reaction overlay the cytoplasm of the epithelial cells at the base of duodenal villi, with the exception of the microvillar border (Fig. 11), but by 4 hours a strong reaction had appeared over this border (Fig. 12). Presumably the proline label was taken up on the ribosomes of the endoplas-mic reticulum to give rise to the polypeptidic chain of glycoproteins, which later migrated to the cell surface to become part of the cell coat.

---

cytoplasm of the columnar cells, a definite line of reaction now appears over the apical cell surface (arrow).

Figures 13 and 14. Radioautographs of rat duodenal columnar cells at the base of a villus after ³H-galactose injection. Epon sections stained with toluidine blue. 33-day exposure. 1050X.

Figure 13. Five minutes after ³H-galactose injection. Heavy clusters of silver grains occur over the Golgi region of the cells located above the nucleus. Only a light reaction appears over the rest of the cytoplasm.

Figure 14. Thirty minutes after ³H-galactose injection. A moderate reaction remains over the Golgi region of the cells, but now in addition a definite reaction occurs over the apical striated border at the top of the picture (arrow).

Figures 15 and 16. Radioautographs of rat duodenal columnar cells at the base of a villus after ³H-fucose injection. Epon sections stained with toluidine blue. 960X. [From G. Bennett, C. P. Leblond, and A. Haddad, *J. Cell Biol.* 1974, **60**, 258.

Figure 15. Two minutes after ³H-fucose injection; 8-week exposure. Above the row of nuclei (N), dense clusters of silver grains overlie the Golgi region of the columnar cells. Very few silver grains occur over other regions of the cytoplasm.

Figure 16. One hour after ³H-fucose injection; 3-week exposure. The columnar cells show a moderate Golgi reaction above the barely visible nuclei (N). The rest of the reaction is seen to be localized as a reactive band along the top of the epithelium (arrow), corresponding to the microvilli of the striated border, and as perpendicular bands along the lateral interfaces between cells (double arrow) corresponding to the lateral cell membranes.

Figures 17 and 18. Stratified epithelia after ³H-fucose injection. Paraffin sections. Hema-toxylin and eosin stains.

Figure 17. Forestomach, 10 minutes after ³H-fucose injection; 6-week exposure. 450X. This stratified squamous epithelium shows paranuclear dotlike reactions (dark arrow) in the stratum spinosum. They should not be confused with the keratohyalin of the granular layer (white arrow).

Figure 18. Oral cavity, 30 hours after ³H-fucose injection; 7-week exposure. 380X. In the stratum spinosum, the reaction outlines the cells precisely. A strong diffuse reaction is also seen at this time over the underlying lamina propria (LP). Below, this reaction outlines the fibers of striated muscle cut in cross section (M).

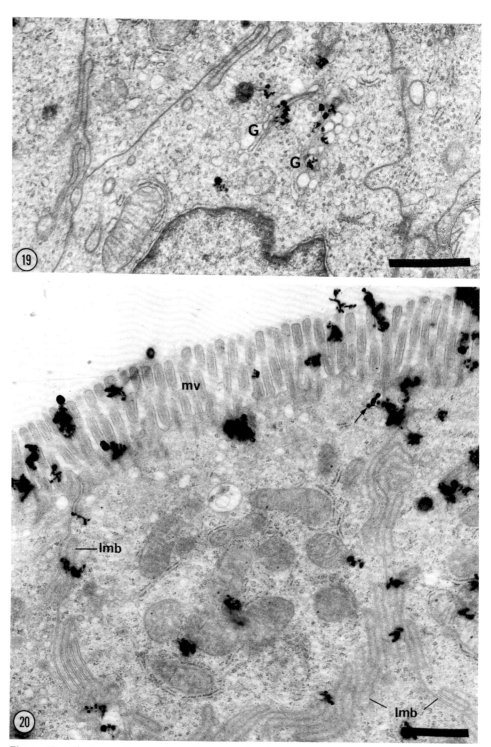

Figure 19. Electron microscopic radioautograph of the Golgi region of a duodenal villus columnar cell, 2 minutes after ^{3}H-galactose injection. Stained with uranyl and lead. 22000X. All of the silver grains are localized over the saccules and vesicles of the Golgi apparatus (G).

When ³H-*mannose* was injected into rats to examine the formation of the carbo-hydrate side chains (unpublished study), radioautographs at the base of duodenal villi showed a pattern similar to that observed with ³H-proline. The diffuse cyto-plasmic radioautographic reaction observed at 10 minutes gave rise to a cell surface reaction which was weak at 1.5 hours and definite by 5 hours after injection. Pre-sumably, the mannose label taken up in the endoplasmic reticulum by the poly-peptide chain migrated with it to the cell surface.

In contrast, ³H-*galactose* at 10 minutes after injection produced, not a diffuse cytoplasmic reaction, but one restricted to the Golgi region, as shown by Neutra and Leblond (1966) in the light microscope (Fig. 13). This was followed by the appearance of a reaction at the cell surface at 30 minutes (Fig. 14). This sequence was confirmed in the electron microscope after ³H-galactose administration *in vitro* using cat intestinal cells (Ito, 1969) or after injection *in vivo* into young rats (Bennett, 1970). The label was initially incorporated into the Golgi saccules (Fig. 19) and subsequently relocated to the apical microvillar border (Fig. 20).

The above evidence demonstrated that some component of the cell coat present at the apical surface of intestinal cells came from the Golgi apparatus. It was known, however, that the cell coat at this site is particularly prominent and may be seen as a "fuzzy" coat in ordinary electron micrographs. We wondered if, on other surfaces of these cells, as well as on other cells in which no "fuzz" is seen in the electron microscope, there might be a different type of cell coat, which arises in a different manner. It was decided, therefore, to re-examine this problem in other locations.

In the meantime a new isotope, ³H-*fucose*, became available. Fucose has proved to be superior to all other sugars as a specific precursor for glycoproteins because, when ingested into the body, it is not broken down or converted to other sugars; rather, its sole fate is to be incorporated as a fucose residue into newly synthesized glycoproteins (Bekesi and Winzler, 1967; Herscovics, 1970) and to a lesser extent into glycolipids (Bosmann et al., 1969). The results with ³H-fucose will therefore be presented in some detail.

When ³H-fucose was injected into young rats, the radioautographic reaction at early time intervals was similar to that previously seen with ³H-galactose in duodenal villus columnar cells, but background was reduced. Thus, at 2 minutes after injection, the light microscope showed a dense accumulation of silver grains in the supranuclear Golgi region (Fig. 15). In the electron miscroscope at that time (Bennett and Leblond, 1970) almost 80% of the silver grains were recorded over Golgi saccules (Fig. 26). Only a negligible number of grains were at the cell sur-faces (Fig. 29). In this regard it is of interest that the lining of Golgi saccules is stained by the techniques for glycoprotein detection (Fig. 10). Thus the fucose is presumed to have been incorporated into glycoproteins at this site.

With time, the distribution of label within the cells changed rapidly. Light micrographs at 1 hour after injection showed some label remaining in the Golgi

Figure 20. Electron microscopic radioautograph of the apical region of a duodenal villus columnar cell. Stained with uranyl and lead. 17250X. A substantial reaction occurs over the apical microvillar cell surface (mv). Several grains occur over lateral cell membranes (lmb). Some grains occur over smooth surface vesicles (arrow) near the cell apex, and other grains appear over lower regions of the cytoplasm, perhaps in association with rough endoplasmic reticulum.

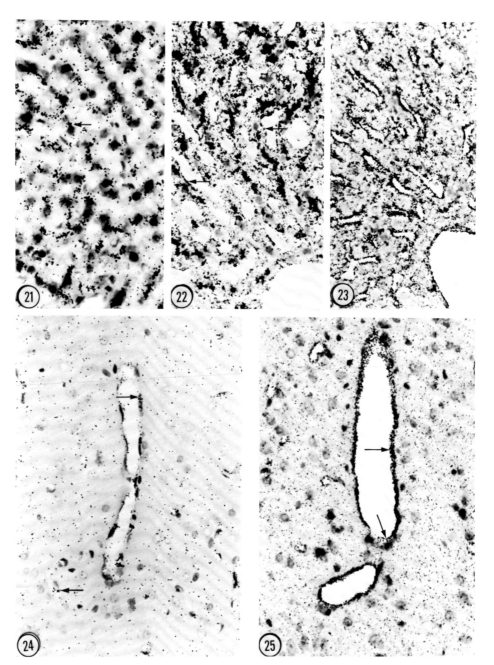

Figures 21–25. Radioautographs of paraffin sections after ³H-fucose injection. [From G. Bennett, C. P. Leblond, and A. Haddad, *J. Cell Biol.* 1974, **60**, 258.]

Figure 21. Liver, 2 minutes after injection. Hematoxylin and eosin stain; 10-week exposure. 390X. Along the liver cords localized reactions (arrow) occur adjacent to hepatocyte nuclei. These reactions are distributed in the same manner as after staining of the Golgi apparatus.

Figure 22. Liver, 20 minutes after injection. Hematoxylin and eosin stain; 10-week exposure. 390X. Localized paranuclear reactions are heavier than in Fig. 7; in addition, some reaction can now be seen along the cell surfaces facing the sinusoids (arrows).

Figure 23. Liver, 4 hours after injection. Hematoxylin and eosin stain; 5-week exposure.

apparatus, but much appeared at the cell surface. In contrast to the previous observations with ^{3}H-galactose, however, labeling appeared not only at the apical microvillar cell surface but also along the lateral cell surfaces (Fig. 16). The electron microscope confirmed these findings: by 20 minutes after injection the percentage of grains localized to the Golgi apparatus had dropped to 30%, while 32% of the grains now occurred over the lateral and apical cell surfaces (Fig. 29). By 60 minutes after injection, the Golgi apparatus contained less than 20% of the label, while the cell surfaces contained almost 50% (Figs. 27 and 29). Meanwhile, several silver grains were observed over the cytoplasm between the Golgi apparatus and cell surfaces (Fig. 29, "Remainder of Cytoplasm"); frequently these grains occurred over smooth-surfaced vesicles measuring about 100 nm or more in diameter (Fig. 27, arrows). (These may be the same vesicles which are stained by carbohydrate detection techniques.) Label also appeared at this and later time intervals in dense and multivesicular bodies scattered throughout the cell (Fig. 27, DB and MVB). Even though over 60% of these bodies became labeled (Bennett and Leblond, 1971), they did not account for more than 5% of the total grain count over the cell (Fig. 29, "Lysosomes").

By 4 hours after injection, reaction over the Golgi apparatus of epithelial cells of the intestine had further decreased, and now only 8% of the silver grains occurred over this organelle (Fig. 29). The lateral and apical cell surfaces, on the other hand, were intensely labeled and together accounted for approximately 70% of the total label. Interestingly enough, the intensity of labeling of the apical microvillar plasma membrane differed markedly from cell to cell. Close inspection showed that the sites at which the radioautographic reaction changed in intensity coincided with the lateral limits of individual cells (Fig. 28). Thus, adjacent cells may differ in their functional status. The fact that the lateral and basal cell surfaces, as well as the apical surface, became labeled indicates that the cell coat material of the entire cell surface, whether visualized as a fuzzy coat in ordinary micrographs or unseen, has a similar origin in the Golgi apparatus.

To summarize, the above results indicate that the protein moieties of cell coat glycoproteins are formed in the rough endoplasmic reticulum. Mannose, a sugar located near the base of the oligosaccharide side chains, is added at this site. Sugars nearer the end or at the end of the side chains, such as galactose and fucose, are incorporated into the cell coat glycoproteins as they pass through the Golgi apparatus. The completed glycoproteins are then transported to the cell surfaces, probably by means of cytoplasmic vesicles pinched off Golgi saccules. The vesicles then open at the cell surface by exocytosis so as to donate their glycoprotein-lined wall to the plasma membrane.

These conclusions are in line with results obtained in investigations of the formation of the thyroid glycoprotein, thyroglobulin, by biochemical methods (Spiro and

---

390X. The Golgi localized reactions are no longer seen, while the reaction along the sinusoidal surfaces of the cells is intense.

Figure 24. Capillaries in cerebral nervous tissue, 10 minutes after injection. Colloidal iron and Feulgen stain; 16-week exposure. 460X. Paranuclear reactions occur over the endothelial cells of large and small capillaries (arrows).

Figure 25. Capillaries in cerebral nervous tissue, 4 hours after injection. Hematoxylin and eosin stain; 10-week exposure. 480X. Surface reaction occurs over the endothelial cells (arrows).

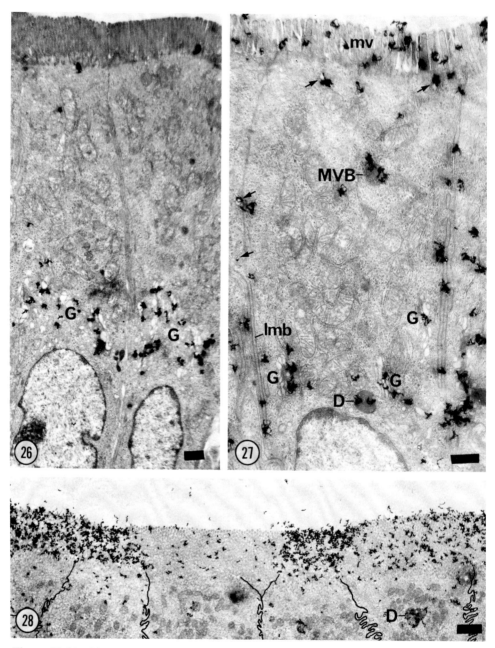

Figures 26–28. Electron microscopic radioautographs of rat duodenal villus columnar cells after ³H-fucose injection. Stained with uranyl and lead.

Figure 26. Two minutes after injection; 7-week exposure. 5000X. The great majority of silver grains are localized over the saccules of the Golgi apparatus (G) located above the nuclei. Only occasional silver grains are seen outside of the Golgi region.

Figure 27. One hour after injection; 7-week exposure. 7000X. Silver grains occur not only over the Golgi apparatus (G) but also over the lateral cell membranes (lmb) and the apical microvillus border (mv). In addition, two lysosomes, a dense body (D) and a multivesicular body (MVB), are both labeled. Of the silver grains in the remaining cytoplasm, several are

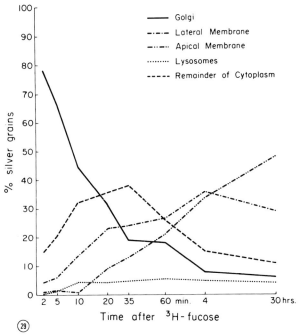

Figure 29. Distribution of silver grains over columnar cells of a duodenal villus at various time intervals after an injection of ³H-fucose.

Spiro, 1966; Spiro, 1970; Herscovics, 1970) and by radioautography (Whur et al., 1969), according to which the polypeptide chain is made first and, as it migrates through the cells, collects the sugar residues of the side chains one by one.

ELABORATION OF CELL COAT GLYCOPROTEINS IN OTHER CELL TYPES

The duodenal villus columnar cell provided an excellent model for cell coat synthesis, since this cell was known to have a thick, well-stained, apical cell coat and it did not actively produce other secretions. However, a cell coat is seen to surround all cell types, and it was of great interest to see if the pattern of cell coat synthesis seen in duodenal columnar cells was a standard one. Light microscopic radioautographic studies showed that cells in most other tissues exhibited the ability to rapidly incorporate exogenous ³H-fucose into glycoprotein in a manner similar to that observed in duodenal columnar cells. Thus, in rats sacrificed early,

associated with small cytoplasmic vesicles (arrows). [From G. Bennett and C. P. Leblond. *J. Cell Biol.* **51**, 875. (1971).]

Figure 28. Four hours after injection; 14-week exposure. The lateral membranes (lmb) of the cells have been darkened with ink for easier visualization. 5000X. The intensity of reaction over the microvilli (mv) is uniform within any given cell but differs dramatically in different cells. (D: dense body.) [From G. Bennett, C. P. Leblond, and A. Haddad, *J. Cell Biol.* 1974, **60**, 258.]

2–10 minutes, after intravenous injection of ³H-fucose, most cells exhibited a discrete paranuclear reaction (Figs. 17, 21, 24), which in the electron microscope was localized over Golgi saccules (Fig. 30). A similar Golgi localization of early reactions after ³H-fucose injection was also reported in thyroid follicular cells (Haddad et al., 1971), hepatocytes (Bennett and Leblond, 1971), and odontoblasts (Weinstock et al., 1972). These observations were interpreted as indicating that, in these various cells, fucose is incorporated into glycoproteins in the Golgi apparatus, a conclusion which received support from the finding in HeLa cells of fucosyltransferase within the smooth microsome fraction (Bosmann et al., 1968), which presumably includes Golgi membranes.

As time elapsed after ³H-fucose injection, all cells examined showed a decrease in the amount of Golgi label, while label appeared in other parts of the cells, indicating that the newly formed glycoproteins had migrated. In many of these cells, the label was relocated along *cell surfaces* in a manner similar to that observed in duodenal columnar cells, indicating that glycoproteins completed in the Golgi apparatus had migrated to the cell coat. There was considerable variation among the different cell types in regard to the amount and migration rate of the label. Thus, the intensity of the paranuclear Golgi reactions was high in some cells (e.g., intestinal epithelium) and weak in others (e.g., lymphocytes) with all intermediates between these extremes. Furthermore, the surface became labeled by 20 minutes after injection in intestinal cells, but only after several hours in some other cells, such as those of the seminiferous epithelium.

A few examples may be presented. In membranes of stratified epithelia small paranuclear grain clusters were observed at early time intervals (Fig. 17), whereas at later time intervals the grains became almost entirely localized at the periphery of the cells (Fig. 18).

In hepatocytes, at the earliest time intervals, grain clusters were located near bile canaliculi (Fig. 21). This location is commonly occupied by the Golgi apparatus, and electron microscopic radioautographic studies (Bennett and Leblond, 1971) revealed a localization of the silver grains at the Golgi saccules. By 20 minutes after injection, a moderate number of grains (Fig. 22) and, by 4 hours, a large number (Fig. 23) occurred over the cell surfaces facing the sinusoids.

Even squamous epithelial cells such as capillary endothelial cells exhibited definite paranuclear localization of silver grains at 10 minutes after injection (Fig. 24); by 4 hours the silver grains covered the entire luminal surface of the cells (Fig. 25).

When the cells of kidney tubules were examined by electron microscopic radioautography, a reaction first appeared over the Golgi saccules in the cells of proximal and distal convulted tubules (Fig. 30). Later, most silver grains occurred over the apical and lateral membranes (Fig. 31); the translocation took about 30 minutes in distal tubules and 4 hours in proximal tubules.

Thus it would appear that, while every cell is surrounded by a cell coat, the rate at which glycoprotein is added to the cell coats covering different cells or even different surfaces of the same cell varies very considerably.

Not all cells exhibited heavy labeling of their surfaces, however. In many secretory cells, the label was relocated in *secretory products*, such as mucus, thyroid colloid, bone and cartilage matrix, and connective tissue ground substance, rather than over the cell surface, so that a fucose-labeled glycoprotein became part of the secretion of these cells (Bennett et al., 1973).

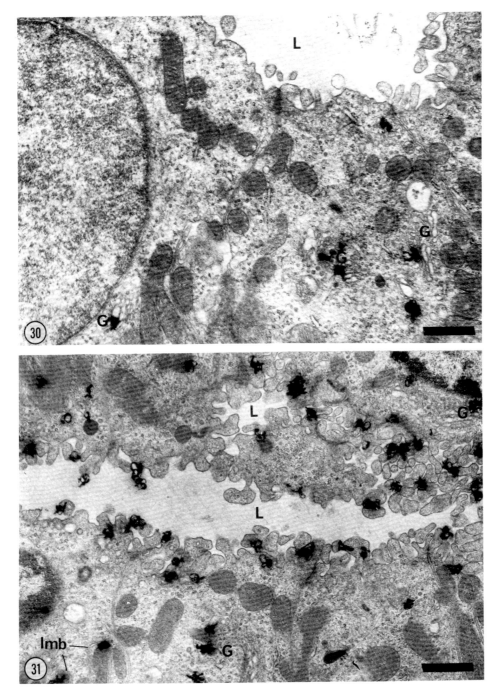

Figures 30–31. Electron microscopic radioautographs of kidney distal convoluted tubule cells after ^{3}H-fucose injection. Stained with uranyl acetate and lead. (L: lumen.)

Figure 30. Ten minutes after injection. 14000X. All but one of the seven silver grains are over, or very close to, Golgi stacks (G). The free surface is unlabeled.

Figure 31. Thirty minutes after ^{3}H-fucose injection. 14000X. Most of the silver grains are associated with the irregular convolutions of the free surface. Some others remain in the Golgi region (G), and two grains are seen over or near a lateral cell membrane (lmb).

TURNOVER OF CELL COAT GLYCOPROTEINS

In the case of a renewing cell population, such as that of the intestinal epithelium, which turns over every 2–3 days, labeling of the cell surface does not necessarily indicate that cell coat components are renewed since, in the young growing cells of these populations, the label may represent a net addition of new material to a growing cell coat. We observed, however, that the intensity of cell surface reaction of intestinal cells at 30 hours after injection was less than that of the corresponding cells at 4 hours; therefore, even during the short life span of these renewing cells, some turnover of cell coat components occurs (Bennett et al., 1974). Biochemical studies have also shown that the glycoproteins of apical microvilli turn over more rapidly than can be accounted for by the turnover of the cells (Alpers, 1972).

In liver, capillaries, and kidney, the cells do not undergo renewal. Thus, even in young animals, most of the cells in these tissues would be mature, so that the growth of their cell coats would have been completed. In such cases, the fact that new glycoproteins are continually added implies a corresponding continual release of the old glycoproteins present. In other words, cell coat components turn over.

Thus, turnover is believed to occur in many renewing as well as nonrenewing cell populations. This conclusion may be related to reports indicating that the membrane components of cultured cells are synthesized at the same rate whether the cells are actively dividing or have stopped doing so (Warren and Glick, 1968; Hughes et al., 1972). This remarkable observation implies that, when cells are growing, membrane components are added mainly for growth; when cells are in a stationary phase, membrane components are still added, but for turnover.

Turnover may be associated with a purposeful secretion of cell coat glycoproteins. Observations of ascites cells in culture have shown that they release glycoproteins to the medium continually (Molnar et al., 1965; Kraemer, 1967). It is possible that in the body too, cells release glycoprotein to the extracellular space or to a lumen when present. Thus, in liver cells, the radioautographic reaction at the surface facing the sinusoids seemed to be less intense at 30 than 4 hours, perhaps because surface glycoproteins are released to the circulation, where they may constitute some of the serum glycoproteins. In the case of distal and proximal tubule cells of rat kidney, the luminal urine may sweep along material from the coat of the microvilli and thus acquire the carbohydrates observed in bladder urine by Maury (1971) and by Lote and Weiss (1971). In the small intestine, the turnover of the glycoprotein hydrolases present at the surface of microvilli (Alpers, 1972) may result from their removal by the passage of chyme. In fact, one of these hydrolases, alkaline phosphatase, has been identified in duodenal juice (Warnes, 1972).

Some loss of glycoproteins may also result from their participation in surface activities, such as intercellular adhesion, uptake of antigens, interaction with viruses and other noxious agents, active transport across the membrane, pinocytosis, and phagocytosis (Winzler, 1970; Hughes, 1973). For instance, adhesion to other cells (or, conversely, the breakdown of the bonds required for the mobility of motile cells) may consume glycoproteins. Pinocytosis and phagocytosis require the formation of vesicles which are derived from the plasma membrane and carry off cell coat glycoproteins. The glycoproteins thus lost have to be replaced. The procedure evolved by nature is to ensure a constant supply of new glycoproteins and, in this way, anticipate their loss or damage. It was recently shown that the rate of formation of surface glycoproteins in ascites cells remains constant, whether the surface sialic acid is intact or has been removed by neuraminidase (Hughes et al., 1972). Thus, regardless of whether or not surface glycoproteins have been damaged or lost, their renewal occurs continually.

SUMMARY

Histochemical stains for carbohydrate, such as periodic acid-Schiff, which specifically reveals 1,2-glycol groups, and colloidal iron, which reveals acidic groups, stain a layer located on the outer surface of the plasma membrane in all examined cells of the rat. Since both 1,2-glycols and acid groups are present in the externally protruding oligosaccharide side chains of the glycoproteins of the plasma membrane, the stained layer, referred to as the cell coat, is attributed to these side chains.

The elaboration of cell coat glycoproteins was examined by radioautography of the intestinal epithelium after injection of labeled precursors: ^{3}H-proline for the tracing of the protein moiety, and ^{3}H-mannose, ^{3}H-galactose, and ^{3}H-fucose for the tracing of the carbohydrate side chains. The evidence suggests that the protein moiety is synthesized on ribosomes and migrates through endoplasmic reticulum, where mannose is collected, and through the Golgi apparatus, where galactose and fucose are collected. The completed glycoprotein is then transported to the cell surface, apparently by vesicles which donate their wall and its glycoprotein lining to the plasma membrane.

A radioautographic survey of most rat tissues after ^{3}H-fucose injection showed uptake of this precursor by the Golgi apparatus and, in many cell types except some secretory cells, passage to the cell surface. These observations show that, in many cells, cell coat glycoproteins are continually supplied to the cell surface from the Golgi apparatus, although at variable rates in different cell types.

The continuous addition of glycoproteins to the cell surface, especially in nonrenewing cells, such as those of kidney and liver, indicates continuous turnover of cell coat glycoproteins.

ACKNOWLEDGMENT

This work was carried out with the support of grants from the Medical Research Council and the National Cancer Institute of Canada.

REFERENCES

Alpers, D. (1972). The relation of size to the relative rates of degradation of intestinal brush border proteins. *J. Clin. Invest.* **51**, 2621–2630.

Bairatti, A., and F. E. Lehmann (1954). Structural and chemical properties of the plasmalemma of *Amoeba proteus. Exp. Cell. Res.* **5**, 220–233.

Bekesi, J. G., and R. J. Winzler (1967). The metabolism of plasma glycoproteins: Studies on the incorporation of L-fucose-l-^{11}C into tissue and serum in the normal rat. *J. Biol. Chem.* **242**, 3873–3879.

Benedetti, E. L., and P. Emmelot (1967). Studies on plasma membranes. IV. The ultrastructural localization and content of sialic acid in plasma membranes isolated from rat liver and hepatoma. *J. Cell. Sci.* **2**, 499–512.

Bennett, G. (1970). Migration of glycoprotein from Golgi apparatus to cell coat in the columnar cells of the duodenal epithelium. *J. Cell Biol.* **45**, 668–673.

Bennett, G., and C. P. Leblond (1970). Formation of cell coat material for the whole surface of columnar cells in the rat small intestine, as visualized by radioautography with L-fucose-³H. *J. Cell Biol.* **46**, 409–416.

Bennett, G., and C. P. Leblond (1971). Passage of fucose-³H label from the Golgi apparatus into dense and multivesicular bodies in the duodenal columnar cells and hepatocytes of the rat. *J. Cell Biol.* **51**, 875–881.

Bennett, G., C. P. Leblond, and A. Haddad (1974). Migration of glycoprotein from the Golgi apparatus to the surface of various cell types as shown by radioautography after labelled fucose injection into rats. *J. Cell Biol.* **60**, 258–284.

Bennett, H. Stanley (1963). Morphological aspects of extracellular polysaccharides. *J. Histochem. Cytochem.* **11**, 14–23.

Bosmann, H. B., A. Hagopian, and E. H. Eylar (1968). Glycoprotein biosynthesis: the characterization of two glycoprotein:fucosyltransferases in HeLa cells. *Arch. Biochem. Biophys.* **128**, 470–481.

Bosmann, H. B., A. Hagopian, and E. H. Eylar (1969). Cellular membranes: the biosynthesis of glycoprotein and glycolipid in HeLa cell membranes. *Arch. Biochem. Biophys.* **130**, 573–583.

Burgos, M. H. (1964). Uptake of colloidal particles by cells of the caput epididymis. *Anat. Rec.*, **148**, 517–525.

Chambers, R. (1940). The relation of the extraneous coats to the organization and permeability of cell membranes. *Cold Spring Harbor Symp. Quart. Biol.*, **8**, 144–153.

Forstner, G. (1969). Surface sugar in the intestine. *Am. J. Med. Sci.*, **288**, 172–180.

Forstner, G. (1971). Release of intestinal surface membrane glycoproteins associated with enzyme activity by brief digestion with papain. *Biochem. J.*, **121**, 781–789.

Gaitonde, M., D. Dahl, and K. Elliot (1965). Entry of glucose carbon into amino acids of rat brain and liver *in vivo* after injection of uniformly ¹⁴C-labeled glucose. *Biochem. J.* **94**, 345–352.

Gasic, G., and L. Berwick (1962). Hale stain for acid containing mucins. Adaptation to electron microscopy. *J. Cell Biol.*, **19**, 223–228.

Haddad, A., M. D. Smith, A. Herscovics, N. J. Nadler, and C. P. Leblond (1971). Radioautographic study of *in vivo* and *in vitro* incorporation of fucose-³H into thyroglobulin by rat thyroid follicular cells. *J. Cell Biol.* **49**, 856–882.

Herscovics, A. (1970). Biosynthesis of thyroglobulin: incorporation of ³H-fucose into proteins by rat thyroids *in vitro*. *Biochem. J.* **117**, 411–413.

Hughes, R. C. (1973). Glycoproteins as components of cellular membranes. *Prog. Biophys. Mol. Biol.* **26**, 191–268.

Hughes, R., B. Sanford, and R. Jeanloz (1972). Regeneration of the surface glycoproteins of a transplantable mouse tumor cell after treatment with neuraminidase. *Proc. Nat. Acad. Sci. U.S.A.* **69**, 942–945.

Ito, I. (1965). Radioactive labelling of the surface coat on enteric microvilli. *Anat. Rec.* **151**, 489 (abstr.).

Ito, S. (1969). Structure and function of the glycocalyx. *Fed. Proc.* **28**, 12–25.

Kraemer, P. M. (1967). Regeneration of sialic acid on the surface of Chinese hamster cells in culture. II. Incorporation of radioactivity from glucosamine-1-¹⁴C. *J. Cell Physiol.* **69**, 199–208.

Kraemer, P. M., and R. A. Tobey (1972). Cell cycle dependent desquamation of heparin sulfate from the cell surface. *J. Cell Biol.* **55**, 713–717.

Leblond, C. P. (1950). Distribution of periodic acid-reactive carbohydrates in the adult rat. *Am. J. Anat.* **86**, 1–25.

Leblond, C. P., R. E. Glegg, and D. Eidinger (1957). Presence of carbohydrates with free 1,2-glycol groups in sites stained by the periodic acid-Schiff technique. *J. Histochem. Cytochem.* **5**, 445–458.

Lote, C. J., and J. B. Weiss (1971). Identification of digalactosylcysteine in a glycopeptide isolated from urine by a new preparative technique. *FEBS Lett.* **16**, 81–85.

Luft, J. H. (1971). Ruthenium red and violet. II. Fine structural localization in animal tissues. *Anat. Rec.* **171**, 369–416.

Maury, P. (1971). Neuraminyl-oligosaccharides of rat urine: Isolation and some structural characteristics. *Biochim. Biophys. Acta* **252**, 48–57.

Molnar, J., D. W. Teegarden, and R. J. Winzler (1965). The biosynthesis of glycoproteins. VI. Production of extracellular radioactive macromolecules by Erlich ascites carcinoma cells during incubation with glucosamine-¹⁴C. *Cancer Res.* **25**, 1860–1866.

Moscona, A. (1952). Cell suspensions from organ rudiments of chick embryos. *Exp. Cell Res.* **3**, 535–539.

Mowry, R. (1963). The special value of methods that color both acidic and vicinal hydroxyl groups in the histochemical study of mucins: With revised directions for the colloidal iron stain, the use of alcian blue G8X, and their combination with the periodic acid-Schiff reaction. *Ann. N.Y. Acad. Sci.* **106**, 402–423.

Neutra, M., and C. P. Leblond (1966). Radioautographic comparison of the uptake of galactose-³H and glucose-³H in the Golgi region of various cells secreting glycoproteins or mucopolysaccharides. *J. Cell Biol.* **30**, 137–150.

Puchtler, H., and C. P. Leblond (1958). Histochemical analysis of cell membranes and associated structures as seen in the intestinal epithelium. *Am. J. Anat.* **102**, 1–24.

Rambourg, A. (1969). Localization ultrastructurale et nature du matériel coloré au niveau de la surface cellulaire par le mélange chromique-phosphotungstique. *J. Microscop.* **8**, 325–342.

Rambourg, A. (1971). Morphological and histochemical aspects of glycoproteins at the surface of animal cells. *Int. Rev. Cytol.* **31**, 57–114.

Rambourg, A., and C. P. Leblond (1967). Electron microscope observations on the carbohydrate-rich cell coat present at the surface of cells in the rat. *J. Cell Biol.* **32**, 27–53.

Rambourg, A., M. Neutra, and C. P. Leblond (1966). Presence of a "cell coat" rich in carbohydrate at the surface of cells in the rat. *Anat. Rec.* **154**, 41–71.

Rambourg, A., W. Hernandez, and C. P. Leblond (1969). Detection of complex carbohydrates in the Golgi apparatus of rat cells. *J. Cell Biol.* **40**, 395–414.

Singer, S. J., and G. L. Nicolson (1972). The fluid mosiac model of the structure of cell membranes. *Science* **175**, 720–731.

Spiro, R. (1970). Glycoproteins. *Ann. Rev. Biochem.* **39**, 599–638.

Spiro, R., and M. J. Spiro (1966). Glycoprotein biosynthesis: Studies of thyroglobulin. Characterization of a particulate precursor and radioisotope incorporation by thyroid slices and particle systems. *J. Biol. Chem.* **241**, 1271–1282.

Warnes, T. W. (1972). Alkaline phosphatase. *Gut* **13**, 926–937.

Warren, L., and M. Glick (1968). Membranes of animal cells. II. The metabolism and turnover of the surface membrane. *J. Cell Biol.* **37**, 729–746.

Weinstock, A., M. Weinstock, and C. P. Leblond (1972). Radioautographic detection of ³H-fucose incorporation into glycoprotein by odontoblasts and its deposition at the site of the calcification front in dentin. *Calc. Tissue Res.* **8**, 181–189.

Whur, P., A. Herscovics, and C. P. Leblond (1969). Radioautographic visualization of the incorporation of galactose-³H and mannose-³H by rat thyroids *in vitro* in relation to the stages of thyroglobulin synthesis. *J. Cell Biol.* **43**, 289–311.

Winzler, R. (1970). Carbohydrates in cell surfaces. *Int. Rev. Cytol.* **29**, 77–125.

Wislocki, G. B., D. W. Fawcett, and E. W. Dempsey (1951). Staining of stratified squamous epithelium of mucous membranes and skin of man and monkey by the periodic acid-Schiff method. *Anat. Rec.* **110**, 359–375.

CHAPTER FOUR

Some Aspects of Cellular Interactions in Development

JEAN-PAUL REVEL

Our understanding of the processes which shape the embryo represents the synthesis of many lines of investigation. These range from such classical approaches as three-dimensional reconstructions from serial sections to more recent studies involving the use of autoradiography and surgical transplantation, of cell dissociation and reaggregation *in vitro*, of cell sorting experiments, or of time lapse cinematography. (See De Haan, 1968; Vaughan and Trinkaus, 1966; Trinkaus, 1969; Bellairs et al., 1969; Bellairs, 1971; Steinberg, 1964; Garber and Moscona, 1971.) It appears clear from these and many other investigations that the carefully orchestrated cell and tissue movements are controlled to some extent by the contacts and junctions that cells have with each other and/or with their substrata. A detailed study of the morphology of these interactions requires a fairly high resolution, often beyond what is available with the light microscope. Yet until recently use of the transmission electron microscope was limited because of the difficulties imposed by the extremely thin sections required. The complex shape of the cells makes it difficult to study the relationships that they have with their neighbors. Even when cell junctions have been found, they have been difficult to identify clearly, because they are often so small that the usual criteria for recognition are difficult to apply.

Two new techniques now available will, I believe, help to overcome these problems. On the one hand, freeze-etching will permit detailed study of cell junctions because it reveals large portions of the cell membrane and unambiguous characterization of the junctions. On the other hand, the scanning electron microscope is coming into more general use and will provide data on the three-dimensional organization of embryos in a way that we are only beginning to appreciate (Waterman, 1972). We will show examples of the use of these techniques in studying the migration of cell sheets and their interactions with their surrounding tissues. We will then go on to describe cell junctions in the young embryo and to compare their genesis with the process of junction formation as seen in regenerating liver and in reaggregating hepatoma cells. Finally, we will report on some studies of adhering junctions in the frog embryo.

SCANNING ELECTRON MICROSCOPY OF YOUNG CHICK EMBRYOS

The scanning electron microscope can be used to particularly good advantage if the embryos are fractured, either before or after critical-point drying (Waterman, 1972). The broken edges reveal internal structures which can easily be related to surface topography (Fig. 1). Much greater detail is obtainable, either in sections or in whole mounts, than with the light microscope. Resolution as good as 100Å can be reached under optimal conditions while retaining a great depth of field. The three-dimensional information conveyed by scanning electron microscope (SEM) pictures (either directly or, better, through stereo pairs) can be appreciated at the light microscope level only by changing focus or through a reconstruction based on transmission electron micrographs. We have found the SEM useful in studies aimed at establishing how much of a role the phenomenon of contact guidance (Weiss, 1961), so well studied in *in vitro* systems, plays in the *in vivo* situation. Cells or sheets of cells could be guided either by features such as ridges, grooves, or other structural attributes of the substrate upon which they move (De Haan, 1968), or, on the contrary, by recognition of molecular features (Konigsberg and Hauschka, 1965), as may be the case for cell-cell recognition (see Garber and Moscona, 1971; Oppenheimer and Humphreys, 1971; Roth et al., 1971). In the latter case, trails guiding cell movement would be detectable only by mapping the distributions of various chemical residues. While the latter information is for the moment just out of reach, the SEM already allows us to make a preliminary test of the first version of the hypothesis. In examining this problem we decided to concentrate on the spreading of the primary mesenchyme between epiblast and hypoblast and on the movement of the future endothelial cells across the back of the young cornea (Hay and Revel, 1969).

Migration of the Mesoderm

After leaving the surface of the embryo at the level of the primitive streak, the future mesodermal cells migrate laterally. Francis Janssen and I have been trying to find out whether structural features of the underside of the epiblast or the inner surface of the hypoblast could serve as guides for this orderly migration. To this end we tried to dissect embryos, removing either the epiblast or the hypoblast, to reveal regions of the zona pellucida where the migrating mesoderm had not yet arrived. On the inner surface of the hypoblast, that is, the surface in contact with the advancing mesenchymal cells, one sees the boundary between hypoblast cells (Fig. 2). No collagen fibrils or basement membrane can be detected. The cells are variable in shape, have many overlapping processes, and form a rather poorly organized epithelial layer. In fact, it is common to find that the cell sheet has large

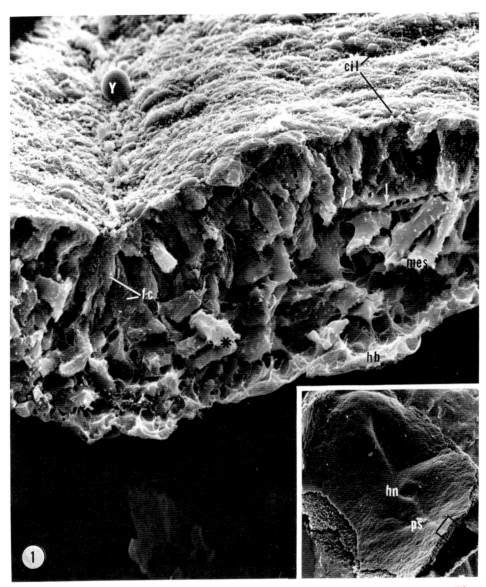

Figure 1. Scanning electron micrographs of a critical-point dried chick embryo, stage 5. The inset shows the whole embryo, including primitive streak (ps), Hensen's node (hn), and head process. The embryo was purposely broken to reveal internal structures, and the fracture can be seen at the lower right. A foreshortened view of an area similar to the one indicated by the square is shown in the main part of the illustration. In spite of difficult optical conditions, one can see that many of the epiblast cells carry a centrally placed cilium (cil). The primitive streak itself is seen as a well marked groove which is typically lined by blebs and ruffles. A yolk droplet has fallen into the groove and is seen as a large sphere (Y). The elongated cells at the level of the primitive streak represent flask cells (fc). They have a relatively narrow neck and a wider cell body. Underneath this region the cells gradually acquire a mesenchymal appearance. A freeze-cleave replica of a cell similar in location to the starred cell is shown in Fig. 7. Most cells have become frankly mesenchymal in appearance at the level of the sharp boundary between the epiblast and the mesoblast (arrows). The hypoblast (hb) forms the very thin cellular layer which limits the underside of the embryo. Note the complexity of its inner surface, and the contacts made by cells of the mesoblast with both epiblast and hypoblast inner surfaces.

53

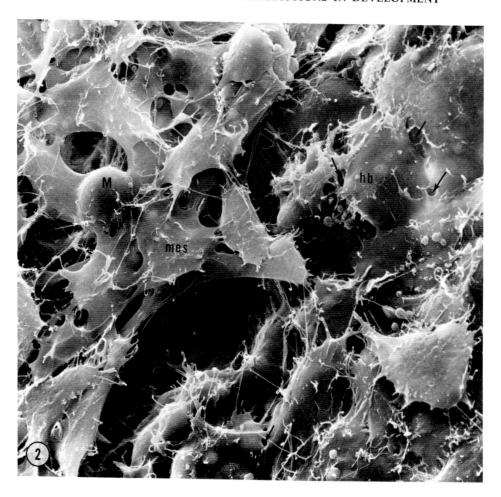

Figure 2. Scanning electron micrograph of the inner surface of the hypoblast, partially covered by the mesoblast. This view was taken in the lateral mesenchyme of a stage 4 embryo and may represent part of the advancing edge of the mesoblast. There are many open spaces (arrow) between hypoblast (hb) cells and much intertwining of processes. The polygonal appearance of the mesenchyme (mes) cells can clearly be recognized, and one can observe many intercellular contacts between cells. The cells appear to be relatively scalelike and rather more "two-dimensional" than had been expected. One of the cells (M) is rounded up, perhaps in preparation for mitosis. No obvious morphological pattern is seen on the hypoblast which could serve to guide the movements of the mesoblast over the hypoblast. Contacts between hypoblast and mesoblast cells, however, can be observed at many places.

holes in it; and when it is viewed from the surface facing the yolk, one can see the mesenchyme through the holes. There are few, if any, obvious structures on the early hypoblast which could act as guides to cellular movements. The underside of the epiblast has more "texture" (Fig. 3). A striking feature here is a gentle undula-

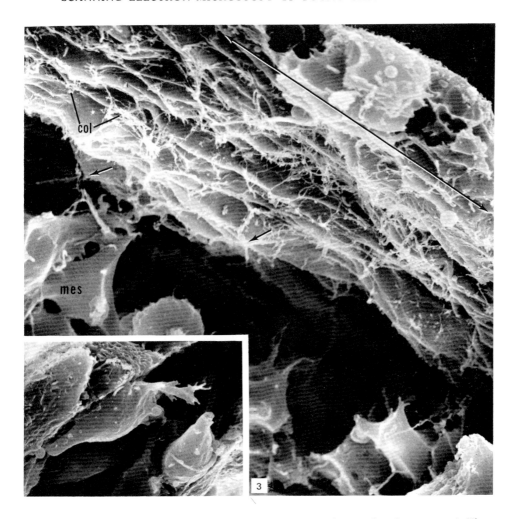

Figure 3. A view of the underside of the epiblast in a cross-fractured embryo, stage 4. The inset shows a face-on view of the cross fracture so that the boundary between the epiblast and the mesoblast can be clearly seen. This preparation shows a mesenchymal cell, apparently firmly anchored to the underside of the epiblast and sending out pseudopodia, presumably as it searches out another site to attach to and migrate laterally. The next attachment point could presumably be either on the inner surface of the hypoblast or epiblast or on another mesenchymal cell. In the main part of the illustration, the embryo has been tilted so as to reveal the underside of the epiblast. One can see that the undersurface is not smooth but shows a series of low rolling hills or ridges. At intervals one observes sharp peaks or promontories (arrows). Numerous fine fibrils (col), presumably collagen, coat the underside of the epiblast. The stellate or polygonal mesenchymal (mes) cells seem to attach to the underside of the epiblast at the level of some of the projections. No pattern emerges which could easily be interpreted as the guiding cells, but examination of many specimens may reveal interactions between the two germ layers at specific points.

tion of the underside of the epithelium, which is broken by sharp conical invaginations where fibroblasts appear to be attached to the base of the epithelium. Whether these sharp invaginations are the result of a mechanical pull where fibroblasts attach, or whether they represent a structural feature of the underside of the epithelium, is not clear at this point. Strands of fine filaments (probably collagen) are applied against the base of the epithelium, and fibroblasts seem at times to associate with these. They may represent the incomplete basement lamina which is seen in sections (Trelstad et al., 1967). The relationship is not so constant, however, as to suggest a definite role of these collagen strands in guiding movements of the mesenchyme.

Examining sheets of migrating mesenchymal cells seen on the surface of the hypoblast or epiblast is the natural counterpart of having a fibroblast explant on a petri dish. Our observations confirm (see Trinkaus, 1967; Bellairs, 1971) that the ruffles which are so prominent a structure *in vitro* also occur at the leading edges of the cells *in vivo*, although often in a less highly developed form. A particularly good example of a ruffle can be seen in the inset of Fig. 3.

Formation of the Corneal Endothelium

The corneal endothelium forms by the migration of mesenchymal cells on the surface of a collagenous layer which has been shown (Hay and Dodson, 1973) to be secreted by the corneal epithelium. Greg Nelson and I prepared the eyes of chick embryos at various stages of development for scanning electron microscopy. After removing the lens of hemisected eyes, we were able to examine the back side of the future cornea. At the early stages the collagenous corneal stroma can be seen as a tightly knit, three-dimensional network of very fine fibrils (Fig. 4, upper inset). There is no indication of the orthogonal array which appears so obvious in sections (Hay and Revel, 1969), probably because the most clearly organized layers, those in proximity to the epithelium, are too deeply situated to be easily observable in the scanning microscope. All the mesenchymal cells are found near the edge of the optic cup, barely distinguishable underneath a network of very fine collagen fibrils. At stage 21 the only cell found on the smooth undersurface of the corneal stroma was an occasional macrophage (Fig. 4, upper inset). At slightly later stages (stages 23–25) the corneal stroma loses its smooth surface to become a wavy structure (Fig. 4), as already observed with the light microscope.

On the basis of the limited number of embryos studied in detail, we believe that the major axis of the waves is parallel to the nasotemporal axis, although the pattern observed seems rather complex. At about stage 23 the mesenchymal cells begin advancing from the edges. It is interesting to speculate that the cells suddenly begin secretion of a collagenase (Gross and Lapiere, 1962), which could allow them to move through a retaining net of collagen. The cells advance as a sheet, converging toward the center of the stroma. The front itself is somewhat irregular in shape, and there are often cells or small clumps of cells which apparently outrun their neighbors (Fig. 4). Ruffles and other indications of membrane activity can be seen in cells behind the advancing front when they do not contact each other. At short distances behind the leaders there seems to be an active secretion of collagen, to judge from the thick net of irregularly arranged fibrils which soon appears. Cells that are at the advancing edge frequently have ruffles or (just as often) lobopodia (Fig. 4, lower inset). The ruffles are much less imposing, however, than those formed *in vitro* by chick heart fibroblasts (Abercrombie et al., 1970). In the latter case, no lobopodia are reported, although they have been described as the organ of locomotion in other systems (Trinkaus, 1973).

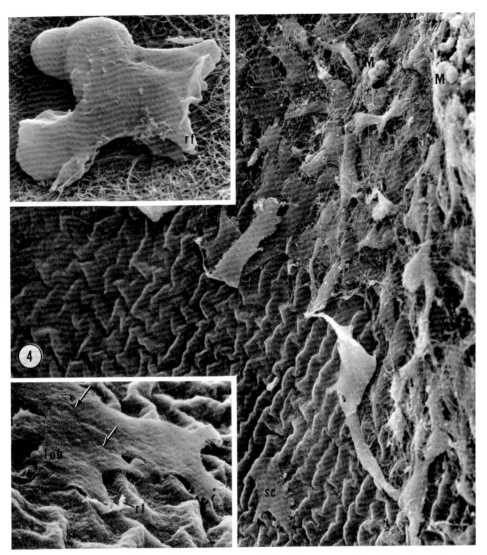

Figure 4. The back surface of the cornea in a chick embryo, stage 23, during the migration of the mesenchymal cells which are destined to form the corneal endothelium. The back surface of the cornea consists of a primary stroma, which is wrinkled in a characteristic pattern of ridges and grooves intersecting at more or less right angles to each other. This waviness cannot be observed at earlier stages, as is illustrated in the upper inset, which shows the primary stroma at a higher magnification in an embryo, stage 20. One can clearly recognize the fine collagen fibrils which form the dense network of the primary stroma. At this stage the back of the cornea is smooth. The only cells found are, like the one illustrated here, macrophages which display anterior ruffles (rf). The migrating mesenchymal cells in the stage 23 embryo, however, have both ruffles and lobopodia. The mesenchymal cells seem to advance as a broad front, but occasionally scouts (sc), either single cells or clumps of closely packed cells, which outrun the front, can be seen. The lower inset shows a highly oblique view of the advancing edge of the endothelial sheet. The cells are in very close contact with each other (arrows). Pseudopodia with ruffles (rf) or lobopodia (lob) appear to feel out the substratum, but the cells do not really follow the grooves or ridges; they often bridge right over the features of the terrain. Although the cells at the very edge of the advancing front appear to be essentially free of collagen, many new collagen fibrils, coarser than those of the primary stroma, are found at short distances behind the advancing edge. Two pairs of dividing cells are seen at (M).

The forward processes of the leading mesenchymal cells seem to search for and follow the grooves of the corneal stroma (Fig. 4a, lower inset). This would suggest that contact guidance may indeed play a role here. Although the cell processes appear to search out the topography of the terrain upon which they migrate, there is no clear orientation or arrangement of the cell bodies with respect to the underlying substratum (Fig. 4, lower inset). We must conclude that, while our evidence suggests that cells *in vivo* recognize roughness in terrain, it is not clear whether this represents true contact guidance.

CELL JUNCTIONS IN THE EMBRYO

The control of cellular movements and the recognition mechanisms which allow the formation of orderly patterns in the embryo are mediated in part by cell interactions. These could take many forms, among which one can count the cell-substrate relationships described in the preceding section, as well as the actual cell-cell contacts that we will now discuss. Our understanding of cell junctions has been greatly advanced in recent years through the use of freeze-cleaving, a technique which has allowed unequivocal characterization of the various types of junctions. In this portion of our paper we will first describe tight junctions in the embryo and follow this by a discussion of gap junctions and of desmosomes.

Embryonic Tight Junctions

Tight junctions in the adult are clearly involved as transepithelial permeability barriers. They seal the juxtaluminal intercellular space in epithelia and are seen as a series of punctate fusions of adjacent membranes (Farquhar and Palade, 1963) in sections. In views of the membrane revealed by freeze-cleaving (Fig. 5) one finds anastomosing "ropes" on the A face of the membrane or a corresponding network of grooves on the B face (Stahelin et al., 1969). We have found that tight junctions are already well developed in the epiblast of embryos as young as stage 4–6 (Revel et al., 1973). It is presumed that in the embryo, as in the adult, such tight junctions play a role in separating the external medium from the internal environment.

There are regions in the embryos where the presence of tight junctions can be expected to cause special problems. These are the regions where the epiblast invaginates to form some of the internal tissues of the embryo, such as in the early Hensen's node or at the level of the primitive streak. Examination in the SEM of cross fractures of early embryos strikingly illustrates the process of invagination (Fig. 1). Near the midline of the embryo the primitive streak appears as a groove. The epithelial cells at the level of the streak are elongated, forming flask cells which leave the surface to penetrate into the inside of the embryo. The lack of obvious processes by which the cells might pull themselves along would suggest that any adhesions necessary to allow them to move are those they already have with their neighbors. There is no obvious formation of pseudopodia, such as are described in sea-urchin embryo gastrulation by Gustafson and Wolpert (1967). In the process of invagination the cells lose contact with the surface with the result that existing zonulae occludentes between erstwhile neighbors are disrupted. Further changes in the appearance of the junctions take place as the cells acquire the typically mesenchymal stellate appearance. The "transformation" of epitheloid cells into "mesenchymal" cells is fairly abrupt and can be seen to occur at the same level as a sharp epithelial lower boundary forms (Fig. 1).

When cells near the midline which have lost their contact with the surface are examined by freeze-etching, one finds that the tight junctions are still present, but now, instead of a continuous network in a well specified locus, islands of disrupted junction are evident on the cell membrane (Fig. 6). As one moves laterally, the remnants of tight junction on the cell membrane of the mesenchymal cells appear to become further simplified and are very often represented only by individual strands (Fig. 7) or "focal tight junctions" (Trelstad et al., 1967). This simplification of tight junction remnants is probably achieved by the separation of previously closely associated cells as they move in different directions or when they divide. Each member of the separating pair would take along with it a certain amount of the junctional material by which the members used to be joined. One would eventually expect only small remnants of junctional material, which would in the end disappear altogether.

If the sole function of the zonula occludens is to act as a permeability barrier, it is clear that incomplete occluding junctions, such as are found in Hensen's node, the primitive streak, and the focal tight junctions in the mesenchyme, cannot perform such a function. Whether the adhesion which is conferred to neighboring cells by the presence of such a junctional remnant is of physiological significance, or whether the tight junction remnants can play a role other than that which is played by the zonula as a whole, cannot be clearly established at this time. Because of the presence of well developed gap junctions in the epiblast and mesenchyme, however, it is believed that the focal tight junctions probably do not play a role in intercellular communication.

Gap Junctions in the Young Embryo

In addition to the focal tight junctions and the zonulae occludentes present in the young embyro (Trelstad et al., 1966), one also finds well developed gap junctions (Fig. 5). In the epiblast of chicks in stages 4–6 these junctions are found near the zonula occludens, as well as elsewhere between the epithelial cells. The epiblast gap junctions have the typical array (Goodenough and Revel, 1970) of close-packed particles on A faces and corresponding pits on B faces. Much smaller gap junctions are also found between mesenchymal cells (Revel, 1972). They are sometimes associated with the remnants of tight junctions described in the preceding paragraph, but this is relatively rare (Fig. 6). In some cases one finds extremely small collections of membrane particles completely independent of other junctional elements. Their appearance suggests forming junctions. There is a typical "halo" around the small clusters like the one that is commonly observed in mature gap junctions (Fig. 8). An interesting feature of these small gap junctions is that one can often find large (10 nm in diameter) membrane particles in their vicinity. Similar membrane particles have now been observed in reaggregating hepatoma cells in regions where the process of gap junction formation can clearly be observed.

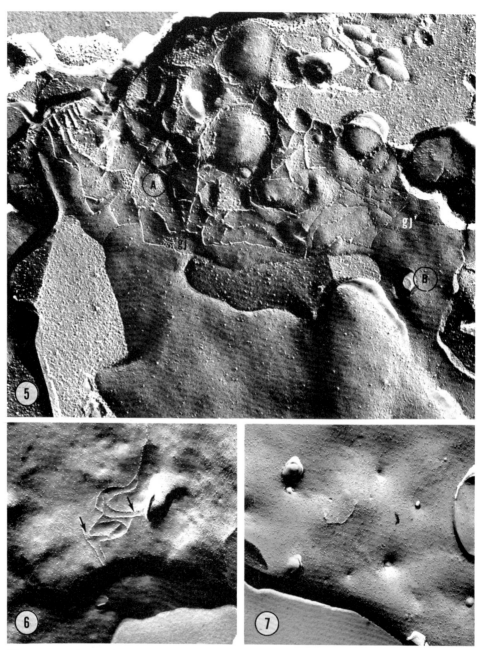

Figure 5. A freeze-cleave preparation of a stage 4 chick embryo, representing the apical tight junction in the epiblast near the primitive streak in a region such as is marked by z.o. in the inset of Fig. 2. The zonula occludens appears as anastomosing ropes on A faces (A) and as corresponding grooves on B faces (B) of the cell membranes. Small gap junctions, seen to be clusters of particles (gj) or pits (gj'), depending on the membrane face observed, are interspersed with the zonula occludens. The appearance of these embryonic junctions is very similar to that seen in adult tissues. The extra embryonic space is seen at the top of the illustration.
Figure 6. An incomplete junction, as seen on the surface of cells which have slipped into the inside of the embryo at the level of the primitive streak or of Hensen's node near the midline of the embryo (star, Fig. 1). One can still recognize the organization of the zonula occludens

Gap Junction Formation

Gap junction formation has been studied in collaboration with Ross Johnson (see Johnson and Sheridan, 1971). Hepatoma cells are dissociated with EDTA and allowed to reaggregate on the bottom of a petri dish. In following the development of electrotonic coupling, one finds that it increases rapidly and reaches control levels in 2 or 3 hours. Freeze-etching at different times during reaggregation reveals the presence of large (10 nm) membrane particles on the A face of the membrane of hepatoma cells. These particles penetrate deeply enough through the membrane to form pits on the B face. The 10-nm membrane particles are at first scattered throughout the cell membrane of the hepatoma cells but later cluster to specific regions of the cell membrane, which become smooth, flat, and devoid of other membrane particles. Aggregates of close-packed particles then appear in the middle of the clusters (Fig. 9). These aggregates are very similar in appearance to small gap junctions and in early stages of formation closely resemble the structures described above in young chick embryos. With time these small gap junctions increase in size, while the number of large particles appears to decrease.

It is not established, however, that there is a precursor-product relationship between the 10-nm particles and the gap junction particles. It is possible to imagine that the 10-nm particles represent individual gap junction particles before they become close packed. Apparent differences in the sizes of individual particles and those that are aggregated in a gap junction may result from geometrical factors which control the disposition of platinum during the shadowing process. On the other hand, they may reflect an actual qualitative discrepancy between the two. There is evidence to suggest that the large particles may play a role in getting cells to contact each other. In fortunate fractures showing neighboring cells, one finds that they are in closest proximity and may even touch in regions where large particles are found. It is tempting to imagine that the 10-nm particles represent the intramembranous locus of a determinant for cell recognition. Only in regions where membrane proximity is established can gap junctions form.

It is of some interest to compare these observations from those that Ann Yee (1972) and I made during a study of gap junction formation in regenerating liver. In this case we did not find the 10-nm particles seen in hepatoma and in chick embryos. The zonula occludens persists during the disappearance of gap junctions which follow hepatectomy, and the newly formed gap junctions appear in close proximity in the zonulae occludentes. Perhaps the latter provide the membrane proximity which is necessary for the establishment of cell junctions. A separate mechanism, such as the appearance of the "large particles," does not have to operate in this case. It would be interesting if one could establish that gap junction

as seen in the epiblast, but it does not form a complete zonule. At the arrows one can just distinguish small gap junctions similar to those seen within the meshes of the intact zonula occludens of the epiblast.
Figure 7. Examination of mesenchymal cells removed from the midline, such as the cells labeled "mes" in Fig. 1, shows remnants of zonulae occludentes in the form of small focal tight junctions. They are similar in appearance to those found in the z.o. but are no longer interconnected. While the examples of incomplete junctions shown in Figs. 6 and 7 represent A faces of the membranes of interiorized cells, examination of B faces would show the corresponding image, that is, cellular grooves of similar morphology.

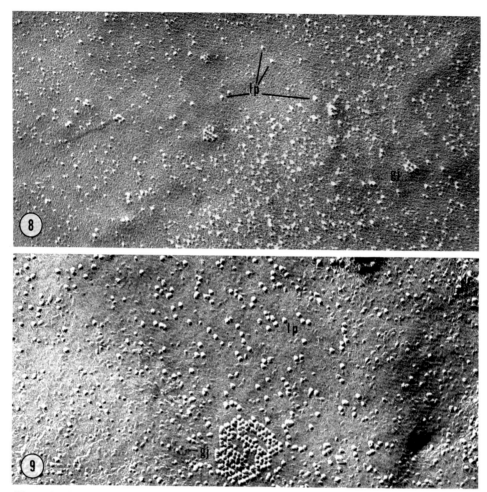

Figure 8. Part of an A face of a cell in the somite region of a young chick embryo, stage 8. Besides the usual membrane particles which are randomly distributed on A faces, one observes some particularly large membrane particles (lp), as well as clusters of close-packed particles surrounded by a particle-poor region of the membrane. These clusters of particles in their surrounding halo are interpreted as gap junctions (gj) in the young embryo. The role and nature of the so-called large particles are discussed in this paper. They may be similar to the 10-nm particles found in reaggregating hepatoma cells.

Figure 9. Portion of an A face of hepatoma cells allowed to reaggregate for 60 minutes. One can see aggregates of 10-nm particles (lp) in regions of the membrane which are poor in the usual membrane particles. The regions where the 10-nm membrane particle aggregates appear are relatively smooth and flat and are postulated to represent zones of adhesion between neighboring cells. Within these zones one sees the formation of clusters of close-packed particles reminiscent of typical gap junctions (gj). It is believed that the processes of gap junction formation observed in hepatoma and in embryos may be very similar.

formation involved a specific recognition mechanism. Since gap junctions are so widely distributed, it might well be advantageous for specificity to be conferred by some other means than the gap junctions themselves, as long as a prerequisite for junction formation was previous recognition and adhesion by a separate mechanism.

Desmosomes in the Embryo

In addition to the mechanisms already mentioned, adhesion and changes in adhesivity between cells during development are believed to play a major role in directing the organization of cells in the embryo (Steinberg, 1964; Wiseman et al., 1972; Trinkaus, 1967). In many cases it is likely that the adhesion (Garber and Moscona, 1971; Roth et al., 1971) is carried out at levels of the cell membrane where no specializations are visible microscopically. One specialized structure which has been known for a long time to be involved in cellular adhesion is the "adhering" junction, particularly the desmosome. Proteolytic enzymes and chelating agents have often been used to dissociate tissues into single cells. The effect of such treatments on embryonic junctions has been discussed, notably by Overton (1968).

Joan Borisenko and I carried out a survey of the stability of desmosomes to different agents (Borisenko and Revel, 1973) and found that, broadly speaking, adhering junctions fall into two categories: those that are sensitive to trypsin or deoxycholate, and those affected by calcium removal. In studying adult tissues, one finds that junctions which are affected in the absence of Ca^{2+} are often those that might be expected to be relatively labile *in vivo*, because they are present in tissues where relationships between cells are constantly changing during cell renewal or migration. Junctions sensitive to the effect of trypsin, on the other hand, are commonly found in tissues where relationships would remain quite stable. Of special interest here are the results obtained in the study of developing stages of *Rana pipiens* larvae. Whereas desmosomes from adult frog skin are sensitive to deoxycholate and trypsin, but not to EDTA, the desmosomes found in the two-cell epithelium of an early larva can be completely dissociated by EDTA but are unaffected by trypsin. The shift to the adult pattern is gradual. Therefore the young larvae have desmosomes which require calcium for their integrity, but at metamorphosis this calcium dependence disappears and is replaced by a trypsin sensitivity. Whether the trypsin-sensitive desmosomes of the adult frog skin are the same desmosomes, modified by the addition or removal of some material, that are found in the embryo, or whether they represent completely different structures, is not clear at this time. It is of interest, however, that the desmosomes display properties which change with the age of the embryo. They could therefore form the basis for some of the changes in intercellular adhesivity which are such prominent features of development.

CONCLUSION

In many instances our perception of the events which take place *in vivo* is the result of extrapolation of data gathered *in vitro*. Only rarely can the same detail be obtained from investigations in the whole animal as is possible with a well controlled model system. This has been an eminently successful approach in our efforts to unravel complex problems. As a corollary, however, we must keep in mind that our understanding of cells is based on an abnormal or extreme behavior, imposed by the artificial conditions prevailing in the experimenter's laboratory. One whose only glimpse of humanity had been gained through the ports of Skylab or Gumdrop would have a rather slanted, if correct, view of a portion of mankind. Some of the techniques that we have used in the work reported here have allowed us to, so to speak, short-circuit the process in a few specific cases. We have been able to "put to the test" the notion that contact guidance may play a role in the young embryo, as it does on a fish scale, and to compare the process of gap junction formation *in vitro* and *in vivo*. While it is too early to make definitive determination of the facts, it is likely that the *in vitro* model, correct in general terms, must not be thought to apply strictly *in vivo*. Obviously refinements of the technique and more intelligent questions could reap rich rewards.

ACKNOWLEDGMENT

I wish to thank my collaborators for all their help and for allowing me to use original unpublished material. This applies particularly to Greg Nelson, Francis Janssen of California Institute of Technology, Ann Yee of Harvard, Joan Borisenko of Tufts, and Ross Johnson of the University of Minnesota. I am also grateful to Pat Koen and Pamela Hoch for their expert help, and to Mike Rees for preparing the photographs. The U. S. Public Health Service supported much of the original work, most recently with Grants GM-06965 and GM-19224.

REFERENCES

Abercrombie, M., J. E. M. Heaysman, and S. M. Pegrum (1970). The locomotion of fibroblasts in culture. *Exp. Cell Res.* **60**, 437–444.

Bellairs, R. (1971). *Developmental Processes in Higher Vertebrates.* Logos Press, London.

Bellairs, R., A. Boyde, and J. E. M. Heaysman (1969). The relationship between the edge of the chick blastoderm and the vitelline membrane. *Arch. Entwicklungs Mech. Org.* **163**, 113–121.

Borisenko, J. Z., and J. P. Revel (1973). Experimental manipulation of desmosome structure. *Am. J. Anat.* (in press).

De Haan, R. L. (1968). Emergence of form and function in the embryonic heart. *Develop. Biol.*, Suppl. 2, 208–250.

Farquhar, M. J., and G. E. Palade (1963). Junctional complexes in various epithelia. *J. Cell Biol.* **17**, 375–412.

Garber, B. B., and A. A. Moscona (1971). Reconstruction of brain tissue from cell suspensions, II. *Develop. Biol.* **27**, 235–243.

Goodenough, D. A., and J. P. Revel (1970). A fine structural analysis of intercellular junctions in the mouse liver. *J. Cell Biol.* **45**, 272–290.

Gross, J. and Lapiere (1962). *Proc. Nat. Acad. Sci. U.S.A.* **48**, 1014.

Gustafson, T., and L. Wolpert (1967). Cellular movement and contact in sea urchin morphogenesis. *Biol. Rev.* **42**, 442–498.

Hay, E. D. and J. W. Dodson (1973). Secretion of collagen by corneal epithelium. *J. Cell Biol.* **57**, 190–213.

Hay, E. D., and J. P. Revel (1969). Fine structure of the developing cornea. In *Monographs in Developmental Biology* (A. Wolsky and P. S. Chen, eds.). S. Karger, Basel, pp. 1–35.

Johnson, R. G., and J. D. Sheridan (1971). Junctions between cancer cells in culture: ultrastructure and permeability. *Science* **174**, 717.

Konigsberg, I., and S. D. Hauschka (1965). Cell and tissue interactions in the reproduction of cell types. In *Reproduction: Molecular, Subcellular and Cellular* (M. Locke, ed.). Academic Press, New York, p. 243.

Oppenheimer, S. B., and T. Humphreys (1971). Isolation of specific macromolecules required for adhesion of mouse tumor cells. *Nature* **232**, 125–127.

Overton, J. (1968). The fate of desmosomes in trypsinized tissues. *J. Exp. Zool.* **168**, 203–214.

Revel, J. P. (1972). In *Comparative Molecular Biology of Extracellular Matrices* (H. C. Slavkin, ed.). Academic Press, New York, pp. 77–139.

Revel, J. P., L. Chang, and P. Yip (1973). Cell junctions in the early chick embryo: A freeze-etch study. *Develop. Biol.* **35**, 302.

Roth, S., E. S. McGuire, and S. Roseman (1971). Evidence for cell-surface glycosyltransferases. *J. Cell Biol.* 51, 536.

Stahelin, L. A., T. M. Mukherjee, and A. W. Williams (1969). Freeze-etch appearance of the tight junctions in the epithelium of small and large intestine of mice. *Protoplasma* **67**, 165–184.

Steinberg, M. S. (1964). The problem of adhesive selectivity in cellular interactions. In *Cellular Membranes in Development* (M. Locke, ed.), Academic Press, New York.

Trelstad, R. D., J. P. Revel, and E. D. Hay (1966). Tight junctions between cells in the early chick embryo as visualized with the electron microscope. *J. Cell Biol.* **31**, C_6-C_{10}.

Trelstad, R. L., E. D. Hay, and J. P. Revel (1967). Cell contact during early morphogenesis in the chick embryo. *Develop. Biol.* **16**, 78–106.

Trinkaus, J. P. (1967). Morphogenetic cell movements. In *Major Problems in Developmental Biology*. Academic Press, New York.

Trinkaus, J. P. (1969). *Cells into Organs: The Forces That Shape the Embryo.* Prentice-Hall, Englewood Cliffs, N.J.

Trinkaus, J. P. (1973). Surface activity and locomotion of *Fundulus* deep cells during blastula and gastrula stages. *Develop. Biol.* **30**, 68–103.

Vaughan, R. B. and J. P. Trinkhaus (1966). Movement of epithelial cell sheets in vitro. *J. Cell Sci.* **1**, 407–413.

Waterman, R. (1972). Use of the scanning electron microscope for observation of vertebrate embryos. *Develop. Biol.* **27**, 276–281.

Weiss, P. (1961). Guiding principles in cell locomotion and cell aggregation. *Exp. Cell Res.* **18** 260–281.

Wiseman, L. L., M. S. Steinberg, and H. M. Philips (1972). Experimental modulation of intercellular cohesiveness: reversal of tissue assembly patterns. *Develop. Biol.* **28**, 498–517.

Yee, A. G. (1972). Gap junctions between hepatocytes in regenerating liver. *J. Cell Biol.* **55**, 294a.

CHAPTER FIVE

Surface Specification of Embryonic Cells: Lectin Receptors, Cell Recognition, and Specific Cell Ligands

A. A. MOSCONA

Next to the genome, the cell surface plays a most important role in the control of embryonic *growth, differentiation,* and *morphogenesis.* The cell surface is involved in practically all aspects of *cell communication* and cell interactions in development, including cell recognition, cell motility, morphogenetic attachment, and aggregation of cells into tissues. Reactions with extracellular factors and with other cells often generate in the cell surface signals, which elicit internally changes in macromolecular synthesis; these changes may, in turn, affect the cell surface and further alter its properties. It is such sequences of reciprocal outside-inside interactions that drive embryonic cells along the course of differentiation and development.

In recent years, the concept of the animal cell surface has undergone profound revision. With few exceptions, past notions about cytomembranes were based essentially on static models which presumed sameness, uniformity, and constancy of membranes, and which considered the erythrocyte "ghost" as a generally valid example. In contrast, the current concept of the cell surface is one of a dynamic three-dimensional mosaic (Singer and Nicolson, 1972), which reflects the genetic and phenotypic diversities of cells and manifests responsiveness to environmental changes and signals. This revision has been prompted by evidence of the cell surface's role in environmental surveillance, control of cell growth, and cell communication, and thus of its importance in morphogenesis and functional integration of cells in the organism. The cell surface is endowed with specificity markers (Boyse and Old, 1969; Edelman et al., 1974; Jones et al., 1974), enzymes (Roseman, 1970), various receptors and transport mechanisms (Pardee, 1968), and junctional specializations (Farquhar and Palade, 1963; Overton, 1973). It undergoes architectural and chemical turnover, modulated by internal and external conditions. Because of the cooperative nature of the cell membrane (Changeux et al., 1967; Lehninger, 1968; Roseman, 1970), localized changes or signals may influence the rest of the cell surface and therefrom affect other cell properties. Considerable information has been forthcoming concerning biosynthesis and assembly of cell surface constituents in normal (Winzler, 1970) and in neoplastic (Roizman and Spear, 1969; Sachs, 1967; Gold et al., 1968; Smith, 1968) cells, as well as the role of intracellular organelles in these processes (Rambourg et al., 1969; Whaley et al., 1972; Winterburn and Phelps, 1972).

This rapidly growing body of information, although based largely on work with nonembryonic cells, has profoundly affected considerations of the role of the cell surface in embryonic differentiation and development and has opened up new approaches to these problems (Anderson and Coggin, 1971; Moscona, 1973). It is of some historical interest that a dynamic concept of the nature of the cell surface, consistent with present views, was recurrently implied by studies on embryonic cell interactions (Weiss, 1947; Sperry, 1951; Grobstein, 1954; Moscona, 1960, 1962a, b) but was sidetracked by the then-prevailing opinions.

This discussion will be concerned with three aspects of our work pertaining to biological properties of embryonic cell surfaces: (1) differentation-related cell surface changes revealed by reactions with lectins; (2) dependence of specific enzyme induction in embryonic cells on cell contacts; (3) morphogenetic cell aggregation and the concept of specific cell ligands.

DIFFERENTIATION-RELATED CHANGES IN CELL REACTIONS WITH LECTINS

Reactions with Concanavalin A

The presence of carbohydrate-containing materials on embryonic cell surfaces has long been known (Moscona, 1952, 1962a, b; Moscona and Moscona, 1952; Grobstein, 1954). The introduction of lectins as molecular probes of sugar-containing cell surface components (reviewed by Sharon and Lis, 1972) led us to examine their usefulness for exploring changes in embryonic cell surfaces during differentiation. Initial studies demonstrated that tissue cells freshly isolated from young chick embryos were readily agglutinated with concanavalin A (Con A) because of the presence of surface sites containing α-D-glucose-like pyranosides (Moscona, 1971). Since Con A had earlier been shown to agglutinate more effectively neoplastic cells than nontransformed adult cells in tissue culture (Burger, 1969; Inbar and Sachs, 1969), the matter of embryonic cell agglutination deserved further study. Specifically, the question arose whether in the course of embryonic cell differentiation the properties of the cell surface became altered so as to result in declining agglutinability of the cells with Con A.

Most of our work (Kleinschuster and Moscona, 1972) was done with cells from embryonic and fetal neural retina, which are particularly suitable for such studies. The retina of chick embryos is a purely neural, avascular cell population, devoid of blood and connective tissue elements; this tissue is readily obtainable in convenient amounts at different embryonic ages and lends itself easily to dissociation into suspensions of single cells. The differentiation program of the retina can be staged with considerable precision in terms of phases of cell replication and growth, and of morphological and biochemical changes. In the chick embryo, replication of neural retina cells ceases almost completely after day 12 of development, and during the following 9 days till hatching the retina is engaged predominantly in differentiation and functional maturation.

In order to react tissue cells with lectin, the tissue must first be disssociated into a cell suspension. Suspensions of retina cells have been obtained in two ways: (1) by mechanical disruption of the tissue, after treatment with EGTA (ethylene bisoxyethylene nitrilotetraacetic acid), a specific calcium chelator (Lamkin and Williams, 1965); (2) by dissociation of the tissue into single cells after treatment of the retina with pure trypsin (Moscona, 1952, 1961a). It was found that these two procedures yielded cells with quite different surface reactivities to lectins. Cell agglutination was studied by adding different concentrations of the lectin to cell suspensions with a constant cell number in small Erlenmeyer flasks on a gyrating shaker (70 rpm; 37°C); the size of agglutinates was scored after 30 minutes (for details see Kleinschuster and Moscona, 1972).

The reactions of retina cells with Con A differed significantly, depending on embryonic age and dissociation procedure of the cells. In general terms, EGTA-dispersed cells from *young* embryos were readily agglutinated, while cells from *late* embryos were not; trypsin-dissociated cells from *late* embryos agglutinated readily, but those from *young* embryos much less so. In more detail, the results were as follows.

EGTA-dispersed cells from young, 8- or 9-day embryos were rapidly and massively agglutinated by Con A (Fig. 1,1); the size of the agglutinates depended on the lectin concentration and was maximal with 50–125 μg ml^{-1}. Agglutination was not affected by neuraminidase, but it was competitively prevented by α-methyl mannose. Thus, carbohydrate-containing sites specific for binding Con A are present and accessible on the surface of these young embryonic cells, either in overall high density or in patches conducive to massive cell agglutination.

Mild trypsinization rapidly reduced the agglutinability of these cells, presumably by degrading Con A receptors or altering their distribution; if such cells were incubated in culture medium for several hours, agglutinability was restored. Therefore, Con A receptors on these young embryo cells are located externally in a way that makes them readily accessible to proteolysis, as well as to reaction with the lectin. In these respects, the early embryonic cell surface resembles that of neoplastic cells (Inbar and Sachs, 1969).

This situation changes as retina cells differentiate and mature (Fig. 1, 3, *4*). Thus, EGTA-dispersed cells from 12- or 16-day embryos formed much smaller agglutinates than 8-day cells, and late fetal, 20-day cells agglutinated only very poorly, if at all, even with massive doses of Con A. However, if these late fetal cells were lightly trypsinized, or if they were dissociated by trypsinization, they agglutinated as well as early embryonic nontrypsinized cells (Fig. 1, *8*).

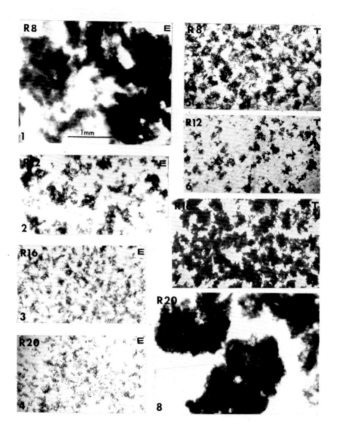

Figure 1. Age-dependent differences in agglutinability with Con A of embryonic neural retina cell suspensions from 8-, 12-, 16-, and 20-day chick embryos (R8-R20). Left row: 30-minute agglutination of cells obtained by dispersing the tissue with EGTA. Right row: 30-minute agglutination of cells dissociated with trypsin. Further explanations in text. (From Kleinschuster and Moscona, 1972.)

Therefore, as cells in the embryo differentiate and mature their surface properties change: young embryonic cells are readily agglutinated by Con A; late fetal cells are not, but are rendered agglutinable by trypsinization. The following interpretations are suggested. As cells differentiate, Con A-binding sites become too sparsely distributed over the surface to effect cell agglutination; trypsinization causes clustering of the sites and thereby renders the cells agglutinable (Nicholson, 1972). Another possibility is that in mature cells these sites are located deeper within the cell surface or are complexed and therefore no longer accessible to the lectin; proteolysis by trypsin exposes the sites and renders them reactive with the lectin (Burger, 1969; Inbar and Sachs, 1969).

These alternatives are not necessarily mutually exclusive, and they can be resolved experimentally. However, present information already shows clearly that, with differentiation, carbohydrate-containing sites on the retina cell surface undergo striking changes in accessibility, topographical distribution, or composition. These changes therefore represent an integral aspect of the overall differentiation program of the cells and contribute to the specification process of their surfaces.

Considering the rapid proliferation of young embryonic retina cells, we examined whether their high reactivity with Con A was related to cell replication, as suggested for other systems. Complete and prolonged inhibition of DNA synthesis in 8-day retinas by treatment with cytosine arabinoside did not significantly reduce cell agglutinability (Kleinschuster and Moscona, 1972). Thus, this experimental approach did not provide a satisfactory explanation of why cells are more agglutinable with Con A in the replicative phase of their developmental program than after cessation of growth.

Another possibility is that developmental changes in Con A receptors on cells are correlated with changes in cell motility and reflect the role of the cell surface in cell movement. In the early embryonic retina there are extensive morphogenetic cell movements related to tissue organization, and these coincide with the phase of cell replication. As morphogenesis is completed, these movements, as well as cell divisions, cease, and this coincides with the gradual reduction of cell agglutinability with Con A; immotile or mature cells are no longer agglutinated. It has been shown that the highly motile gastrula cells of sea-urchin embryos also are highly reactive with Con A (Krach et al., 1973; O'Dell et al., 1973). Conceivably, the display of Con A receptors on malignant cells may also be related to their motility and invasiveness. In this connection it has been suggested that cell reactions with Con A may involve microtubules and microfilaments (Edelman et al., 1973), which are thought to play a role in cell motility. Finally, Con A immobilizes embryonic cells and prevents morphogenetic cell movements, as will be described below. Thus, a strong case can be made for a significant correlation between cell motility and the topography or accessibility of Con A receptors on the cell surface.

In summmary, these correlations between changes in reactivity of the cell surface with Con A, on the one hand, and cell motility and replication, on the other, clearly suggest that the condition of Con A receptors is indicative, in a broad sense, of the state of cell differentiation as reflected in the properties of the cell surface. The nature and significance of these correlations remain to be established. For example, practically nothing is known about possible effects of Con A on metabolic processes in embryonic cells. Nor is it known how many kinds of Con A receptors exist in the cell surface and whether they have similar or different functions. Finally, it is to be assumed that Con A receptors are structurally-functionally related to other constituents of the cell surface and therefore localized effects may trigger "chain reactions" within the cell surface.

Reactions with Wheat Germ Agglutinin

Wheat germ agglutinin (WGA) is a lectin which binds selectively to N-acetylglucosamine residues. It agglutinates certain neoplastic cells, but not nontransformed cells, except after trypsinization (Burger, 1969; Inbar and Sachs, 1969). In contrast to Con A, WGA agglutinates only trypsinized embryonic retina cells regardless of their developmental age (Kleinschuster and Moscona, 1972); EGTA-dispersed cells are not agglutinated. Since WGA binds to nontrypsinized retina cells, absence of agglutination suggests that the topography of the binding sites precludes it, unless the cell surface is first altered with trypsin. Thus, while with respect to

Con A receptors early embryonic cells resemble neoplastic cells, with respect to WPA receptors they resemble normal adult cells, which also are not agglutinated by WGA, unless trypsinized.

Agglutination of trypsinized retina cells with WGA is hapten-inhibited, and it is dependent on the concentration of the lectin (Fig. 2). In general, retina cell agglutinability with WGA increases with embryonic age. Unlike Con A, WGA does not immobilize retina cells; in fact, embryonic retina cells agglutinated with WGA resume morphogenetic movements, reaggregate, and reconstruct tissue (see below).

Developmental changes in cell surface reactions with Con A and WGA similar to those described here have recently been observed also in cells from other tissues and embryos ranging from human (Weiser, 1972) to sea-urchin (Krach et al., 1973). Thus, programed changes in cell surface glycoproteins with different re-activities to lectins may turn out to be a rather general consequence of differentiation and of cell surface specification in development. The details of these changes will undoubtedly vary from system to system, and these differences should be useful in establishing correlations with specific developmental processes and cell functions.

Stimulation of DNA Synthesis in Retina Cells

Reaction with lectins modify the cell surface and thereby can alter cell functions. Stimulation of mitogenesis in lymphocytes by lectins is a typical example (Edelman et al., 1974). We have asked if lectins could stimulate DNA synthesis also in nonlymphoid, embryonic cells, and if embryonic cell susceptibility to such stimu-

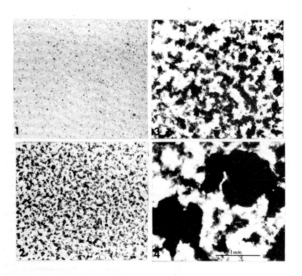

Figure 2. Agglutination (20 minutes) by wheat germ agglutinin (WGA) of trypsin-dissociated neural retina cells from 10-day chick embryos. Effects of different concentrations of WGA. *1:* no WGA; *2, 3,* and *4:* 60, 125, and 250 μg ml⁻¹. Further explanations in text. (From Klein-schuster and Moscona, 1972.)

lation would vary with differentiation. This question was examined in suspensions and monolayer cultures of neural retina cells of the chick embryo. The results (reported in detail elsewhere: Kaplowitz and Moscona, 1974) have shown that reaction of Con A or WGA with the cell surface stimulates DNA synthesis in these cells. Furthermore, the developmental stage of the cells is a decisive factor in their responsiveness to this stimulation: only between the embryonic ages of 9 and 12 days are these cells reactive to this effect of the lectins, even though lectin receptors are present on their surfaces at all ages. It should be pointed out that cell agglutination by lectins is not prerequisite for the stimulation of DNA synthesis; however, it is not known if the same receptor sites are involved in both these effects.

The above results suggest, as a concept for future exploration, that the ability of the retina cell surface to perceive or generate signals for DNA replication changes with differentiation. Accordingly, this change may be said to represent still another aspect of the specification program of the cell surface, and it may coincide with the onset of other conditions which restrict DNA replication. Only during a limited period in their ontogeny do retina cells possess surface properties such that when perturbed with lectins they elicit signals for DNA replication. This contrasts with lymphocyte differentiation, which results in persistent responsiveness to stimulation of mitogenesis. In the neural retina, differentiation results in termination of mitotic activities and "phasing out" of reactivity to mitogenic stimulation by lectins; it also coincides with cessation of cell motility and decline in agglutinability with Con A. These transitions demonstrate the changing properties of the embryonic cell surface during differentiation; they also show the usefulness of lectins in revealing such changes.

CELL SURFACE INTERACTIONS IN EMBRYONIC ENZYME INDUCTION

The embryonic cell surface has long been thought to have a critical role in the regulation of inductive processes in differentiation. The studies summarized here provide decisive evidence for this; they are based on an experimental approach which shows that, when the histological organization of retina cells is perturbed, their responsiveness to a specific enzyme induction is markedly reduced.

The induction of the enzyme *glutamine synthetase* (GS) is uniquely characteristic for the developmental program of the embryonic neural retina (for a review and description of this system, see Moscona, 1972). In the embryonic chick retina, GS is inducible by hydrocortisone; the steroid elicits differential gene expression which results in accumulation of stable, active RNA templates for GS, and in rapid increase in the rate of enzyme synthesis and enzyme accumulation. Neither DNA replication nor cell multiplication is involved. Glutamine synthetase can be readily induced *in vitro* in organ cultures of isolated retina tissue from 10- to 12-day chick embryos. Figure 3 shows the steroid-induced increase in the rate of GS synthesis measured by immunoprecipitation of radioactively labeled nascent enzyme (Moscona et al., 1972) and the corresponding increase in GS specific activity.

The hormonal induction of GS in the embryonic chick retina is stringently dependent on histotypic cell contacts, that is, on the cells being associated in a neuroretinal tissue pattern (Morris and Moscona, 1970, 1971). If retina tissue

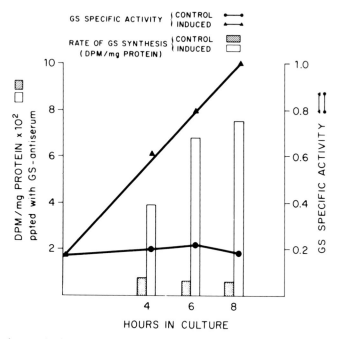

Figure 3.   Induction by hydrocortisone of glutamine synthetase (GS) in cultures of embryonic neural retina tissue from 12-day chick embryos. Accumulation of GS specific activity (solid lines) in induced ($\triangle$–$\triangle$) and control ($\bigcirc$–$\bigcirc$) cultures. White bars show increases in the rate (per 15 minutes) of enzyme synthesis in induced retina; these measurements were obtained by radioimmunoprecipitation procedures, using antiserum against GS, to precipitate newly synthesized enzyme; the rate data are expressed as DPM precipitated specifically by the antiserum, per milligram protein. (From Moscona et al., 1972.)

is dispersed into single cells and these are maintained as monolayer cultures, GS induction is low or fails to take place (Fig. 4). Although in such monolayers the cells make contacts, their mutual relationships are quite different from those within the intact tissue, and consequently the conditions of their surfaces are altered.

This absence of GS induction in monolayer cultures of retina cells is not due to cell proliferation, leakage of the enzyme from cells, or failure of the steroid inducer to enter the dispersed cells. Furthermore, cytoplasmic receptors capable of binding the steroid hormone are present in these cells (Koehler and Moscona, article in preparation). A clue to the failure of induction was provided by immunoprecipitation measurements of the enzyme; these have shown that in the monolayered cells the inducer did not elicit an increase in the rate of enzyme synthesis above the basal level present in noninduced cells. As explained above, such an increase requires differential gene expression; therefore, this gene expression evidently does not occur in dispersed and misarranged retina cells, even though the inducer is taken up by the cells. Accordingly, under these conditions, either the appropriate genes are not activated, or their products do not become functional.

The situation is quite different if cells freshly dissociated from retina tissue are made to reaggregate into multicellular clusters. Under suitable culture conditions, the cells in suspension can reassemble rapidly into compact aggregates within

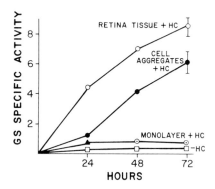

Figure 4. Induction of glutamine synthetase (GS) by hydrocortisone (HC) in cultures of neural retina tissue, and in monolayers and aggregates of retina cells. Tissue and cells were obtained from 10-day chick embryos. The inducer was added at the beginning of culture (except in experiments marked −HC). Monolayer cultures were prepared from freshly trypsinized retina cells that were plated at a cell density and under conditions preventing cell reaggregation. Cell aggregates were prepared by rotation in culture flasks of freshly prepared cell suspensions. (For details see Morris and Moscona, 1970, 1971.)

which they reform histotypic cell contacts, reconstruct retina tissue patterns, and continue to differentiate (Moscona 1961a; Sheffield and Moscona, 1969, 1970). In such histotypically organized cell reaggregates, GS is inducible (Fig. 4) and the enzyme level increases as morphogenesis in the aggregates progresses. Moreover, when cells that have been precultured in monolayer for 24 hours are redispersed and reaggregated, they again become inducible (Moscona and Moscona, original results). Thus, resumption of histotypic cell contacts by reaggregation of the cells restores responsiveness to GS induction.

Another way of testing the dependence of GS induction on histotypic cell contacts is to prevent aggregating retina cells from recombining correctly and reestablishing retinotypic associations. This was achieved by adding to suspensions of dissociated cells a low concentration of Con A, which agglutinates and immobilizes the cells. Since histological organization of cells in aggregates requires cell movements, treatment with Con A caused the cells to remain randomly connected, thereby preventing much of the retinotypic reconstruction that occurs in the absence of lectin. In such cell clusters, GS induction was considerably lower than in normal cell aggregates. This effect was reversible: if after 5 hours the agglutinated cells were redispersed and Con A was washed out (using α-methyl mannose), normal aggregation and tissue reconstruction ensued and the cells became inducible for GS. Agglutination of trypsin-dissociated retina cells with WGA did not durably immobilize these cells and allowed their subsequent histotypic reaggregation; accordingly, WGA did not prevent GS induction in aggregates of retina cells (Moscona and Moscona, original results).

Further evidence of correlation between responsiveness to GS induction and cellular organization in the retina was provided by yet another experimental approach. Instead of disorganizing the cells in vitro by dissociation of the tissue, they were caused to become misarranged in situ, that is, within the tissue framework. This was accomplished by treating early embryonic (5-day) undifferentiated retinas with 5-bromodeoxyuridine (BrdU), an analog of thymidine which is incorporated into DNA in these cells. At this early stage of retina development the cells proliferate rapidly and are in the processs of becoming organized into a retinotypic pattern. The BrdU prevented their organization; instead of the orderly neuroretinal architecture the cells formed grossly aberrant and chaotic structures (Fig. 5). Such misorganized retina cells were nonresponsive to GS induction (Moscona and Moscona, original results). Treatment with BrdU of older retinas,

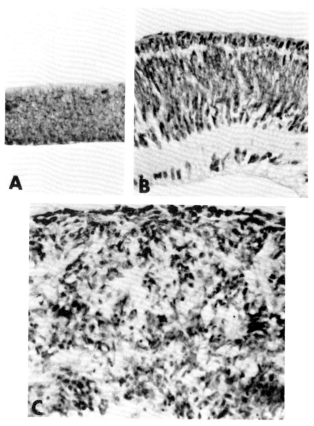

Figure 5. The effect of BrdU on morphogenesis in embryonic chick neural retina. Same magnification in the three pictures. (A) Histological section through neural retina of 5-day chick embryo. (B) Section through a retina from a 10-day embryo, showing the resulting growth and characteristic cell stratification. (C) Section through a 10-day retina which was treated with BrdU between 5th and 6th days of embryonic development; it shows growth but chaotic cell organization and is not responsive to GS induction.

which had already acquired a retinotypic organization, did not reduce responsiveness to GS induction (Moscona, 1972).

Although a detailed interpretation of the BrdU effect requires further analysis, these and the other results obtained demonstrate that in embryonic retinas responsiveness to GS induction is correlated with, or dependent on, the cells being retinotypically organized. Therefore, specific cell contacts and cell surface conditions are evidently critical factors in the ability of the steroid inducer to elicit in the neural retina gene expressions essential for GS induction. It should be pointed out that specific cell contacts are an outstanding aspect of neural differentiation, and that GS induction is integrally associated with neural retina differentiation. Thus, this system may be particularly favorable for revealing what may be a generally valid distinction between induction of enzymes that characterize phenotypic differentiation of embryonic cells and induction of enzymes involved in homeostatic adaptations, the latter being less stringently dependent than the first on histotypic cell organization and specific cell surface contacts. For example, tyrosine aminotransferase of rat hepatoma is readily inducible in monolayer cultures of these cells (Thompson et al., 1966).

In speculating about the need for histotypic cell contacts in GS induction, one might assume that such contacts generate conditions in the cell surface which are signaled to the interior and determine cell responsiveness to inducers of differentiation. In view of the involvement of specific cell ligands in histotypic linking of cells, the state of such ligands and their interactions could be highly important with respect to cell surface conditions required for inducibility of GS. Uncoupling or miscoupling of cell ligands, such as probably occurs when tissue cells are dispersed in monolayer cultures, could result in modification of the cell surface and, in turn, in altered cell responsiveness to developmental inductions. Such assumptions suggest obvious directions for future study. The problem is clearly of major importance for understanding differentiation: specific cell contacts are involved in many other embryonic inductions; disruption of normal cell contacts in neoplasia may be related to "dedifferentiation" of various tumor cells.

CELL AGGREGATION

Introduction

Cell aggregation *in vitro* and tissue reconstruction from cells in suspension constitute an experimental method for studying mechanisms of cell recognition, selective cell adhesion, cell interactions, and histogenesis. It was employed in studies on amphibian cells by the great embryologist J. Holtfreter some 30 years ago. The introduction of enzymic methods for preparing single cell suspensions from mammalian and avian tissues and organs (Moscona, 1952), and of procedures for "cell aggregation by rotation" (Moscona, 1961a), provided a controllable, quantitative approach to analysis of cellular and molecular aspects of cell aggregation. As is now generally known, when trypsin-dissociated embryonic tissue cells are maintained under suitable culture conditions, they reassemble into multicellular aggregates, within which they reconstruct their original tissue patterns and continue to differentiate (Moscona and Moscona, 1952; Moscona, 1957a,b). Such experimental construction of tissues from discrete cells makes it possible to study mechanisms which mediate cell association into collective systems. These methodologies and basic findings have been corroborated, extended, and employed in diverse studies by many investigators (for reviews, see Moscona, 1962b, 1965, 1972).

It should be emphasized that experimental cell reaggregation *in vitro* reflects morphogenetic events that are fundamental to embryonic development. In order to construct tissues and organs, cells in the embryo must become assorted and aggregated selectively in appropriate locations. Many organs, including the heart, adrenal, cortex, gonad, and various parts of the nervous system, arise in the embryo from cells that originate away from their final locations. These cells emigrate from their sites of origin and "home" toward their prospective locations, reaggregate there, and cooperatively establish the primordia of future tissues or organs. Experimental cell aggregation provides a relatively simplified model system for studying the component processes involved in these morphogenetic phenomena. By making it possible, in effect, to "synthesize" *in vitro* tissues from suspensions of discrete cellular units under controlled conditions, cell aggregation has been especially useful in studies on cell recognition and mechanisms involved in attachment and assembly of cells into tissues.

It has long been assumed that embryonic cells possess on their surfaces mechanisms for mutual recognition and selective adhesion (Holtfreter, 1939; Tyler, 1946; Weiss, 1947; Moscona, 1952, 1962b, 1963). In contrast to immunological recognition, which serves in safeguarding the genetic sovereignty of the organism from foreign intrusion, embryonic cell recognition functions in integrating and ordering different cells within the embryo; it enables cells with different specificities and affinities to sort out and aggregate into tissue-forming groupings.

Embryonic cell specificity and recognition can be categorized as follows, on the basis of cell surface properties:

1. *Cell-type specificity* reflects surface identities and affinities of individual cells; it is involved in *self-recognition* between homotypic cells, that is, cells of the same phenotype; and in *allorecognition*, that is, between different cell phenotypes which have complementary affinities and cooperative functions (e.g., in associations of neuron-muscle, neuron A–neuron B, and specific mesenchyme-epithelium interactions).

2. *Tissue specificity* refers to cell surface affinity shared by the disparate cells that make up a single tissue; this common surface identity distinguishes the cell population of one tissue from that of another tissue in the same organism (e.g., retina cell specificity, liver cell specificity, cerebrum cell specificity). Tissue specificity has been studied more than cell-type specificity and will be the main subject of this discussion.

These specificities, residing in the cell surface, arise in the course of embryonic differentiation; as embryo cells multiply and diversify, different cell surfaces acquire disparate characteristics resulting in differential cell affinities and adhesivenesses; thus, depending on their surface complementarities, adjacent cells recognize whether they are morphogenetically matched and, accordingly, either adhere and aggregate into a tissue, or separate and move on until the appropriate contact is made. Hence, embryonic *cell recognition* and *selective cell adhesion* represent closely related aspects of a fundamental cell-linking mechanism based on interactions of specific constituents of cell surfaces (see below).

Tissue Reconstruction

Treatment of an embryonic tissue with trypsin cleaves intercellular bonds, removes from the cell surface protein–carbohydrate-containing materials, and makes it possible to release cells into suspension. Reaggregation of the cells can be obtained by gently swirling flasks with cell suspensions on a gyrating shaker (Moscona, 1961a); the cells recombine into clusters, within which they sort out and reconstruct tissue. A 24-hour aggregation sequence in a suspension of embryonic retina cells is shown in Fig. 6. To further illustrate this process, Fig. 7 shows several stages in the reaggregation of embryonic myocardial cells, as seen by scanning electron microscopy (Shimada et al., 1974). The dispersed cells (Fig. 7a) rapidly form primary clusters (Fig. 7b), which increase in size by accretion of cells and small clusters; the emergent aggregates become increasingly compact (Fig. 7c) as

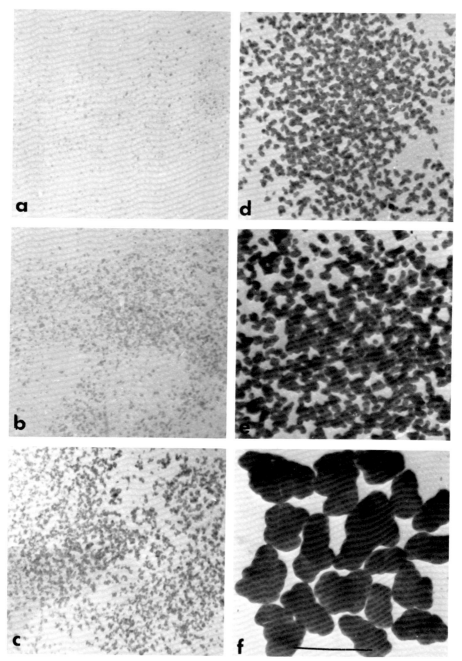

Figure 6. Twenty-four-hour sequence of cell aggregation by rotation in culture flasks. Cell suspension obtained by trypsin dissociation of neural retina from 7-day chick embryo. Rotation in 25-ml Erlenmeyer flasks, 70 rpm, ¾-inch diameter of rotation, 38°C; 6×10^6 cells per 3 ml medium (Eagle's basal medium with 10% horse serum). Approximate times (from top, left): 0.5, 1, 3, 4, 6, and 24 hours. Scale bar=0.5 mm. (Original.)

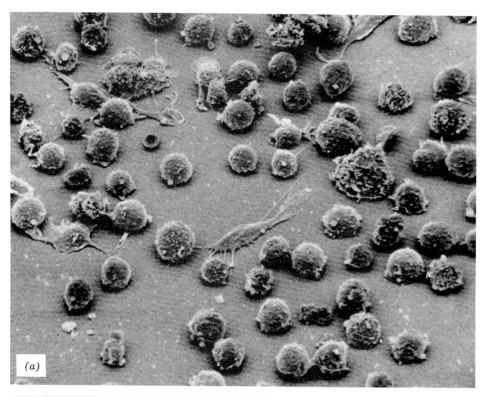

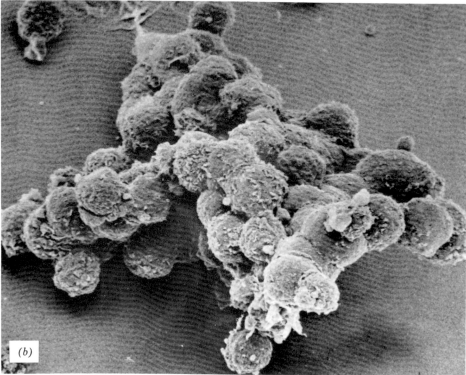

Figure 7. Scanning electron micrographs of aggregation of myocardial cells from 7-day chick embryo. (For technical and other details, see Shimada et al., 1974.) (a) Low-magnification micrograph of dispersed cells (X1100). (b) Three-hour cell aggregate (X2000). (c)

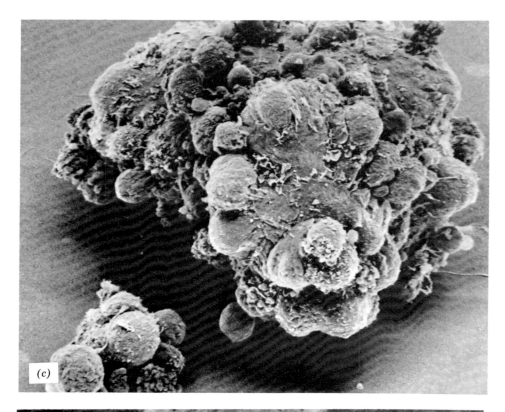

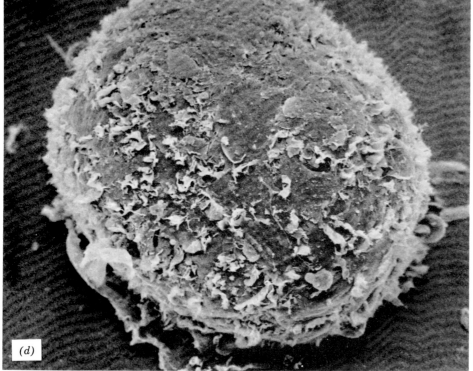

Seven-hour cell aggregate (X2000). (d) Forty-eight-hour cell aggregate. (Figs. 7a, b, d from Shimada et al., 1974; Fig. 7c, original, courtesy Dr. Y. Shimada.)

the cells are arranged into a tissue; within 24 hours such cell aggregates become spherical, pulsating "cardioids" (Fig. 7d). Myocardial tissue consists of several kinds of cells which become coordinately associated in the cardioids; thus, reconstruction in aggregates results in differential grouping, assortment, and selective attachments of the different cells present in the population, in accordance with their affinities and identities. Cell multiplication is not causally involved in this process.

Retina cell aggregation and reconstruction of neuroretinal tissue from suspensions of embryonic retina cells provide an even more striking example of cellular allorecognition, which is reflected in the sorting out and patterning of the several cell types that comprise this tissue. Electron miscroscopic analysis has shown that in early aggregates of retina cells the various cell types adhere mostly indiscriminately. However, they rapidly sort out and become aligned in distinct layers, reforming the characteristic pattern of neuroretina tissue (Sheffield and Moscona, 1969). The sorting of the cells coincides with the reappearance on the cell surface of glycoproteins which stain with ruthenium red, and which had been removed by the trypsinization of the cells. With further development of the aggregates, synaptic junctions form and eventually the reconstructed tissue attains advanced structural and biochemical retinotypic differentiation (Sheffield and Moscona, 1970).

A similarly striking cell sorting out is seen in the aggregation of brain cells. For example, aggregates of cerebrum cells obtained from embryonic mouse or chick reconstruct with remarkable fidelity their characteristic tissue patterns (Fig. 8) and pursue morphologic and biochemical programs that are typical of their respective brain regions (Garber and Moscona, 1972a, b; DeLong, 1970; DeLong and Sidman, 1970; Seeds, 1971).

In regard to tissue-specific cell recognition, this can be demonstrated by combining in the same suspension cells from two or more different tissues of the same embryo (heterotypic cell combinations). The initial cell clusters formed in such composite mixtures usually contain the different cells interspersed indiscriminately, but this rapidly gives way to sorting out of cells according to tissue type, resulting in formation of histotypic cell groupings which then undergo characteristic differentiation (Moscona and Moscona, 1952; Moscona, 1956, 1957a,b, 1965; Garber and Moscona, 1972a, b; Roth, 1968). For example, mixtures of embryonic liver and cartilage cells form composite aggregates in which the cells sort out in accordance with their tissue origins and reconstruct hepatic tissue and cartilage (Fig. 9). Tissue-specific embryonic cell recognition has recently been demonstrated in an experimental system using isolated plasma membranes from embryonic cells (Merrell and Glaser, 1973).

The overall architecture of heterotypic cell aggregates reflects their cellular composition. Depending on the kinds of cells in the combination, cells from a given tissue may become localized in the center of the composite aggregate, on its surface, or in an intermediate location; or the different cells may form completely separate aggregates. Examination of various cell combinations suggested (Moscona, 1956, 1961b) that cells from different tissues might display, with respect to their inside-outside localization in aggregates, a "hierarchical order" which might prove useful in studies on sorting out, interactions, and selective adhesiveness of cells. This suggestion was subsequently borne out by studies of Steinberg (1963) and others on topographies of cell distribution in composite aggregates.

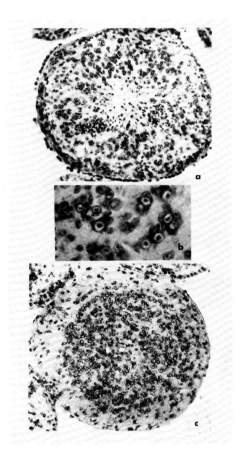

Figure 8. Sections of cell aggregates derived from suspensions of brain cells from 6-day chick embryos. (*a*) Cerebrum cell ggregate, 14 days in culture. (*b*) Same, at higher magnification to show neurons. (*c*) Optic tectum. (Garber and Moscona, 1972a.)

Requirements for Macromolecular Synthesis

In order to reaggregate histotypically, trypsinized embryonic cells require synthesis of proteins, mediated by RNA with a half-life of approximately 2 hours. Complete inhibition of protein synthesis in cells freshly dissociated by trypsin reversibly sup· presses histotypic reaggregation; however, it does not prevent the initial, random clustering of the cells which is due to the largely nonspecific "stickiness" of freshly trypsinized cell surfaces (Moscona and Moscona, 1963, 1966). The requirement in cell aggregation for protein synthesis varies, depending on kinds of cells, embryonic age of cells (i.e., their state of differentiation), dissociation procedure, and culture conditions. With respect to dissociation procedure, diverse proteases alter the cell surface differently, resulting in different cell aggregation kinetics and patterns. Cells dispersed by mechanical disruption or after treatment with calcium chelators tend to flocculate rapidly (Kleinschuster and Moscona, 1972); such flocculation is distinct from histotypic aggregation, and its nature is not yet clear.

Protein synthesis is needed for the regeneration of cell surface constituents removed by treatment with trypsin and essential for histotypic cell reaggregation.

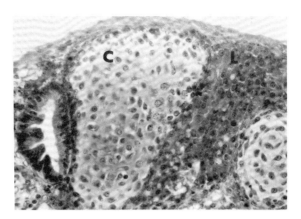

Figure 9. Tissue reconstruction in composite aggregate, produced in a suspension containing embryonic cartilage-forming and liver-forming cells; 48 hours in culture. Histological section through part of an aggregate. Cells from the two tissues had segregated into separate regions and reconstructed nodules of cartilage (C) and masses of hepatic tissue (L) with a bile duct on far left. (Original.)

As mentioned above, restoration of specific cell adhesiveness coincides with the reappearance on the cell surface of glycoproteins which stain with ruthenium red; their reappearance is suppressed in the absence of protein synthesis. Glucosamine is involved in the formation of cell surface glycoproteins (Glaeser et al., 1968), and it has been found that inhibition in dissociated cells of glucosamine synthesis by DON (an analog of glutamine) suppresses reaggregation of cells in which glutamine is limiting (Oppenheimer et al., 1969). Consistent with the requirements for macromolecular synthesis is the fact that reaggregation of embryonic cells is temperature sensitive (Moscona, 1962b; Moscona and Moscona, 1966); at temperatures inhibitory to biosynthetic processes, trypsin-dissociated cells in suspension do not undergo histotypic reaggregation and form only small, randomly and loosely constructed cell clumps.

The requirement for specific biosynthesis in reaggregation of trypsinized embryonic cells is further evident from the suppression by proflavine of cell aggregation. Proflavine, an acridine which differentially blocks the formation of the enzyme glutamine synthetase in neural retina cells (Wiens and Moscona, 1972), reduces preferentially (and reversibly) the synthesis of glycoproteins in these cells and inhibits their reaggregation (Hausman and Moscona, 1973). It is of interest that such proflavine-inhibited cells are agglutinated by Con A; this suggests that Con A-binding sites are not primarily involved in histotypic adhesion of cells, and that treatment with proflavine preferentially hinders regeneration of sites on the cell surface which are specifically involved in ligating cells histotypically to each other.

The Cell-Ligand Hypothesis

On the basis of aggregation studies with embryonic and other cells, a working hypothesis was evolved (Moscona, 1960, 1962a, 1963, 1968) which postulated that cell recognition and selective cell adhesion are mediated by interactions of specific components on the cell surface that function as cell-receptor sites. These cell-linking components were assumed to be protein-carbohydrate complexes and were referred to by the general term "cell ligands" (Moscona, 1967, 1968). The hypothesis suggested that cells from different embryonic tissues differed in the properties of their cell ligands. Cells with a high degree of ligand complementarity would tend to become histotypically linked through ligand interactions, thereby displaying *positive* recognition; cells with low ligand complementarity would adhere less effectively, transiently, or not all all, thus displaying *negative* recognition. The hypothesis postulated that the characteristics of cell ligands on cell surfaces evolve in the course of embryogenesis, resulting in *ligand patterns* which distinguish cells from different tissues (tissue specificity) and various kinds of cells (cell specificity). Such ligand patterns undergo continuous diversification and "tuning" as differentiation progresses, thereby specifying cell surfaces with increasing subtlety. Thus, qualitative, quantitative, and architectural differences in ligand patterns would confer on cells differential adhesiveness and affinity, causing them to sort out selectively according to tissue type and cell kind. The characteristics of ligand patterns and their "tuning" during differentiation may depend not only on the cell's genetic regulatory mechanisms, but also on the cell surface's responses to contact with other cells and to factors in the extracellular milieu.

The cell-ligand concept had a twofold purpose. It aimed to consider the problem of embryonic cell adhesion and recognition in the broader biological context of cell surface interactions mediated by complementarities of receptor mechanisms (Moscona, 1963); examples of such mechanisms are the reaction of sperm with egg, mating contact in microorganisms (Brock, 1959; Sneath and Lederberg, 1961; Wiese and Shoemaker, 1970), and hormone binding to the cell membrane (Cuatrecasas, 1971). However, its main purpose was to aid in formulating testable questions concerning the nature of differences between embryonic cell surfaces, and of cell recognition and specific cell contact. Thus, one of the immediate questions raised by the ligand hypothesis with respect to tissue specificity of cell surfaces was whether there are qualitative differences between surfaces of cells from different embryonic tissues. The existence of such tissue-specific cell surface disparities has, in fact, been confirmed immunologically (Goldschneider and Moscona, 1972) by demonstrating the presence of antigenically different determinants on cell surfaces from different embryonic tissues. Thus, antiserum prepared in rabbits against suspensions of live embryonic neural retina cells, when thoroughly absorbed with nonretina cells, agglutinated only neural retina cells and was shown by immunofluorescence tests to react specifically with the surface of retina cells; similarly prepared antiserum against liver cells reacted specifically with the surface of liver cells (Fig. 10). In conferring tissue specificity on cell surfaces, such antigenic differences are consistent with the disparities postulated for tissue-specific cell ligands.

A crucial test of the cell-ligand hypothesis was the feasibility of isolating from the exterior of live embryonic cells materials with tissue-specific cell-ligand activity. It was assumed that addition of such materials to homologous cells in suspension should enhance cell reaggregation and therefore result in larger-sized aggregates than occurred in controls. The first positive evidence of this effect was obtained in work on embryonic retina cells (Moscona, 1962a), and it led to the isolation in this laboratory of highly specific cell aggregating factors from several systems (Moscona, 1963, 1968; Humphreys, 1963; Lilien and Moscona, 1967; Garber and

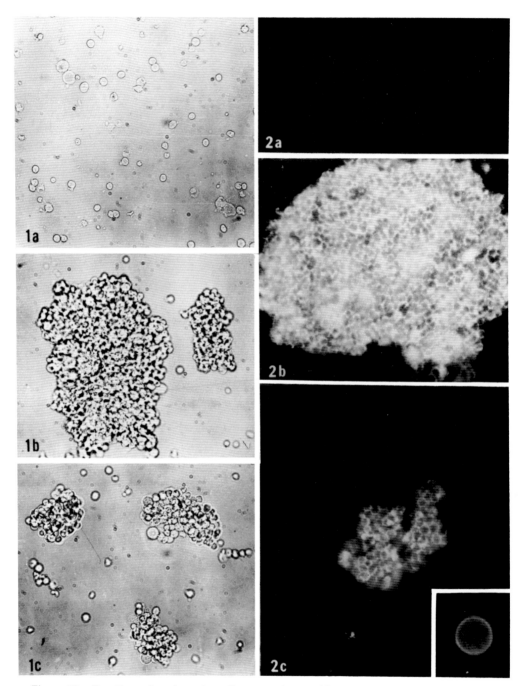

Figure 10. Reaction of embryonic cell suspensions with tissue-specific antisera. (From Goldschneider and Moscona, 1972.)

1a–c. Suspension of neural retina cells from 10-day chick embryo exposed to rabbit serum and observed for agglutination. X380. (*a*) Retina cells plus preimmunization rabbit serum, diluted 1:10. No agglutination. Small, medium, and large cell types are normally present in this suspension. (*b*) Retina cells plus rabbit antiretina serum (absorbed with embryonic liver cells), diluted 1:10. Massive agglutination of the retina cells. (*c*) Retina cells plus rabbit antiretina

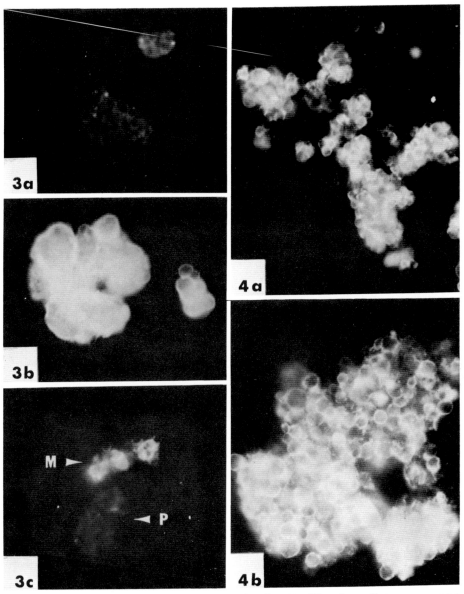

3a

3b

3c

M ▷

◁ P

4a

4b

serum (absorbed with embryonic liver cells), diluted 1:80. All retina cell types are represented in the agglutinates.

2a–c. Suspension of neural retina cells from 10-day chick embryo exposed to rabbit serum, washed, and reacted with fluorescein-conjugated goat antirabbit IgG. X380. (*a*) Retina cells plus preimmunization rabbit serum, diluted 1:20. No fluorescence or agglutination. (*b*) Retina cells plus rabbit antiretina serum (absorbed with embryonic liver and cerebrum cells), diluted 1:20. Marked agglutination and intense fluorescence of all cells. (*c*) Retina cells plus rabbit antiretina serum (absorbed with embryonic liver and cerebrum cells), diluted 1:80. Fluorescent staining is restricted to surface of retina cells (see insert).

3a–c. Suspension of viable cells from 10-day chick embryonic liver exposed to rabbit serum (1:40 dilution), washed, and reacted with fluorescein-conjugated goat antirabbit IgG. X380 (*a*) Liver cells plus preimmunization rabbit serum. Two clusters of liver parenchymal

Moscona, 1972b). These findings were confirmed and extended to other systems by work in several laboratories (McClay, 1971; Kondo and Sakai, 1971; Daday and Creaser, 1970; Oppenheimer and Humphreys, 1971; Tonégawa, 1973). The recent work of Merrell and Glaser (1973) demonstrating adhesive specificity with isolated plasma membranes is consistent with the results obtained in these studies.

To obtain these cell aggregating factors, advantage was taken of the turnover of cell surface components and their release from dispersed cells into the culture medium. It was found tht dispersed embryonic retina cells, maintained in suspensions or in monolayer cultures in a serum-free medium, synthesize and release a factor which, when added to aggregating retina cells, strikingly enhances their reaggregation (Fig. 11). Most importantly, the effect of this retina factor is tissue specific; it does not enhance aggregation of cells from other tissues (Lilien and Moscona, 1967; Lilien, 1968; Hausman and Moscona, 1973). This specificity distinguishes cell aggregating factors from other substances that may be present in the medium and may cause nonspecific or atypic cell clumping or cell flocculation (Pessac and Defendi, 1972). Within the massive aggregates produced in the presence of the aggregating factor, the cells become organized into retinotypic tissue; that is, the factor does not simply lump or agglutinate the cells, but also enhances their assembly and reattachment into morphogenetic patterns. Thus, the overall effect of this factor is consistent with the predicted activity of the postulated specific cell ligands.

In the early experiments (Moscona, 1962b) the effect of retina cell factor, obtained from suspensions of 10-day embryonic retina cells, was tested on reaggregation of trypsin-dissociated cells from 10-day retinas dispersed in Tyrode's physiological salt solution and aggregated by rotation. The critical tests were conducted at 25°C because, at this temperature, normal cell reaggregation is minimal because of suppression of metabolism and therefore also of regeneration of cell ligands on the cell surface. Thus, these experiments tested whether addition to such cells of the factor would supply them with components necessary for cell adhesion and aggregation. To eliminate the possibility of nonspecific cell clumping by DNA that might have been released from damaged cells, DNAse was added to all the cultures. The size of aggregates in control and experimental cultures was scored after 24 hours.

In control cultures only small cell clusters and free cells were present after 24 hours of rotation at 25°C. Cultures with the factor were strikingly different; in these, the cells formed numerous large masses, in addition to the smaller clusters

---

cells (hepatocytes) are shown. Hepatocytes tend to clump in the course of these tests and show slight internal autofluorescence. No specific surface fluorescence is present. (b) Liver cells reacted with antiliver serum (absorbed with embryonic neural retina and muscle cells). Intense fluorescence of the cell surface of hepatocytes, but not of mesenchymal cells. (c) Reaction of liver cell suspensions with antiheart serum (absorbed with neural retina cells). Mesenchymal cells (M) show bright surface fluorescence; parenchymal cells (P) do not fluoresce specifically.

4a-b. Agglutination and specific surface staining by antiheart serum (absorbed as above) of embryonic heart cells (a) and skeletal muscle cells (b). X400. (Modified from Goldschnieder and Moscona, 1972).

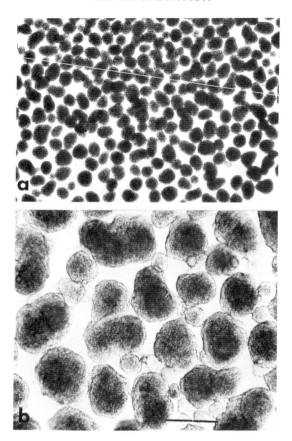

Figure 11. Effect of retina-specific cell aggregating factor on aggregation of suspension of neural retina cells from 10-day chick embryo. Twenty-four-hour aggregates obtained by rotation of culture flasks (25-ml Erlenmeyer) with 3 ml of cell suspension at 37°C (70 rpm, 3/4-inch diameter of rotation). (a) Control without factor (in balanced salt solution only). (b) With the factor added at zero hour. Scale bar = 0.5 mm. (Original.)

(Fig. 12). The effect of the retina factor was specific for retina cells; no such enhancement of cell aggregation was obtained when this factor was tested on suspensions of embryonic kidney or limb-bud cells.

It is of interest that no such factor could be obtained from retina cells of much older embryos (19 days). This is consistent with the fact that the ability of trypsin-dissociated cells to reaggregate histotypically declines with embryonic age and advancing specialization (Moscona, 1962b); such fully differentiated retina cells also failed to respond to aggregating factor obtained from younger cells. It is thus conceivable that, as cells complete their morphogenesis and differentiation and become committed to specialized functions, they are no longer capable of reforming specific surface components which are required for morphogenetic processes in earlier stages of embryonic development. Accordingly, it would be of interest to determine if "dedifferentiation" in some cases of neoplasia or in regenerating tissues results in reversal of this situation.

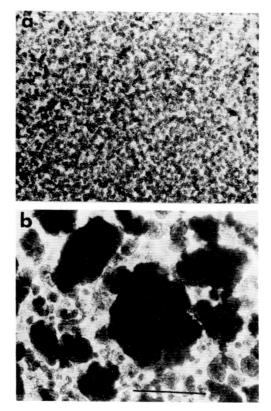

Figure 12. Effect of retina-specific cell aggregating factor on aggregation of suspensions of neural retina cell at 25°C; 24-hour cultures. For other details, see Fig. 11 (a) Control (without factor). (b) With the factor added at zero hour. Scale bar = 0.5 mm. (From Moscona, 1962a.)

In the presently employed procedure, cell aggregating factors are obtained from primary monolayer cultures of cells. Formation of the retina cell aggregating factor is prevented at suboptimal temperatures, suggesting that simple leakage of materials from cells is not the reason for its accumulation in the supernatant medium. Formation of the factor is also prevented by agents which block protein synthesis and inhibit cell reaggregation (Lilien, 1968). Of special interest is the preferential inhibition by proflavine of factor formation since, as mentioned above, proflavine reversibly suppresses retina cell reaggregation. These two effects are causally related; this was demonstrated by experiments in which cells treated with proflavine, and therefore unable to aggregate, were caused to reaggregate by supplying them with the retina-specific factor (Fig. 13). The likely interpretation is that proflavine hinders metabolically the effective regeneration of cell ligands on trypsin-dissociated cells, but it does not interfere with the binding to the cell surface of exogenously supplied cell ligands or with their ability to mediate cell attachment. Thus, the deficiency in the cell aggregation mechanism caused by proflavine can be "repaired" by providing cells with the specific cell aggregating factor ((Hausman and Moscona, 1973). It follows that this factor, exteriorized by dispersed cells, contains specific cell-ligating components which correspond in

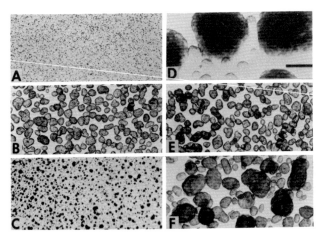

Figure 13. Inhibition by proflavine of retina cell aggregation and restoration of aggregability by retina-specific cell aggregating factor. (A) A freshly prepared suspension of neural retina cells (trypsin-dissociated retina tissue from a 10-day chick embryo). (B) Twenty-four-hour cell aggregates obtained in rotation cultures of cell suspension. (C) Twenty-four-hour rotation culture of cells in the presence of 12μM proflavine. (D) Twenty-four-hour aggregates obtained in the presence of the retina-specific cell aggregating factor. (E) Twenty-four-hour cell aggregates obtained in the presence of factor preparation collected from cultures which were treated with proflavine to inhibit formation of the specific cell ligands. (F) Effect of addition of active, retina-specific cell aggregating factor on retina cell aggregation in 12 μM proflavine (24-hour cultures). Scale bar = 0.5 mm. (From Hausman and Moscona, 1973.)

function to cell ligands postulated to be normally associated with the cell surface.

The binding to the cell surface of cell ligands in the factor preparation and their mechanism of cell linking are currently being studied. Earlier evidence suggested that the ligands bind to anchoring sites or acceptor sites in the cell membrane (Moscona, 1968). Additional arguments for such a two-component ligand-acceptor system have been recently reported for sponge cell aggregation by Weinbaum and Burger (1973). The nature of the acceptor site is unknown.

Specific cell aggregating factors were prepared also from embryonic mouse and chick cerebrum cells (Garber and Moscona, 1972b); they enhanced selectively the aggregation of cerebrum cells, but not of cells from other regions of the brain or from other tissues (Fig. 14). Evidence is forthcoming that cell aggregating factors are obtainable also from other regions of the embryonic brain, and these appear to be equally specific for the homologous cells (Garber, in preparation).

Thus, cell aggregating factors obtained from cells of different embryonic tissues contain cell-ligand activity which reflects in its characteristic effects the tissue specificity of the cells, in terms of differential adhesiveness and recognition of cells. Of special interest are brain cell aggregating factors in view of the fundamental importance of cell recognition and selective cell contacts in the primary morphogenesis of the nervous system during embryonic development (Sperry, 1951, 1965; Barbera et al., 1973). The specification of neuronal surfaces (Hunt and Jacobson, 1974) which has been postulated to be involved in the formation of neuronal associations may thus reflect the appearance of specific cell-ligand patterns on neuronal surfaces and their conduciveness to the sorting out and specific

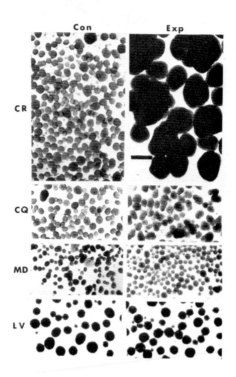

Figure 14. Specific enhancement of aggregation of cells from cerebrum of 13-day mouse embryos by cerebrum cell aggregating factor; 24-hour aggregates. Left row, controls (Con); right row, with factor (Exp). CR: cerebrum, CQ: corpora quadrigemina, MD: medulla, LV: liver. Scale bar = 0.5 mm. (From Garber and Moscona, 1972b.)

associations of neurons. It may be speculated that disturbances in the development of such ligand patterns on neuronal surfaces, due to mutational or congenital defects, would deprive the cell surface of characteristic specificity and thus give rise to missorting and misarrangement of brain cells such as are described for the "reeler" mutant in mice (DeLong and Sidman, 1970).

Cell-ligand preparations are now being purified and characterized. The initial evidence that the characteristic function of sponge cell aggregating factors resides in carbohydrate–protein complexes (Margoliash et al., 1965; Moscona, 1968) has been confirmed (Turner and Burger, 1973), and the evidence points to their being proteoglycans with unique physical characteristics (Henkart et al., 1973). Analysis of a purified cell-ligand preparation from embryonic chick neural retina cells indicates that its activity is associated with protein-carbohydrate complexes which form a single band on polyacrylamide gel electrophoregrams; electrophoresis in SDS-gels resolves a double band in the 50,000 M.W. region (McClay and Moscona, 1974; Hausman and Moscona, 1973). Labeling experiments with amino acids and sugars have shown that the materials banding in this region are synthesized and exteriorized by the retina cells in monolayer cultures from which the factor is obtained (Hausman and Moscona, 1973). Amino acid analyses show a relatively high content of glutamic and aspartic acid residues. The major carbohydrates are mannose, galactose, glucosamine, and sialic acid.

Much further work will be required to determine in detail the chemistry and structure of cell ligands from various tissues, the differences responsible for their specificities, their precise mechanisms of linking cells to each other, the control of their biosynthesis, and their architecture and distribution on cell surfaces. Similarly, the matter of acceptor sites for ligands on the cell surface and of the two-component ligand-acceptor systems remains to be investigated. Finally, it is necessary to consider the problems of cell ligands in the context of their relationships to other constitutents of the cell membrane, such as various other receptors, histocompatibility antigens, junctional specializations, and processes involved in cell motility, as well as structures on the inner side of the plasma membrane.

If a glycoprotein nature of the cell ligands is assumed, the mechanism by which ligands on adjacent cells interact to link cells could involve protein-carbohydrate, protein-protein, or carbohydrate-carbohydrate interactions. Preliminary information does not favor the last of these possibilities, but it does not firmly exclude any. In this context, of interest are the glycosyltransferases of cell surfaces (Roseman, 1970). Accepting the general concept of cell ligands described above and their role in cell recognition and selective adhesion, Roth et al. (1971) suggested that glycosyltransferases are structural components of the cell-ligand system and that they mediate cell adhesions by binding to specific carbohydrate residues on the opposite cell surface. This provocative suggestion remains to be fully tested. The possibility exists that glycosyltransferases may have other functions on the cell surface: they may be involved in termination of the carbohydrate portion of ligands, in cross-linking and stabilizing components within the cell membrane, in transport of sugars across the cell membrane, in binding ligands to acceptors in the cell membrane, or in degradation and turnover of ligands.

The tissue specificity of cell aggregating factors isolated so far is consistent with the tissue-specific category of cell recognition and cell adhesion, as discussed above; it supports the suggestion that cells in the population of a given tissue share surface determinants that distinguish them from cell populations of other tissues. However, such tissue-specificity does not readily explain the cell-type specific recognition and adhesion involved in the detailed assortment and organization of different kinds of cells within a single tissue. Thus, in addition to tissue specificities of cell ligands one must postulate the existence of still another, subtler order of differences which distinguishes diverse kinds of cells and determines their sorting out, association, and orientation within the tissue pattern. In view of the multiplicity of cell kinds, the subtleties of cell recognition, and the variety of mulicellular patterns, it is debatable whether the required diversity of cell surfaces could be explained solely on the basis of simple chemical differences between ligands. A more likely model is suggested by the heterogeneity of immunoglobulin molecules, which derives from their consisting of constant and variable portions (Edelman and Gall, 1969).

Another attractive possibility, not exclusive of the one above, is that the detailed diversity of cell surface specificities may be generatd by unique topographical-temporal distribution patterns of ligands on cell surfaces, superimposed on their chemical disparities. This kind of multidimensional diversity would provide a high degree of heterogeneity, versatility, and modifiability even within a relatively limited range of chemical permutations of ligands; this could result in unique and dynamic surface mosaics (for a comparable viewpoint concerning histocompatibility antigens, see Boyse and Old, 1969). Thus, interactions between cell surfaces might be based not only on chemical complementarities, but on matching topographies as well. Such architectural diversity would confer on cell surfaces also positional and polarity information for the ordering of cells into tissues. It can easily be imagined how a relatively limited chemical code of cell ligands deployed in various topographical-temporal permutations might generate on cell surfaces a highly versatile lexicon for morphogenetic cell recognition and organization.

These considerations, derived from the cell-ligand hypothesis, though speculative and oversimplified, point to future research aims in the study of cell surface specification, cell recognition, and selective adhesion. In this context, of prime interest is the genetic-developmental regulation of tissue- and cell-specific ligands in various normal and neoplastic states of differentiation. Recent work shows that this difficult problem is already being effectively approached. Thus, Artzt et al. (1973) have demonstrated surface antigens common to cells of mouse cleavage embryos and to primitive teratocarcinoma cells, but undetectable on differentiated cells (see also Bennett et al., 1971, and Edidin et al., 1971). Of particular importance are studies from T. T. Puck's laboratory on the use of surface antigens as genetic markers of mammalian cells (Wuthier et al., 1973; Jones et al., 1974), in that they may lead to identification of genes controlling the operation of cell surface antigens, including specific cell ligands and histocompatibility antigens, in various normal and abnormal states of differentiation (see also McDevitt and Banaceraff, 1969).

Thus, elucidation of the chemical, architectural, and temporal characteristics of cell-ligand patterns holds the key to detailed understanding of cell organization in embryonic morphogenesis. Such information would undoubtedly be relevant also to certain aspects of neoplasia and congenital malformations. While these are admittedly far-reaching expectations, they are encouraged by the progress already achieved in studies on cell aggregation and tissue reconstruction from cells *in vitro*.

ACKNOWLEDGMENTS

The work reported here has been supported by Research Grant HD-01253 from the National Institute of Child Health and Human Development, and in part by the Louis Block Fund, The University of Chicago. Collaboration with Drs. D. Fischman, B. Garber, I. Goldschneider, R. Hausman, P. Kaplowitz, S. Kleinschuster, D. McClay, M. Moscona, J. Morris, and Y. Shimada and the technical assistance of Barbara Schulak are gratefully acknowledged.

REFERENCES

Anderson, N. G., and J. H. Coggin, Jr. (1971). Models of differentiation, retrogression, and cancer. In *Embryonic and Fetal Antigens in Cancer* (N. G. Anderson and J. H. Coggin, Jr., eds.), Oak Ridge National Laboratory, Conf-710527, pp. 7–38. Oak Ridge, Tenn.

Artzt, K., P. Dubois, D. Bennett, H. Condamine, C. Babinet, and F. Jacob (1973). Surface antigens common to mouse cleavage embryos and primitive teratocarcinoma cells in culture. *Proc. Nat. Acad. Sci. U.S.A.* **70**, 2988–2992.

Barbera, A. J., R. B. Marchase, and S. Roth (1973). Adhesive recognition and retinotectal specificity. *Proc. Nat. Acad. Sci. U.S.A.* **60**, 2482–2486.

Bennett, D., E. A. Boyse, and L. J. Old (1971). Cell surface immunogenetics in the study of morphogenesis. In *Lepetit Symposium,* North-Holland London (L. G. Silvestri, ed.), pp. 247–263.

Boyse, E. A., and L. J. Old (1969). Some aspects of normal and abnormal cell surface genetics. *Ann. Rev. Genet.* **3**, 269–289.

Brock, T. D. (1959). Biochemical basis of mating in yeast. *Science* **129**, 960.

Burger, M. M. (1969). A difference in the architecture of the surface membrane of normal and virally transformed cells. *Proc. Nat. Acad. Sci. U.S.A.* **62**, 994–998.

Changeux, J. P., J. Thiery, Y. Tung, and C. Kittel (1967). On the cooperativity of biological membranes. *Proc. Nat. Acad. Sci. U.S.A.* **57**, 335–341.

Cuatrecasas, P. (1971). Properties of the insulin receptor of isolated fat cell membranes. *J. Biol. Chem.* **246**, 7265–7274.

Daday, H., and E. H. Creaser (1970). Isolation of a protein responsible for aggregation of avian embryonic cells. *Nature* **226**, 970–972.

DeLong, G. R. (1970). Histogenesis of fetal mouse isocortex and hippocampus in reaggregating cell cultures. *Develop. Biol.* **22**, 563–575.

DeLong, G. R., and R. L. Sidman (1970). Alignment defect of reaggregating cells in cultures of developing brains of reeler mutant mice. *Develop. Biol.* **22**, 584–599.

Edelman, G. M., and W. E. Gall (1969). The antibody problem. *Ann. Rev. Biochem.* **38**, 415–466.

Edelman, G. M., I. Yahara, and J. L. Wang (1973). Receptor mobility and receptor-cytoplasmic interactions in lymphocytes. *Proc. Nat. Acad. Sci. U.S.A.* **70**, 1442–1447.

Edelman,, G. M., P. G. Spear, U. Rutishauser, and I. Yahara (1974). Receptor specificity and mitogenesis in lymphocyte populations. Chapter 8 in this volume.

Edidin, M., H. L. Patthey, E. J. McGuire, and W. D. Sheffield (1971). An antiserum to "embryoid body" tumor cells that reacts with normal mouse embryos. In *Embryonic and Fetal Antigens in Cancer* (N. G. Anderson and J. H. Coggin, Jr., eds.). Oak Ridge National Laboratory, Conf-710527, pp. 239–248. Oak Ridge, Tenn.

Farquhar, M. G., and G. E. Palade (1963). Junctional complexes in various epithelia. *J. Cell Biol.* **17**, 375–398.

Fischman, D. A., and A. A. Moscona (1971). Reconstruction of heart tissue from suspensions of embryonic myocardial cells: Ultrastructural studies on dispersed and reaggregated cells. In *Cardiac Hypertrophy* (N. R. Alpert, ed.). Academic Press, New York, pp. 125–139.

Garber, B. B., and A. A. Moscona (1972a). Reconstruction of brain tissue from cell suspensions. I. Aggregation patterns of cells dissociated from different regions of the developing brain. *Develop. Biol.* **27**, 217–234.

Garber, B. B., and A. A. Moscona (1972b). Reconstruction of brain tissue from cell suspensions. II. Specific enhancement of aggregation of embryonic cerebral cells by supernatant from homologous cell cultures. *Develop. Biol.* **27**, 235–271.

Glaeser, R. M., J. E. Richmond, and P. W. Todd (1968). Histotypic self-organization by trypsin-dissociated and EDTA-dissociated chick embryo cells. *Exp. Cell Res.* **52**, 71.

Gold, P., M. Gold, and S. O. Freedman (1968). Cellular location of carcinoembryonic antigens on the human digestive system. *Cancer Res.* **28**, 1331–1339.

Goldschneider, I., and A. A. Moscona (1972). Tissue-specific cell surface antigens in embryonic cells. *Cell Biol.* **53**, 435–449.

Grobstein, C. (1954). Tissue interactions in the morphogenesis of mouse embryonic rudiments *in vitro.* In *Aspects of Synthesis and Order in Growth* (D. Rudnick, ed.). Princeton University Press, Princeton, N.J., pp. 233–267.

Hausman, R. E., and A. A. Moscona (1973). Cell surface interactions: differential inhibition by proflavine of embryonic cell aggregation and production of specific cell aggregating factor. *Proc. Nat. Acad. Sci. U.S.A.* **70**, 3111–3114.

Henkart, P., S. Humphreys, and T. Humphreys (1973). Characterization of sponge aggregation factor: a unique proteoglycan complex. *Biochemistry* **12**, 3045–3050.

Holtfreter, J. (1939). Gewebeaffinitat, ein Mittel der embryonalen Formbildung. *Arch. Exp. Zellforsch.* **23**, 169–209.

Humphreys, T. (1963). Chemical dissolution and *in vitro* reconstruction of sponge cell adhesions. I. Isolation and functional demonstration of the components involved. *Develop. Biol.* **8**, 27–47.

Hunt, R. K., and M. Jacobson (1974). Neuronal specificity revisited. *Current Topics Develop. Biol.* **8**, 203.

Inbar, M., and L. Sachs (1969). Interaction of the carbohydrate-binding protein concanavalin A with normal and transformed cells. *Proc. Nat. Acad. Sci. U.S.A.* **63**, 1418–1425.

Jones, Carol, P. Wuthier, and T. T. Puck (1974). Application of somatic cell genetics to surface antigens. Chapter 16 in this volume.

Kaplowitz, Paul B., and A. A. Moscona (1974). Stimulation of DNA synthesis by concanavalin A in cultures of embryonic neural retina cells. *Biochem. Biophys. Res. Commun.* (in Press).

Kleinschuster, S. J., and A. A. Moscona (1972). Interactions of embryonic and fetal neural retina cells with carbohydrate-binding phytoagglutinins: cell surface changes with differentiation. *Exp. Cell Res.* **70**, 397–410.

Kondo, K., and H. Sakai (1971). Demonstration and preliminary characterization of reaggregation promoting substances from embryonic sea urchin cells. *Develop. Growth Differ.* **13**, 1–14.

Krach, S. W., A. Green, G. L. Nicolson, and S. B. Oppenheimer (1973). Cell surface changes occurring during sea urchin embryonic development monitored by quantitative agglutination with plant lectins. *Exp. Cell. Res.* (in press).

Lamkin, E. G., and M. B. Williams (1965). Spectrophotometric determination of calcium and magnesium in blood serum with arsenate and EGTA. *Anal. Chem.* **37**, 1029–1031.

Lehninger, A. L. (1968). The neuronal membrane. *Proc. Nat. Acad. Sci. U.S.A.* **60**, 1069–1080.

Lilien, J. E. (1968). Specific enhancement of cell aggregation *in vitro*. *Develop. Biol.* **17**, 657–678.

Lilien, J. E., and A. A. Moscona (1967). Cell aggregation: its enhancement by a supernatant from cultures of homologous cells. *Science* **157**, 70–72.

Margoliash, E., J. R. Schenck, N. P. Hargie, S. Burokas, W. R. Richter, G. H. Barlow, and A. A. Moscona (1965). Characterization of specific cell aggregating materials from sponge cells. *Biochem. Biophys. Res. Commun.* **20**, 383–388.

McClay, D. R. (1971). An autoradiographic analysis of the species specificity during sponge cell reaggregation. *Biol. Bull.* **141**, 313–327.

McClay, D. R., and A. A. Moscona (1974). Purification of the specific cell-aggregating factor from embryonic neural retina cells. *Exp. Cell Res.* (in press).

McDevitt, H. O., and B. Benaceraff (1969). Genetic control of specific immune response. *Adv. Immunol.* **11**, 31–74.

Merrell, Ronald, and Luis Glaser (1973). Specific recognition of plasma membranes by embryonic cells. *Proc. Nat. Acad. Sci. U.S.A.* **70**, 2784–2798.

Morris, John E., and A. A. Moscona (1970). Induction of glutamine synthetase in embryonic retina: its dependence on cell interactions. *Science* **167**, 1736–1738.

Morris, John E., and A. A. Moscona (1971). The induction of glutamine synthetase in aggregates of embryonic neural retina cells: correlation with differentiation and multicellular organization. *Develop. Biol.* **25**, 420–444.

Moscona, A. A. (1952). Cell suspensions from organ rudiments of chick embryos. *Exp. Cell Res.* **3**, 535–539.

Moscona, A. A. (1956). Development of heterotypic combination of dissociated embryonic chick cell. *Proc. Soc. Exp. Biol. Med.* **92**, 410–416.

Moscona, A. A. (1957a). The development *in vitro* of chimeric aggregates of dissociated embryonic chick and mouse cells. *Proc. Nat. Acad. Sci. U.S.A.* **43**, 184–194.

Moscona, A. A. (1957b). Reconstruction of tissues from dissociated cells. In "Discussion on morphogenesis in animal and tissue cultures" (P. J. Gaillard, Chairman). *J. Nat. Cancer Inst.* **19**, 602–605 (Decennial Tissue Culture Conference, Woodstock, Vt., October 1956).

Moscona, A. A. (1960). Patterns and mechanisms of tissue reconstruction from dissociated cells. In *Developing Cell Systems and Their Control* (Society for the Study of Development and Growth, 18th Growth Symposium) (D. Rudnick, ed.). Ronald Press, New York, pp. 45–70.

Moscona, A. A. (1961a). Rotation-mediated histogenetic aggregation of dissociated cells: a quantifiable approach to cell interactions *in vitro. Exp. Cell Res.* **22**, 455–475.

Moscona, A. A. (1961b). How cells associate. *Sci. Am.* **205**, 142–162.

Moscona, A. A. (1962a). Analysis of cell recombinations in experimental synthesis of tissues *in vitro. J. Cell Comp. Physiol.* **60**, Suppl. 1, 65–80.

Moscona, A. A. (1962b). Cellular interactions in experimental histogenesis. *Int. Rev. Exp. Pathol.* **1**, 371–529.

Moscona, A. A. (1963). Studies on cell aggregation: demonstrations of materials with selective cell-binding activity. *Proc. Nat. Acad. Sci. U.S.A.* **49**, 742–747.

Moscona, A. A. (1965). Recombination of dissociated cells and the development of cell aggregates. In *Cells and Tissues in Culture* (E. N. Willmer, ed.). Academic Press, New York, pp. 489–529.

Moscona, A. A. (1967). Aggregation of sponge cells: cell-linking macromolecules and their role in the formation of multicellular systems. *In Vitro* **3** (1968) (21st Annual Meeting of Culture Association, Philadelphia, Pa., June, 1967).

Moscona, A. A. (1968). Cell aggregation properties of specific cell-ligands and their role in the formation of multicellular systems. *Develop. Biol.* **18**, 250–277.

Moscona, A. A. (1971). Embryonic and neoplastic cell surfaces: availability of receptors for concanavalin A wheat germ agglutinin. *Science* **171**, 905–907.

Moscona, A. A. (1972). Induction of glutamine synthetase in embryonic neural retina: A model for the regulation of specific gene expression in embryonic cells. In *Symposium on Biochemistry of Cell Differentiation*, Vol. 24 (7th Meeting, Federation of European Biochemical Societies, Varna, Bulgaria) (A. Monroy, ed.). Academic Press, London, pp. 1–23.

Moscona, A. A. (1973). Cell aggregation. In *Cell Biology in Medicine* (E. E. Bittar, ed.). John Wiley, New York, pp. 571–591.

Moscona, A. A., and M. H. Moscona (1952). The dissociation and aggregation of cells from organ rudiments of the early chick embryo. *J. Anat.* **86**, 287–301.

Moscona, A. A., and M. H. Moscona (1967). Comparison of aggregation of embryonic cells dissociated with trypsin or versene. *Exp. Cell Res.* **45**, 239, 243.

Moscona, M. H., and A. A. Moscona (1963). Inhibition of adhesiveness and aggregation of dissociated cells by inhibitors of protein and RNA synthesis. *Science* **142**, 1070–1071.

Moscona, M. H., and A. A. Moscona (1966). Inhibition of cell aggregation *in vitro* by puromycin. *Exp. Cell Res.* **41**, 703–706.

Moscona, M., N. Frenkel, and A. A. Moscona (1972). Regulatory mechanisms in the induction of glutamine synthetase in the embryonic retina: immunochemical studies. *Develop. Biol.* **28**, 229–241.

Nicolson, G. L. (1972). Topography of membrane concanavalin A sites modified by proteolysis. *Nature New Biol.* **239**, 193–197.

O'Dell, D. S., G. Ortolani, and A. Monroy (1973). Increased binding of radioactive concanavalin A during maturation of *Ascidia* eggs (in press).

Oppenheimer, S. B., and T. Humphreys (1971). Isolation of specific macromolecules required for adhesion of mouse tumor cells. *Nature* **232**, 125–127.

Oppenheimer, S. B., M. Edidin, C. W. Orr, and S. Roseman (1969). An L-glutamine requirement for intercellular adhesion. *Proc. Nat. Acad. Sci. U.S.A.* **63**, 1385–1404.

Overton, J. (1973). Experimental manipulation of desmosome formation. *J. Cell Biol.* **56**, 636–646.

Pardee, A. B. (1968). Membrane transport proteins. *Science* **162**, 632–637.

Pessac, B., and V. Defendi (1972). Cell aggregation: role of acid mucopolysaccharides. *Science* **175**, 898.

Rambourg, A., W. Hernandez, and C. P. LeBlond (1969). Detection of complex carbohydrates in the Colgi apparatus of rat cells. *J. Cell Biol.* **40**, 395–415.

Roizman, B., and P. G. Spear (1969). Macromolecular biosynthesis of animal cells infected with cytolytic viruses. *Current Topics Develop. Biol.* **4**, 79–108.

Roseman, S. (1970). The synthesis of complex carbohydrates by multiglycosyltransferase systems and their potential function in intercellular adhesion. *Chem. Phys. Lipids* **5**, 270–297.

Roth, S. (1968). Studies on intercellular adhesive selectivity. *Develop. Biol.* **18**, 602–612.

Roth, S., E. S. McGuire, and S. Roseman (1971). Evidence for cell-surface glycosyltransferases: their potential role in cellular recognition. *J. Cell Biol.* **51**, 536–551.

Sachs, L. (1967). An analysis of the mechanism of neoplastic cell transformation by polyoma virus, hydrocarbons, and X-irradiation. *Current Topics Develop. Biol.* **2**, 129–148.

Seeds, N. W. (1971). Biochemical differentiation in reaggregating brain cell culture. *Proc. Nat. Acad. Sci. U.S.A.* **68**, 1858–1861.

Sharon, N., and H. Lis (1972). Lectins: cell-agglutinating and sugar-specific proteins. *Science* **177**, 949–959.

Sheffield, J. B., and A. A. Moscona (1969). Early stages in the reaggregation of embryonic chick neural retina cells. *Exp. Cell Res.* **57**, 462–466.

Sheffield, J. B., and A. A. Moscona (1970). Electron microscopic analysis of aggregation of embryonic cells: the structure and differentiation of aggregates of neural retina cells. *Develop. Biol.* **23**, 36–61.

Shimada, Y., A. A. Moscona, and D. A. Fischman (1974). Scanning electron microscopy of cell aggregation: cardiac and mixed retina-cardiac cell suspension. *Develop. Biol.* (in press).

Singer, S. J., and G. L. Nicolson (1972). The fluid mosaic model of the structure of cell membranes. *Science* **175**, 720–731.

Smith, R. T. (1968). Tumor-specific immune mechanisms. *New Engl. J. Med.* **278**, 1207–1212.

Sneath, P. H. A., and J. Lederberg (1961). Inhibition by periodate of mating in *Escherichia coli* K-12. *Proc. Nat. Acad. Sci. U.S.A.* **47**, 86–90.

Sperry, R. W. (1951). Developmental patterning of neural circuits. *Chicago Med. School Quart.* **12**, 66–73.

Sperry, R. W. (1965). Embryogenesis of behavioral nerve nets. In *Organogenesis* (R. L. DeHaan and H. Ursprung, eds.). Holt, Rinehart, and Winston, New York, pp. 161–186.

Steinberg, M. S. (1963). Reconstruction of tissues by dissociated cells. *Science,* **141**, 401–408.

Thompson, E. B., G. M. Tomkins, and J. F. Curran (1966). Induction of tyrosine α-ketoglutarate transaminase by steroid hormones in a newly established tissue culture cell live. *Proc. Nat. Acad. Sci. U.S.A.* **56**, 296–303.

Tonégawa, Y. (1973). Isolation and characterization of a particulate cell-aggregation factor from sea urchin embryos. *Develop. Growth Differ.* **14**, 327–351.

Turner, R. S., and M. M. Burger (1973). Involvement of a carbohydrate group in the active site for surface guided reassociation of animal cells. *Nature* **244**, 509–510.

Tyler, A. (1946). An auto-antibody concept of cell structure, growth and differentiation. *Growth* **10**, 7–19.

Weinbaum, G., and M. M. Burger (1973). Two component system for surface guided reassociation of animal cells. *Nature* **249**, 510–512.

Weiser, M. M. (1972). Concanavalin A agglutination of intestinal cells from the human fetus. *Science* **177**, 525–528.

Weiss, P. (1947). The problem of specificity in growth and development. *Yale J. Biol. Med.* **19**, 235–278.

Whaley, W. G., M. Dauwalder, and J. E. Kephart (1972). Golgi apparatus: influence on cell surfaces. *Science* **175**, 596–599.

Wiens, A. W., and A. A. Moscona (1972). Preferential inhibition by proflavine of the hormonal induction of glutamine synthetase in embryonic neural retina. *Proc. Nat. Acad. Sci. U.S.A.* **69**, 1504–1507.

Wiese, L., and D. W. Shoemaker (1970). On sexual agglutination and mating type substances (gamones) in isogamous heterothallic chlamydomonads. II. The effect of concanavalin A upon the mating-type reaction. *Biol. Bull.* **138**, 88–95.

Winterburn, P. J., and C. F. Phelps (1972). The significance of gycosylated protein. *Nature* **236**, 147–151.

Winzler, R. J. (1970). Carbohydrates in cell surfaces. *Int. Rev. Cytology* **29**, 77–125.

Wuthier, P., C. Jones, and T. T. Puck (1973). Surface antigens of mammalian cells as genetic markers, II. *J. Exp. Medicine* **138**, 229–244.

CHAPTER SIX

Differentiation of Cartilage

DANIEL LEVITT, PEI-LEE HO, AND ALBERT DORFMAN

The process of chondrogenesis represents an example of differentiation which may be reproducibly observed *in vitro*. The recent progress in the elucidation of the chemical nature and the pathways of biosynthesis of components of cartilage matrix permits the use of biochemical parameters in the study of chondrogenesis.

For purposes of this study it is useful to define differentiation as the process which results in conversion of mesenchyme cells (before overt specialization or stable determination) to a particular phenotype, for example, fibrocyte, chondrocyte, or myocyte. The process is depicted as irreversible in Fig. 1 since it appears to be undirectional (Weiss, 1973). The term "dedifferentiation," as widely used

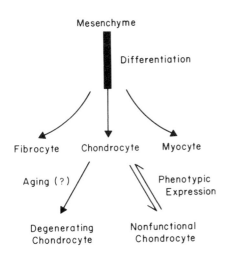

Figure 1. A schematic representation of the differentiation of limb-bud mesenchyme cells to chondrocytes, fibrocytes, and myocytes.

101

in recent literature, is confusing since it has been employed to encompass the reversible modulation of phenotypic expression of differentiated cells. Such reversible modulation is displayed with respect to the synthesis of cartilage matrix in tissue culture by mature chondrocytes (Fig. 1). Growth conditions strongly influence the expression of the differentiated phenotype by such cells (Holtzer and Abbott, 1968; Nameroff and Holtzer, 1967; Lavietes, 1970; Bryan, 1968a, b). Modification of the environment results in increase or decrease of the percentage of cells that show chondrocyte morphology and the quantity of chondroitin sulfate proteoglycan synthesized. Decrease of proteoglycan synthesis does not necessarily indicate dedifferentiation, but instead may result from modification of expression based on many different intercellular metabolic changes. In order to demonstrate dedifferentiation it is necessary to show that a differentiated cell reverts to some more primitive cell type, which in turn differentiates to some other phenotype. No such demonstration has appeared.

Prior definitions of chondrocytes have relied heavily on cell morphology, formation of characteristic nodules, and metachromatic staining. With improved biochemical techniques, it is now possible to delineate cartilage cells on the basis of capacity to synthesize specific macromolecules. At present, a chondrocyte may be defined as a cell which synthesizes large amounts of chondroitin sulfate proteoglycan and collagen of the composition $[\alpha 1 (II)]_3$. Preliminary evidence suggests that the protein core of cartilage chondroitin sulfate proteoglycan may be unique (Palmoski and Goetinck, 1973; Levitt and Dorfman, 1973).

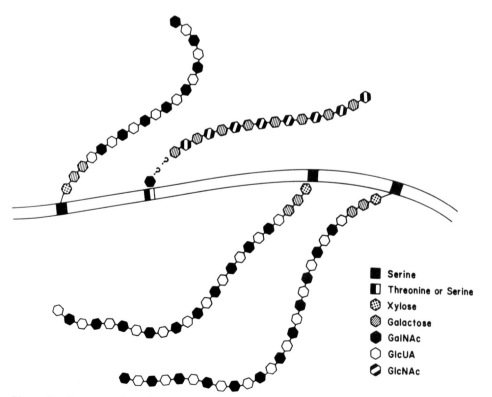

Figure 2.   Structure of cartilage proteoglycan.

The principal glycosaminoglycans of cartilage are chondroitin 4-SO_4, chondroitin 6-SO_4, and keratan sulfate. Chondroitin sulfate chains are attached by way of a galactosylgalactosylxylosyl trisaccharide to the hydroxyl groups of serine in the protein core, as illustrated in Fig. 2 (Rodén, 1970). Dissociated preparations of proteoglycan subunits appear to have molecular weights of approximately 2.5×10^6 daltons (Hascall and Sajdera, 1969, 1970).

Chondroitin sulfate chains are formed by the coordinated action of six distinct glycosyltransferases and appropriate sulfotransferases (Fig. 3). All of the glycosyl residues derive ultimately from glucose by way of a series of uridine nucleotide sugars. Biosynthesis of the proteoglycan appears to be initiated on the rough endoplasmic reticulum, where the protein core is probably translated. Addition of xylose, two galactose residues, and the first glucuronic acid moiety occurs by the action of four distinct enzymes (enzymes 1–4 in Fig. 3). Enzymes 5 and 6 then catalyze the addition of alternating GalNAc and GlcUA residues, and enzyme 7 transfers sulfate from adenylyl phosphosulfate to appropriate hydroxyl groups. It seems likely that the requisite enzymes are organized in a multienzyme system on the endoplasmic reticulum since the synthesis of a single chondroitin sulfate proteoglycan molecule requires more than 10^4 separate transferase reactions. On the basis of kinetic considerations, the possibility of accomplishing this polymerization in a random solution of enzymes and intermediates seems remote.

Chondrogenesis has been investigated most frequently in embryonic chick somites. Many of these investigations have been reviewed elsewhere (Thorp and

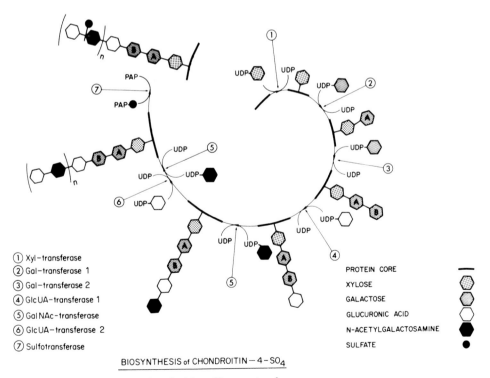

① Xyl−transferase
② Gal−transferase 1
③ Gal−transferase 2
④ GlcUA−transferase 1
⑤ GalNAc−transferase
⑥ GlcUA−transferase 2
⑦ Sulfotransferase

PROTEIN CORE
XYLOSE
GALACTOSE
GLUCURONIC ACID
N-ACETYLGALACTOSAMINE
SULFATE

BIOSYNTHESIS of CHONDROITIN − 4 − SO_4

Figure 3. Biosynthesis of chondroitin sulfate proteoglycan.

Dorfman, 1967). Early work indicated that the ventral spinal cord and notochord were necessary for the induction of cartilage in somite sclerotome. Specific inducers produced by spinal cord-notochord were sought with little success (Lash et al., 1962). Subsequent studies have shown that somite chondrogenesis occurs in culture in the absence of spinal cord or notochord under appropriate growth conditions (Strudel, 1962; Lash, 1968a, b; Ellison and Lash, 1971; Ellison et al., 1969).

DIFFERENTIATION OF LIMB BUD CELLS IN CULTURE

Because of difficulty in defining levels of differentiation in somite cultures, a more reproducible system which would also afford increased quantities of tissue was sought. Early embryonic limb-bud cells grown *in vitro* seemed to furnish such a system. Chondrogenic and myogenic phenotypes are not stably determined in the limb until stage 25, the period during which the first metachromatic extracellular matrix can be detected (Fell and Canti, 1934; Searls, 1967; Searls and Janners, 1969).

When limb-bud cells derived from stage 23-24 chick embryos are grown at high densities (greater than a confluent monolayer) in tissue culture dishes, large amounts of cartilage are formed. However, at lower densities, only cells with a fibroblastic morphology are recognizable.

Horwitz and Dorfman (1970) demonstrated that cartilage cells divide in soft agar more readily than do fibroblasts and that prior growth over agar enriches a culture with respect to chondrocytes, probably by diminishing the overgrowth of fibroblasts. Limb-bud cells (stage 23-24) were grown over agar at densities which do not permit chondrogenesis on plastic dishes. After 2 days, the suspension was plated on plastic dishes, again at low density, and formation of cartilage was monitored by morphology, metachromasia, and incorporation of $^{35}SO_4$ into glycosaminoglycans. The results presented in Table 1 indicate that initial culture over agar permits differentiation of limb-bud cells to cartilage when subsequently plated at low density (Levitt and Dorfman, 1972).

The mechanism whereby growth over agar or growth at high density promotes subsequent chondrogenesis is not clear. Limb-bud cells aggregate when maintained over agar, and aggregation of limb-bud cells enhances cartilage formation (Moscona, 1952). However, growth of limb-bud cells in gyratory shaking flasks for 2 days at low density followed by subculture on plastic dishes results in only minimal cartilage development. Cells initially grown in spinner bottles do not subsequently differentiate to chondrocytes when subcultured on plastic at high density, when grown over agar, or when grown in gyratory shaking flasks. Some type of cell contact appears to be prerequisite for subsequent chondrogenesis by limb-bud mesenchyme. These interactions are probably specific for limb-bud mesenchyme since if cell densities are increased by adding a heterotypic cell type, such as rat glial tumor cells, to cultured limb-bud cells, chrondrogenesis is drastically reduced, not enhanced.

Nevo and Dorfman (1972) have shown that a number of highly charged polyanionic substances stimulate synthesis of chondroitin sulfate proteoglycan by chondrocytes previously grown in suspension culture. Since agar is a sulfated polysaccharide, it is possible that the augmented differentiation of limb-bud cells may

TABLE 1. EFFECT OF GROWTH ON AGAR DIFFERENTIATION OF LIMB-BUD MESENCHYME

Experiment No.	A Cpm/10^6 Cells	B Cpm/10^6 Cells	C Cpm/10^6 Cells
1	16,500	348	—
2	6,200	590	—
3	4,630	595	428
4	53,800	5760	7,050
5	245,000	4950	—
6	26,300	—	1,610

Six hours before harvest, 3.3 μCi H_2 $^{35}SO_4$ ml^{-1} of medium was added to each culture. On the 9th day of culture, chondroitin sulfate was isolated (Dorfman and Ho, 1970). The descriptions of cells used in A, B, and C are as follows. (A) Dissociated cells plated at 10^7 cells in 10 ml media for 48 hours in 100-mm dishes containing 0.5% agar bases, then redissociated and placed on 60-mm plates at a density of 0.5 $\times$ 10^6 cells per dish for 7 days. (B) Dissociated cells plated for 48 hours at a density of 10^7 cells in 10 ml of medium on 100-mm dishes without agar, then redissociated and placed on 60-mm dishes at a density of 0.5 $\times$ 10^6 cells per dish for 7 days. (C) Dissociated cells plated at a density of 3.6 $\times$ 10^6 cells in 3 ml of medium on 60-mm tissue culture dishes and grown for 9 days. All three cell densities are below confluency.

result from this property. Some support for this idea was obtained by the observation that agarose and pectin were less effective in promoting chondrogenesis than was agar. However, addition of chondroitin 4-SO_4, chondroitin sulfate proteoglycan, keratan sulfate, heparan sulfate, or agar hydrolysate did not promote chondrogenesis by limb-bud cells grown at low density on plastic dishes either coated with soluble rat tail collagen or uncoated. Evidence has been accumulating that proteoglycan and/or collagen may play a role in the differentiation of other systems (Strudel, 1972; Ruggeri, 1972; Cohen and Hay, 1971). It is possible that some critical interaction of highly charged polyanions and collagens may be necessary to create local conditions which stimulate chondrogenesis.

EFFECTS OF BUdR TREATMENT ON DIFFERENTIATION OF LIMB BUD CELLS

The nucleoside 5-bromo-2'-deoxyuridine (BUdR), an analog of thymidine, has been demonstrated in a variety of systems to interfere selectively with the expression of processes characteristic of specialized cells. Many of these effects have been recently reviewed (Levitt and Dorfman, 1974). Earlier studies were concerned primarily with the influence of this drug on differentiated cells. For example, myoblast fusion, chondroitin sulfate synthesis by mature chondrocytes, hemoglobin synthesis by early erythroblasts, and hyaluronate production by amnion cells are all inhibited by treatment with BUdR. However, these phenotypic properties are re-expressed after several days in the absence of BUdR. This finding indicates that the effect of the drug on the expression of specialized properties in differentiated cells does not result from mutagenic properties.

Since differentiation of mesenchyme to chondrocytes might be an irreversible process consisting of one or more critical steps, it was of interest to examine the effects of BUdR on cells before either stable determination or overt phenotypic expression.

If, during the brief period of culture over agar, limb-bud cells are exposed to BUdR, subculture at low density on plastic dishes (in the absence of BUdR) is accompanied by marked inhibition of cartilage formation. No cartilage is evident 40 days after repeated subculture in the absence of BUdR. A similar irreversible inhibition of chondrogenesis occurs if high-density cultures of limb-bud mesenchyme are treated with BUdR during the initial 48 hours of culture. The BUdR effect is dose dependent and can be prevented by simultaneous but not subsequent addition of excess molar concentrations of thymidine.

Detailed study of the nature and quantities of glycosaminoglycans produced by cultured limb-bud cells and progeny of BUdR-treated mesenchyme reveals a tenfold decrease in the amount of chondroitin 4/6-SO$_4$ produced by the BUdR progeny. Separation of chondroitinase ABC digests by paper chromatography indicates that chondroitin 6-SO$_4$ constitutes 70–75% of the chondroitin sulfate fraction in both control cells and BUdR-treated progeny. The hyaluronic acid content is also diminished in BUdR progeny but to a lesser extent than chondroitin sulfate. Small amounts of dermatan sulfate and heparan sulfate were present in both types of cells; no keratan sulfate could be detected.

EFFECTS OF BUdR TREATMENT ON SYNTHESIS OF CHONDROITIN SULFATE PROTEOGLYCAN

Partial inhibition of chondroitin sulfate synthesis could be due either to incomplete suppression of synthesis or to differential suppression of two different types of proteoglycan. It became of interest, therefore, to investigate the nature of the proteoglycan produced by progeny of BUdR-treated cells, as compared with that produced by control cells cultured for 1 day (before cartilage formation) and 9 days (after cartilage formation). Figure 4 shows elution patterns on Bio-Gel A-50M of chondroitin sulfate proteoglycan extracted from these three types of cultures. Three peaks are apparent, one located near the void volume, one significantly retarded, and one in the salt fraction. The first two peaks appear to be high molecular weight species, whereas the smallest entity could be a product of abortive synthesis or degradation.

The major difference between the products of these three cultures is the quantity of chondroitin sulfate proteoglycan in the void volume peak. This chondroitin sulfate-containing material appears to be present in large quantities in cultures which show cartilage formation and will be designated as CSPG-2. The second peak (CSPG-1) is present in similar quantities in all three cultures. Both CSPG-1 and CSPG-2 are digested more than 85% by chondroitinase ABC. When papain-digested or 0.5 N KOH-treated proteoglycan from either peak was chromatographed on Sephadex G-200, the distribution of chain lengths was similar in the material obtained from all three types of cultures. The smaller size of CSPG-1 proteoglycan molecule might be due to fewer chains per protein core or to a different core protein, perhaps containing fewer serine residues available for xylosylation. Similar results have been obtained for cultured chondrocytes treated with BUdR and sternal cartilage from nanomelic chick embryos (Palmoski and Goetinck, 1972).

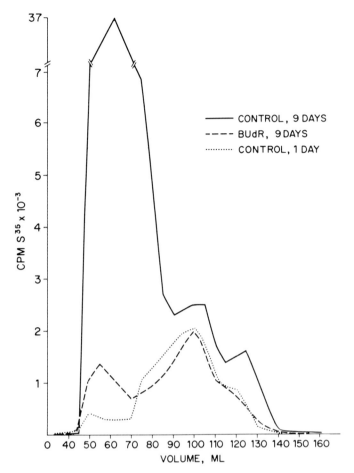

Figure 4.   Limb-bud cells cultured for 1 day, 9 days, and 2 days with $32\mu M$ BUdR, then 7 days in the absence of BUdR, were labeled with 5 μCiml^{-1} H$_2^{35}$SO$_4$ for 6 hours. Cells and media were separated. Cells (including adherent matrix) were extracted overnight at room temperature with 4.0 M guanidine HCl. Soluble material was dialyzed against distilled water and then redissolved in 4.0 M guanidine HCl. Material was separated on a 1.3 $\times$ 100 cm column, Bio-Gel A-50M, and eluted with 0.5 M NaCl at room temperature.

EFFECTS OF BUdR TREATMENT ON COLLAGEN SYNTHESIS

Chonodrocytes synthesize a unique type of collagen composed solely of $\alpha 1$ chains, which differ in primary structure from the $\alpha 1$ collagen chains of skin, bone, or cornea (Miller and Matukas, 1969; Trelstad et al., 1970). Cartilage-type collagen is designated as $[\alpha 1 (II)]_3$, whereas dermal collagen is indicated as $[\alpha 1 (I)]_2 \alpha 2$. Since treatment of limb-bud cells before determination with BUdR prevents subsequent differentiation, it became of interest to examine collagen synthesis of such cultures. The capacity of progeny of BUdR-treated cells to incorporate [^{3}H] + proline into hydroxyproline of protein is shown in Table 2. No inhibition of total collagen synthesis was observed by this method; in fact, a slight but reproducible

TABLE 2. BIOSYNTHESIS OF COLLAGEN BY CULTURED LIMB-BUD CELLS

| Ascorbate | Hydroxyproline/Proline/10^6 Cells | | | |
| | Control | | BUdR | |
	Cells	Media	Cells	Media
—	0.8×10^{-2}	2.1×10^{-2}	1.2×10^{-2}	4.0×10^{-2}
+	3.4×10^{-2}	10.5×10^{-2}	4.5×10^{-2}	17.2×10^{-2}

Cells were grown over agar for 48 hours with or without 32 μM BUdR. Cells were subcultured onto plastic dishes, grown for 7 more days, and then incubated for 16 hours with 2 μCi ml^{-1} proline[^{14}C] in the presence or absence of 50 μg ml^{-1} ascorbic acid. Incorporation of label into hydroxyproline was determined according to Lukens (1965).

stimulation occurred. In collaboration with Barbara D. Smith and George R. Martin, National Institute of Dental Research, experiments were initiated to determine the nature of the collagen produced by cultured limb-bud cells. These results demonstrate that 9-day control cultures produce collagen with an $\alpha1/\alpha2$ chain ratio of 10–1, typical of chondrocytes. Limb-bud cells grown for 1 day (before differentiation to cartilage) and progeny of BUdR-treated mesenchyme synthesize collagen with an $\alpha2$ ratio of 2:1. The detailed structure of the $\alpha1$ chains formed has not yet been determined. These findings indicate that treatment of predifferentiated limb mesenchyme with BUdR prevents subsequent expression of the cartilage phenotype with respect to collagen synthesis.

MECHANISM OF ACTION OF BUdR TREATMENT ON CHONDROITIN SULFATE PROTEOGLYCAN SYNTHESIS

As mentioned previously, exposure to BUdR results in a striking diminution in the quantity of chondroitin sulfate proteoglycan synthesized, presumably CSPG-2. Such a decrease could result from a decrease in the rate of synthesis or an increase in the rate of degradation. To test the latter possibility, the rates of turnover of sulfated glycosaminoglycans were compared in control cells and cells previously exposed to BUdR. The data illustrated in Fig. 5 indicate slightly increased turnover rates in the BUdR progeny. However, the magnitude of this change was not sufficient to account for the large difference in glycosaminoglycan content in the two types of cultures.

Since the effects of BUdR on polysaccharide production could not be explained by changes in degradation, a more detailed analysis of glycosaminoglycan synthesis was undertaken. On the basis of the known pathways of synthesis illustrated in Fig. 3, diminished synthesis of chondroitin sulfate proteoglycan could result from (a) diminished pools of nucleotide precursors, (b) diminished levels of activity of enzymes involved in polysaccharide chain synthesis, or (c) diminished synthesis of core protein such as occurs after exposure to puromycin (de la Haba and Holtzer, 1965; Telser et al., 1965).

In order to determine the effect of exposure to BUdR on nucleotide sugar synthesis in progeny of cells treated with BUdR, the activities of two enzymes critical to the synthesis of nucleotide sugar precursors were measured. The results shown

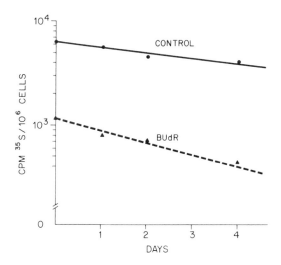

Figure 5. Turnover of sulfated glycosaminoglycans in limb-bud cell cultures. Cultures were labeled with 5 μCi ml⁻¹ H₂³⁵SO₄ for 6 hours on the 9th day of culture. Each dish was washed twice with media; then cells plus fresh media were assayed for sulfate-labeled glycosaminoglycans on the 9th, 10th, 11th, and 13th days of growth (0, 1, 2, and 4 days of chase).

in Table 3 indicate only minor depression of levels of activity of UDPglucose dehydrogenase (E.C.1.1.1.22) and UDP-N-acetylhexosamine-4-epimerase. Additionally, the activity of glucose-6-phosphate dehydrogenase (E.C.1.1.1.49), an enzyme not involved in this pathway, was not decreased in limb-bud cells exposed to BUdR during the first 2 days of culture. Pool sizes of UDP-glucuronic acid, UDP-N-acetylgalactosamine, and UDP-N-acetylglucosamine showed no significant differences between control cells and progeny of BUdR-treated cells (Table 4).

The levels of activity of two of the glycosyltransferases involved in chondroitin sulfate chain synthesis were measured. Xylosyltransferase (enzyme 1, Fig. 3) catalyzes the transfer of D-xylose from UDP-xylose to core portein acceptor, initiating

TABLE 3. ACTIVITY OF NUCLEOTIDE SUGAR SYNTHETIC ENZYMES

Assay	Control	BUdR
UDP-N-Acetylhexosamine-4-epimerase[a]	27,650	15,600
UDPGlucose dehydrogenase[b]	56.5	51.0
Glucose-6-phosphate dehydrogenase[c]	5.4	5.0

[a] Counts per minute [C¹⁴]-N-acetylglucosamine mg⁻¹ protein hour⁻¹.

[b] Micromoles NADH mg⁻¹ protein minute⁻¹.

[c] Micromoles NADPH mg⁻¹ protein minute⁻¹.

Determination of UDP-N-acetylhexosamine epimerase activity involved measurement of the conversion of UDP-N-acetylgalactosamine[¹⁴C] to UDP-N-acetylglucosamine [¹⁴C] by the 20,000 g supernatant. The UDP sugars were then hydrolyzed in 0.2 N HCl at 100°C for 30 minutes and neutralized with 0.4 N NaOH, and N-acetylgalactosamine was separated from N-acetylglucosamine by descending paper chromatography (Davidson, 1966). Dehydrogenase activities were determined by the reduction of either NAD or NADP, measured spectrophotometrically as increased absorbance at 340 nm.

TABLE 4. ESTIMATES OF UDP-SUGAR POOLS IN LIMB-BUD CELL CULTURES

UDP-Sugar	CPM C^{14}/10^8 Cells	
	Control	BUdR
UDP-glucuronic acid	2525	2375
UDP-N-acetylglucosamine	2240	2930
UDP-N-acetylgalactosamine	2110	2025

UDP-sugar pools were estimated on 9-day high-density cultures. Cells were labeled for 1 hour with either [^{14}C]galactose or sodium acetate[^{14}C], then sonicated for 10 sec in cold 5% Cl$_3$ CCOOH. After 1 hour, the TCA supernatant was neutralized and applied to either Dowex 1 × 8 (Cl$^-$ form, 200-400 mesh) for separation of UDP-GlcUA or Dowex 1 × 2 (Cl$^-$ form, 200-400 mesh) for UDP-GlcNAc and UDP-GalNAc. UDP-GlcNAc and UDP-GalNAc were hydrolyzed for 30 minutes in 0.5 M HCl at 100°, then separated on Whatman No. 1 paper according to Davidson (1966).

the formation of chondroitin sulfate chains. N-Acetylgalactosamine transferase (enzyme 5, Fig. 3) adds N-acetylgalactosamine from UDP-N-acetylgalactosamine to oligosaccharides containing terminal glucuronic acid residues. The results shown in Table 5 indicate that the levels of xylosyl and N-acetylgalactosamine transferases are reduced only 35–40%, whereas glycosaminoglycan synthesis is decreased 85–90% in progeny of BUdR-treated cells.

Since these data indicate that the progeny of BUdR-treated cells contain at least part of the enzymic machinery necessary for chondroitin sulfate chain synthesis, the possibility that the marked diminution in proteoglycan synthesis was due to deficient core protein formation was examined. Because no method is yet available for direct measurement of core protein, an indirect procedure was adopted. Brett and Robinson (1971) had previously shown that high concentrations of D-xylose

TABLE 5. CHONDROITIN SULFATE SYNTHESIS AND TRANSFERASE ACTIVITIES

Chondroitin Sulfate Synthesis	CPM/10^6 Cells	
	Control	BUdR
^{35}SO$_4$ incorporation	13,600	1,680
[^{14}C]Acetate incorporation	7,840	1,160
Transferase Activities	CPM Mg^{-1} Protein Hour^{-1}	
	Control	BUdR
Xylosyltransferase	2,060	1,390
N-Acetylgalactosamine transferase	24,200	13,800

Chondroitin sulfate synthesis was determined as previously described. Xylosyltransferase activity was assayed by measuring the transfer of xylose from UDP-xylose[^{14}C] to Smith-degraded proteoglycan according to Stoolmiller et al. (1972). N-Acetylhexosamine transferase activity was determined by measuring the addition of N-acetylgalactosamine from UDP-N-acetylgalactosamine[^{14}C] to a hexasaccharide acceptor possessing a terminal β-glucuronic acid residue (Horwitz and Dorfman, 1968).

partially overcome the inhibition by puromycin of chondroitin sulfate synthesis in minced cartilage. A series of experiments was conducted to measure the capability of xylose to increase chondroitin sulfate synthesis in progeny of BUdR-treated limb-bud cells.

The data summarized in Table 6 show that addition of xylose stimulates glycosaminoglycan synthesis in control cells by only 40% but that cells previously treated with BUdR are stimulated 240–300%. D-Xylose appears to act as a chondroitin sulfate chain initiator, eliminating the need for core protein and xylosyltransferase. In the presence of puromycin, addition of xylose partially overcomes the severe suppression of chondroitin sulfate formation (Table 6).

After these experiments were completed, a report by Hatanaka (1973) appeared, indicating that D-xylose, L-xylose, and L-arabinose derepressed the uptake of D-glucose by NIH Swiss mouse fibroblasts. In order to be certain that the stimulation of polysaccharide synthesis by D-xylose was not due to permeability changes, the effect of L-xylose and L-arabinose on BUdR progeny and puromycin-inhibited cells was studied. No stimulation by either of these pentoses was observed.

Furthermore, it was demonstrated, utilizing [^{14}C]xylose, that xylose is incorporated directly into polysaccharide chains that lack core protein (Levitt and Dorfman, unpublished results). Gel filtration indicated that these chains were approximately the same size as chondroitin sulfate chains synthesized by control cultures.

These results suggest that progeny of BUdR-treated mesenchyme are competent to synthesize chondroitin sulfate at a reasonable rate if a chain initiator is present.

Because previous studies had demonstrated that BUdR treatment also depresses chondroitin sulfate synthesis in differentiated chondrocytes, it was necessary to determine whether xylose also reverses BUdR effects in such cells. When cultured chondrocytes from sternae of 15-day-old embryonic chicks were treated with BUdR, chondroitin sulfate synthesis was diminished to 10–20% of control levels. Addition of xylose resulted in no increase in glycosaminoglycan production by control cells; however, a sevenfold increase in chondroitin sulfate synthesis by BUdR-treated cultures occurred (Table 7).

TABLE 6. EFFECT OF XYLOSE ON GLYCOSAMINOGLYCAN SYNTHESIS IN CONTROL AND BUdR-TREATED LIMB-BUD CELLS

D-Xylose (mM)	Puromycin	^{35}SO$^{2-}_4$ CPM/10^6 Cells		^{3}H CPM/10^6 Cells	
		Control	BUdR	Control	BUdR
0	—	10,000	2220	5050	970
10.6	—	14,400	5060	7820	2250
42.4	—	13,700	6000	7180	2210
0	+	840	340	260	160
10.6	+	3,100	2720	920	700
42.4	+	4,380	2780	1320	890

High-density cultures of limb-bud cells were labeled with 5 μCi mi^{-1}H$_2$^{35}SO$_4$ and 10 μCi ml^{-1}[^{3}H]acetate plus the indicated concentration of D-xylose with or without 10 μg ml^{-1} puromycin for 6 hours on the 9th day of culture. Glycosaminoglycans were isolated from pooled cells plus media according to Dorfman and Ho (1970).

TABLE 7. EFFECT OF D-XYLOSE ON GLYCOSAMINOGLYCAN SYNTHESIS IN CULTURED CHONDROCYTES

D-Xylose (mM)	Puromycin	$^{35}SO_4$ CPM/10^6 Cells		^3H CPM/10^6 Cells	
		Control	BUdR	Control	BUdR
0	—	89,400	13,400	63,200	10,200
10.6	—	85,900	87,400	67,800	47,000
42.4	—	81,700	95,300	54,400	57,600
0	+	4,860	2,420	2,540	1,380
10.6	+	29,620	60,750	17,600	35,800
42.4	+	34,000	71,000	25,800	43,600

On the 10th day of culture, chondrocytes growing in the presence or absence of 32 μM BUdR were labeled with 5 $\mu Ciml^{-1}H_2{}^{35}SO_4$ and 10 $\mu Ciml^{-1}$[^3H]acetate plus the indicated concentration of D-xylose with or without 10 μgml^{-1} puromycin for 6 hours. Glycosaminoglycans were isolated from pooled cells plus media as in Table 6.

An unusual reproducible observation in these experiments was that chondrocytes treated with both BUdR and puromycin were stimulated by xylose to produce larger amounts of glycosaminoglycans than were chondrocytes treated with only puromycin. The interpretation of this observation is not entirely obvious.

The effects of xylose on differentiated cartilage conflict with previous interpretations of experiments measuring enzymes involved in the synthesis of nucleotide sugars in BUdR-treated chondrocytes (Schulte-Holthausen et al., 1969; Marzullo, 1972). A 75–90% diminution of UDP-N-acetylhexosamine-4-epimerase and UDPglucose dehydrogenase activities was found. It was concluded that these reduced enzyme activities were responsible for decreased levels of glycosaminoglycan synthesis in BUdR-treated cartilage cells. We have examined the activities of xylosyltransferase and N-acetylgalactosaminyltransferase and find only a 40–50% decrease in activity in cartilage cells exposed to BUdR.

Treatment of differentiated chondrocytes with BUdR apparently does result in some suppression of enzyme activity. It seems likely that the extent of suppression may vary with conditions, such as time of exposure or concentration of BUdR. However, despite some inhibition of enzyme activity in BUdR-treated chondrocytes, these transferases are able to synthesize high levels of glycosaminoglycans when xylose is present as a chain initiator.

The results discussed so far are summarized in Fig. 6. They suggest that treatment of mesenchyme cells with BUdR before differentiation irreversibly prevents acquisition of the ability to synthesize cartilage-specific collagen and cartilage-specific proteoglycan. More information is required to verify the existence of the cartilage-specific proteoglycan. In contrast to the effects of BUdR on mesenchyme, suppression of chondroitin sulfate proteoglycan synthesis in chondrocytes was found to be reversible (Abbott and Holtzer, 1968; Levitt and Dorfman, unpublished results).

The results summarized so far imply that BUdR inhibits differentiation by maintaining mesenchyme cells in their primitive predifferentiated state and irreversibly prevents maturation to differentiated phenotypes. Although true for the parameters discussed, such a conclusion appears to be a misleading oversimplification.

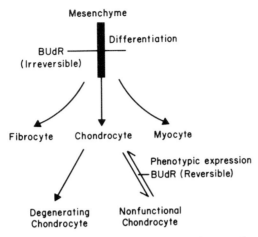

Figure 6. A summary of BUdR effects on differentiation of mesenchyme and/or expression of differentiated function of chondrocytes.

Previous workers have shown that early embryonic tissue is capable of synthesizing small quantities of chondroitin sulfate (Franco-Browder et al., 1963) and that some of the enzymes necessary for glycosaminoglycan biosynthesis are present before overt cartilage production (Medoff, 1967). Both xylosyltransferase and N-acetylgalactosaminyltransferase activities are easily detectable in 1-day cultures of stage 23–24 limb-bud mesenchyme at levels approximately 50% of those in cells grown for 9 days *in vitro*. However, somewhat surprisingly, the synthesis of chondroitin sulfate in 1-day cultures of chick limb-bud mesenchyme was not stimulated by addition of xylose (Table 8).

The reason for the failure of mesenchyme cells, in contrast to the progeny of BUdR-treated mesenchyme, to utilize xylose for chain initiation is not clear. Whatever the explanation, these findings indicate that mesenchyme cells, even after BUdR treatment, undergo changes of biochemical capacity.

TABLE 8. EFFECT OF XYLOSE ON 1-DAY CULTURES OF LIMB-BUD CELLS

D-Xylose (mM)	Puromycin	CPM ^{35}S 10^6 Cells	CPM [^{3}H] acetate 10^6 Cells
0	—	1120	675
10.6	—	1300	735
21.2	—	1100	630
42.4	—	1130	645
0	+	106	75
10.6	+	145	100
21.2	+	194	120
42.4	+	188	135

High-density cultures were incubated for 6 hours with 5 μCi ml^{-1} H$_2$^{35}SO$_4$, 10 μCi ml^{-1} [^{3}H] acetate, and the indicated concentration of D-xylose $\pm$10 μg ml^{-1} puromycin after 1 day of growth. Glycosaminoglycans were then isolated, and radioactivity was determined.

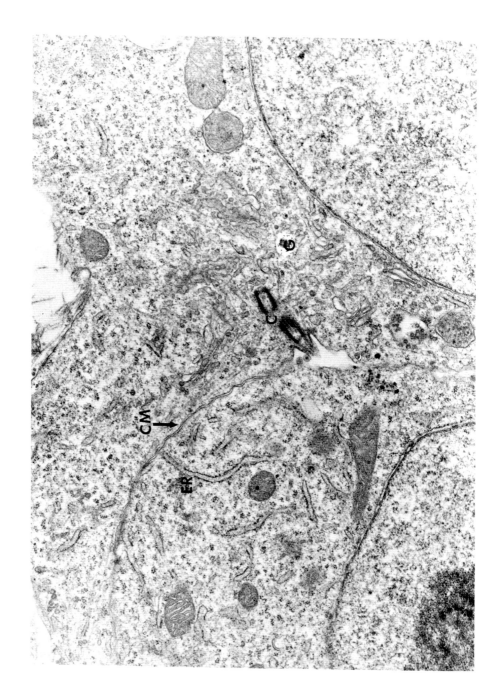

ULTRASTRUCTURAL STUDIES

Ultrastructural studies were carried out on 7-day control cells, 1-day control cells, and 7-day BUdR progeny. After 1 day *in vitro,* limb-bud mesenchyme cells approximate each other closely (Fig. 7). Extracellular matrix is scant and consists of few collagen fibers and fine fibrils. The cytoplasm contains a large proportion of free ribosomes with little rough endoplasmic reticulum. The Golgi apparatus is well formed but is present minimally throughout the cytoplasm and exhibits few excretion vacuoles forming at the mature face. Cilia are frequently observed emerging from the cells. In contrast, after 7 days in culture, limb mesenchyme cells possess all the characteristics of mature chondrocytes (Fig. 8*a, b*), including copious amounts of widely distended rough endoplasmic reticulum, an extensive Golgi apparatus, and a well developed matrix consisting of fine collagen fibrils and discrete granules.

The progeny of BUdR-treated cells possess many characteristics of mature chondrocytes (Fig. 9*a, b*). The rough endoplasmic reticulum is extensive and often distended and contains a fibrogranular material similar to that seen in the differentiated control cells. It would be of great interest to determine the nature of this material. The Golgi apparatus is well developed and extensive despite the low level of proteoglycan synthesis. The matrix of BUdR progeny contains large collagen fibers with a characteristic banding pattern but is almost totally devoid of granules.

Table 9 summarizes the biochemical and ultrastructural characteristics of predifferentiated mesenchyme, BUdR progeny, and chondrocytes. While predifferentiated mesenchyme cells possess several biochemical characteristics in common with BUdR progeny, they do not yet contain the large quantities of rough endoplasmic reticulum and Golgi associated with synthesis of extracellular products, nor do they exhibit the potential to produce chondroitin sulfate chains in the presence of D-xylose. Treatment of early limb mesenchyme with BUdR inhibits differentiation to cartilage by preventing the expression of at least *two specific genes,* one involved in the formation of cartilage-specific core protein and the other required for the formation of cartilage-specific collagen. Synthesis of the nonspecific chondroitin

TABLE 9. EFFECT OF XYLOSE ON 1-DAY CULTURES OF LIMB-BUD CELLS

	Mesenchyme	Chondrocyte	BUdR-Mesenchyme Progeny
Collagen	$[\alpha1\,(I)]_2\alpha2$	$[\alpha1\,(II)]_3$	$[\alpha1\,(I)]_2\,\alpha2$
Proteoglycan	CSPG-1	CSPG-1	CSPG-1
		CSPG-2	
Potential for CS Synthesis	Low	High	High
Endoplasmic Reticulum	Poorly developed	Extensive	Extensive

Figure 7.   An electron micrograph (X18,750) of limb-bud mesenchyme grown for 1 day in culture. G: Golgi apparatus, C: cilia, M: mitochondrion, CM: cell membrane, ER: endoplasmic reticulum, GR: granules, COL: collagen.

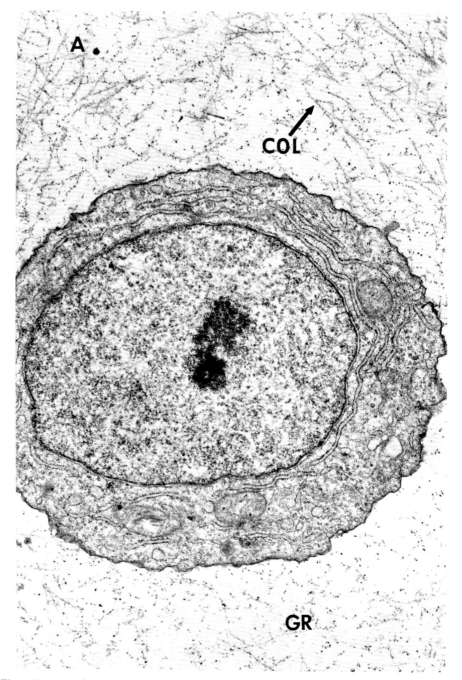

Figure 8a. An electron micrograph of limb-bud cells grown for 7 days, showing typical chondrocyte structure (X7000). Abbreviations as in Fig. 7.

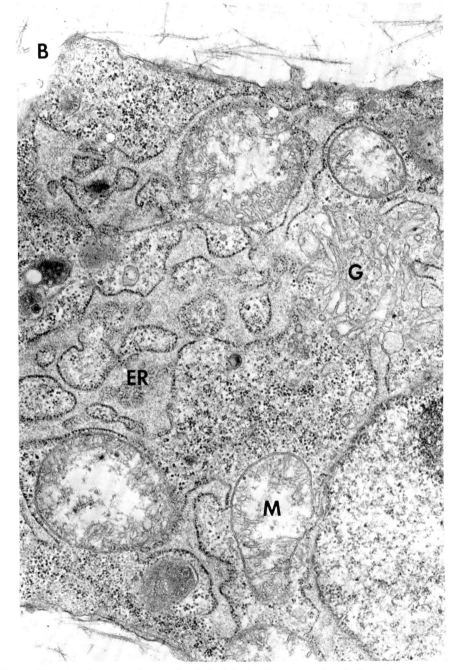

Figure 8b. An electron micrograph of limb-bud cells grown for 7 days, showing typical chondrocyte structure (X24,000). Abbreviations as in Fig. 7.

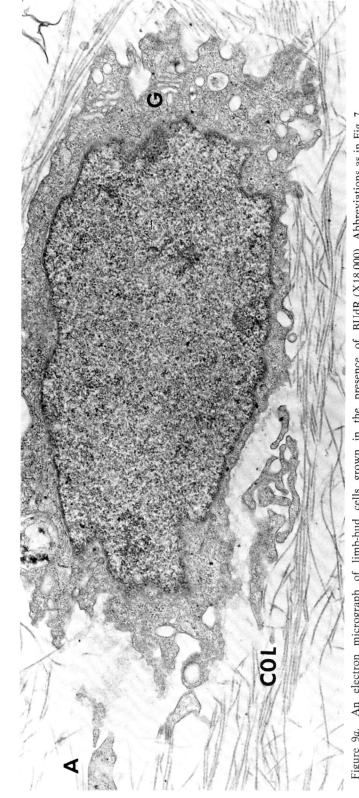

Figure 9a. An electron micrograph of limb-bud cells grown in the presence of BUdR (X18,000). Abbreviations as in Fig. 7.

118

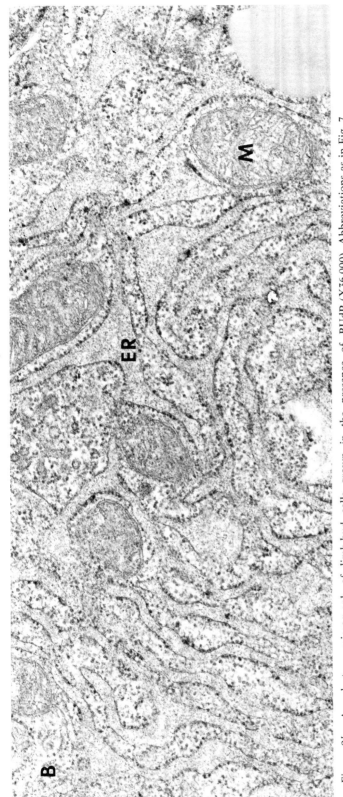

Figure 9b. An electron micrograph of limb-bud cells grown in the presence of BUdR (X36,000). Abbreviations as in Fig. 7.

sulfate proteoglycan (CSPG-I) and collagen [$\alpha 1\,(I)_2\,\alpha 2$] is not inhibited by treating limb mesenchyme with BUdR.

METABOLISM OF BUdR IN LIMB BUD CELLS

It appears that BUdR exerts its irreversible inhibition of chondrogenesis in limb-bud cells by substituting for thymidine in cellular DNA. Autoradiographs after exposure of mesenchyme to [³H]BUdR localize this compound in the nucleus. It is not extracted by treatment with cold Cl_3CCOOH.

In order to determine whether the behavior of progeny of BUdR-treated cells is due to persistence of the analog, the rate of disappearance of [³H]BUdR was studied. An abnormally rapid loss of BUdR from DNA of cultured limb-bud cells was observed after removal of the drug and continued culture in its absence for 6–8 days. As demonstrated in Fig. 10, BUdR is diluted from limb-bud DNA more rapidly than can be accounted for by simple exponential loss due to cell division. Thymidine is also removed at a more rapid rate than expected. Guanidine, however, is diluted at a rate corresponding to the number of cell divisions.

Since cell cultures with all three labeled bases were initiated in the presence of an excess of unlabeled BUdR and radioautography indicatd that more than 90% of all cells incorporated [³H]BUdR into DNA, it is unlikely that the abnormal removal of BUdR in limb-bud cells is due to selective death of cells after incorporation of the analog. Similar experiments using mature sternal chondrocytes incubated

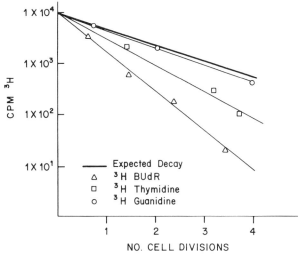

Figure 10. Turnover of DNA in BUdR-treated limb-bud cell cultures. Freshly dissociated limb-bud cells were cultured over agar for 48 hours in the presence of 32 μM BUdR and 1 μCi ml⁻¹ of either [³H]BUdR, [³H]thymidine, or [³H]guanidine. Cells were trypsinized with 0.25% trypsin, washed 3 times with media, and subcultured in the absence of label on 60-mm plastic culture dishes. Radioactivity in cold 5% Cl_3CCOOH-material was measured after digestion with T1 and pancreatic ribonuclease at 0, 2, 4, 6, and 8 days of chase. Cell number was determined in a Coulter counter, Model B. Plating efficiency on subculture was greater than 95%.

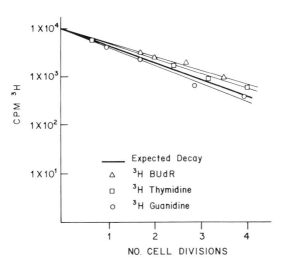

Figure 11. Turnover of DNA in BUdR-treated chondrocytes. Sternal chondrocytes were treated with 32 μM BUdR and 1 μCi ml^{-1} of either [^{3}H]BUdR, [^{3}H]thymidine, or [^{3}H]guanidine 4 days after cultures were initiated. Nonradioactive BUdR and labeled nucleotides were present for 2 days, after which cells were washed 3 times with media. Label in DNA and cell counts were determined as in Fig. 8 after 0, 2, 4, 6, and 8 days of chase.

with both unlabeled BUdR and either [^{3}H]BUdR, [^{3}H]thymidine, or [^{3}H]guanidine do not demonstrate abnormally rapid loss of these bases (Fig. 11).

DISCUSSION

Although BUdR strikingly affects a variety of eukaryotic cells, no simple, clear-cut explanation for the mechanism of action of this drug has yet been proposed. A preliminary attempt to account for the known facts is presented in Fig. 12. In the case of limb-bud cells, exposure to BUdR at a critical period of determination irreversibly prevents differentiation to cartilage. Although there is some decrease in activity of certain enzymes involved in chondroitin sulfate synthesis, the results presented here suggest that the major lesion appears to be failure to acquire the capacity to synthesize cartilage-specific collagen and a specific core protein of chondroitin sulfate proteoglycan. Fig. 12 shows the genes for these specific products on the left and indicates their activation by a control mechanism during the process of conversion of limb-bud mesenchyme cells to chondrocytes. Since BUdR treatment exerts similar effects on acquisition of the capacity to form certain other phenotypically specific proteins (e.g., globin, pancreatic zymogens, actin, and myosin) but does not result in a general decrease in protein synthesis, it is difficult to attribute the effect of the drug to random substitution for thymidine in structural genes. It seems more likely that some particular portion of the genome which controls the production of specialized molecules is particularly sensitive to BUdR incorporation. This does not preclude the incorporation of BUdR into other parts of the genome. Indeed, some decrease in activity of certain enzymes probably results from more general random effects.

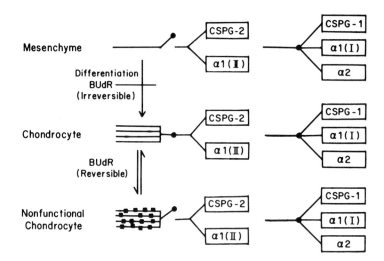

Figure 12. Proposed mechanism of action of BUdR in differentiation of cartilage. Products shown on right are those produced by mesenchyme, while those in center are produced only by chondrocytes. Control mechanism on right is fully functional in mesenchyme, while control mechanism in center becomes activated during differentiation. BUdR interferes irreversibly in amplification of control mechanism for synthesis of cartilage-specific products and reversibly in its function.

Two distinct functions of equal importance, replication and transcription, are served by DNA. Differentiation of mesenchyme to cartilage may involve a selective replication of regions concerned with the control of genes responsible for specialized macromolecule synthesis. These regions could be rich in thymidine or contain thymidine at certain strategic locations, rendering them extraordinarily sensitive to BUdR substitution. If BUdR then prevented normal amplification at a particular point during cell development, selective replication of these regions could never again occur. This proposal would explain both the irreversibility of BUdR treatment in predifferentiated cells and the abnormal metabolism of BUdR in the DNA in these cells.

Differentiated cells would be expected to already contain this amplified special control of the genome. Incorporation of BUdR in this region, however, would be expected to disrupt the synthesis of special molecules by altering normal regulatory relationships, but would not destroy the already amplified control process. Subsequent substitution of thymidine for BUdR would allow reversion to normal levels of expression. Such control regions in eukaryotic genomes have been previously suggested (Britten and Davidson, 1969, 1971).

The observation that progeny of BUdR-treated cells continue to produce collagen of the $[\alpha 1 (I)]_2 \alpha 2$ type but do not form the $[\alpha 1 (II)]_3$ type of collagen and continue to synthesize a non-cartilage-specific proteoglycan (CSPG-1) suggests that synthesis of so-called luxury molecules (Holtzer and Abbott, 1968) may be under controls different from those for other cell proteins (Fig. 12). The genes for the latter proteins may act constitutively. This conclusion has interesting evolutionary implications. It might be postulated that certain highly specialized molecules arise as a result of duplication of genes coding for more ubiquitous molecules. The examples of collagen and chondroitin sulfate proteoglycan may be likened to the relationship of actin- and myosin-like molecules present in cells as diverse as slime molds and fibroblasts to muscle actin and myosin. In the course of evolution, gene duplication resulted in the production of new structural genes which not only have mutated independently but also have come under new and different control mechanisms. It is possible that similar control mechanisms are responsible for maintaining the integration of viral genomes, which are consequently released as a result of BUdR treatment. Such a release by BUdR has now been demonstrated in many tissue culture systems (Lowy et al., 1971; Aaronson et al., 1971).

The model proposed does not afford a detailed mechanism for differentiation but does offer certain possibilities for experimental exploration. The striking effects of BUdR indicate that this analog may offer a powerful tool for elucidation of this fundamental biological problem.

ADDENDUM

After presentation of this paper it became known to the authors that Okayama, Kimata, and Suzuki (1973) found that p-nitro-β-D-xyloside acts as an initiator for chondroitin sulfate chains in cartilage slices. Okayama and Lowther (1973) found similar effects of 4-methyl-umbelliferyl-β-D-xyloside. We have now shown that both xylosides strikingly stimulate chondroitin sulfate synthesis in cultured chondrocytes, the progeny of BUdR-treated limb buds, puromycin-treated limb bud cells and chondrocytes grown in the presence of BUdR. The xylosides are active in concentrations as low as 0.05 mM. Contrary to the findings reported in the text for D-xylose in limb bud cells cultured for one day, the xylosides strikingly stimulate chondroitin sulfate synthesis. This result indicates that limb bud mesenchyme (stage 24) have a high capacity for synthesis of chondroitin sulfate chains.

ACKNOWLEDGMENTS

The original investigations described herein were supported by Grant AM-05996, HD-04583, and HD-00001 and the Chicago and Illinois Heart Associations.

REFERENCES

Aaronson, S. A., G. J. Todaro, E. M. Scolnick (1971). Induction of murine C-type viruses from clonal lines of virus-free BALB/3T3 cells. *Science* **174**, 157–159.

Abbott, J., and H. Holtzer (1968). The loss of phenotypic traits by differentiated cells. V. The effect of 5-bromodeoxyuridine on cloned chondrocytes. *Proc. Nat. Acad. Sci. U.S.A.* **59**, 1144–1151.

Baker, J. R., L. Roden, and A. C. Stoolmiller. Biosynthesis of the chondroitin sulfate proteoglycan: xylosyl transfer to Smith-degraded cartilage proteoglycan and other exogenous acceptors. *J. Biol. Chem.* **247**, 3838–3847.

Brett, M. J., and H. C. Robinson (1971). The effect of xylose on chondroitin sulphate biosynthesis in embryonic cartilage. *Proc. Aust. Biochem. Soc.* **4**, 92 (abstr.).

Britten, R. J., and E. H. Davidson (1969). Gene regulation for higher cells: A theory. *Science* **165**, 349–357.

Britten, R. J., and E. H. Davidson (1971). Repetitive and non-repetitive DNA sequences and a speculation on the origins of evolutionary novelty. *Quart. Rev. Biol.* **46**, 111–133.

Bryan, J. (1968a). Studies on clonal cartilage strains. I. Effect of contaminant non-cartilage cells. *Exp. Cell Res.* **52**, 319–326.

Bryan, J. (1968b). Studies on clonal cartilage strains. II. Selective effects of different growth conditions. *Exp. Cell Res.* **52**, 327–337.

Cohen, A. M., and E. D. Hay (1971). Secretion of collagen by embryonic neuroepithelium at the time of spinal cord-somite interaction. *Develop. Biol.* **26**, 578–605.

Davidson, E. A. (1966). UDP-*N*-acetyl-D-glucosamine-4-epimerase from embryonic cartilage. In *Methods in Enzymology*, Vol. 8 (E. F. Neufeld and V. Ginsburg, eds.). Academic Press, New York, pp. 277–281.

De la Haba, G. and H. Holtzer (1965). Chondroitin sulfate: inhibition of synthesis by puromycin. *Science* **149**, 1263–1265.

Dorfman, A., and P.-L. Ho (1970). Synthesis of acid mucopolysaccharides by glial tumor cells in tissue culture. *Proc. Nat. Acad. Sci. U.S.A.* **66**, 495–499.

Ellison, M. L., and J. W. Lash (1971). Environmental enhancement of *in vitro* chondrogenesis. *Develop. Biol.* **26**, 486–496.

Ellison, M. L., E. J. Ambrose, and G. C. Easty (1969). Chondrogenesis in chick embryo somites *in vitro*. *J. Embryol. Exp. Morphol.* **21**, 331–340.

Fell, H. B., and R. G. Canti (1934). Experiments on the development *in vitro* of the avian knee-joint. *Proc. Roy. Soc.* **B116**, 316–351.

Franco-Browder, S., J. DeRydt, and A. Dorfman (1963). The identification of a sulfated mucopolysaccharide in chick embryos, stages 11–23. *Proc. Nat. Acad. Sci. U.S.A.* **49**, 643–647.

Hascall, V. C., and S. W. Sajdera (1969). Protein-polysaccharide complex from bovine nasal cartilage. The function of glycoprotein in the formation of aggregates. *J. Biol. Chem.* **244**, 2384–2396.

Hascall, V. C., and S. W. Sajdera (1970). Physical properties and polydispersity of proteoglycan from bovine nasal cartilage. *J. Biol. Chem.* **245**, 4920–4930.

Hatanaka, M. (1973). Sugar effects on murine sarcoma virus transformation. *Proc. Nat. Acad. Sci. U.S.A.* **70**, 1364–1367.

Holtzer, H., and J. Abbott (1968). Oscillations of the chondrogenic phenotype *in vitro*. In *The Stability of the Differentiated State*, Vol. 1 (H. Ursprung, ed.). Springer-Verlag, New York, pp. 1–16.

Horwitz, A. L., and A. Dorfman (1968). Subcellular sites for synthesis of chrondromucoprotein of cartilage. *J. Cell Biol.* **38**, 358–368.

Horwitz, A. L., and A. Dorfman (1970). The growth of cartilage cells in soft agar and liquid suspension. *J. Cell Biol.* **45**, 434–438.

Lash, J. (1968a). Chondrogenesis: genotypic and phenotypic expression. *J. Cell Physiol.* **72**, 35–46.

Lash, J. W. (1968b). Somitic mesenchyme and its response to cartilage induction. In *Epithelial Mesenchymal Interactions* (R. Fleischmajer and R. E. Billingham, eds.). Williams and Wilkins, Baltimore, Chapter 10, pp. 165–72.

Lash, J. W., F. A. Hommes, and F. Zilliken (1962). Induction of cell differentiation. I. The *in vitro* induction of vertebral cartilage with a low molecular weight tissue component. *Biochim. Biophys. Acta* **56**, 313–319.

Lavietes, B. B. (1970). Cellular interaction and chondrogenesis *in vitro*. *Develop. Biol.* **21**, 584–610.

Levitt, D., and A. Dorfman (1972). The irreversible inhibition of differentiation of limb-bud mesenchyme by bromodeoxyuridine. *Proc. Nat. Acad. Sci. U.S.A.* **69**, 1253–1257.

Levitt, D., and A. Dorfman (1973). Control of chondrogenesis in limb-bud cell cultures by bromodeoxyuridine. *Proc. Nat. Acad. Sci. U.S.A.* **70**, 2201–2205.

Levitt, D., and A. Dorfman (1974). Concepts and mechanisms of cartilage differentiation. In *Current Topics in Developmental Biology*, Vol. 8 (A. A. Moscona and A. Monroy, eds.). Academic Press, New York, pp. 103–145.

Lowy, D. R., W. P. Rowe, N. Teich, and J. W. Hartley (1971). Murine leukemia virus: high-frequency activation *in vitro* by 5-iododeoxyuridine and 5-bromodeoxyuridine. *Science* **174**, 155–156.

Lukens, L. N. (1965). Evidence for the nature of the precursor that is hydroxylated during the biosynthesis of collagen hydroxyproline. *J. Biol. Chem.* **240**, 1661–1669.

Marzullo, G. (1972). Regulation of cartilage enzymes in cultured chondrocytes and the effect of 5-bromodeoxyuridine. *Develop. Biol.* **27**, 20–26.

Medoff, J. (1967). Enzymatic events during cartilage differentiation in the chick embryonic limb bud. *Develop. Biol.* **16**, 118–143.

Miller, E. J., and V. J. Matukas (1969). Chick cartilage collagen: a new type of α1 chain not present in bone or skin of the species. *Proc. Nat. Acad. Sci. U.S.A.* **64**, 1264–1268.

Moscona, A. A. (1952). The dissociation and aggregation of cells from organ rudiments of the early chick embryo. *J. Anat.* **86**, 287–301.

Nameroff, M., and H. Holtzer (1967). The loss of phenotypic traits by differentiated cells. IV. Changes in polysaccharides produced by dividing chondrocytes. *Develop. Biol.* **16**, 250–281.

Nevo, Z., and A. Dorfman (1972). Stimulation of chondromucoprotein synthesis in chondrocytes by extracellular chondromucoprotein. *Proc. Nat. Acad. Sci. U.S.A.* **69**, 2069–2072.

Okayama, M., K. Kimata, and S. Suzuki (1973). The influence of *p*-nitrophenyl-β-D-xyloside on synthesis of proteochondroitin sulfate by slices of embryonic chick cartilage. *J. Biochem.* **74**, 1069–1073.

Okayama, M., and D. Lowther (1973). Effects of β-xylosides on the synthesis of chondroitin-sulfate protein complexes by cartilage slices. *Proc. Aust. Biochem. Soc.* **6**, 75.

Palmoski, M. J., and P. F. Goetinck (1972). Synthesis of proteochondroitin sulfate by normal, nanomelic, and 5-bromodeoxyuridine-treated chondrocytes in cell culture. *Proc. Nat. Acad. Sci. U.S.A.* **69**, 3385–3388.

Rodén, L. (1970). Biosynthesis of acidic glycosaminoglycans (mucopolysaccharides). In *Metabolic Conjugation and Metabolic Hydrolysis*, Vol. 2 (W. Fishman, ed.). Academic Press, New York, pp. 346–432.

Ruggeri, A. (1972). Ultrastructural histochemical and autoradiographic studies on the developing chick notochord. *Z. Anat. Entwicklungs Gesch.* **138**, 20–33.

Schulte-Holthausen, H., S. Chacko, E. A. Davidson, and H. Holtzer (1969). Effect of 5-bromodeoxyuridine on expression of cultured chondrocytes grown *in vitro*. *Proc. Nat. Acad. Sci. U.S.A.* **63**, 864–870.

Searls, R. L. (1967). The role of cell migration in the development of the embryonic chick limb bud. *J. Exp. Zool.* **166**, 39–50.

Searls, R. L., and M. Y. Janners (1969). The stabilization of cartilage properties in the cartilage-forming mesenchyme of the embryonic chick limb. *J. Exp. Zool.* **170**, 365–376.

Strudel, G. (1962). Induction de cartilage *in vitro* par L-extrait de tube nerveux et de chorde de l-embryon de poulet. *Develop. Biol.* **4**, 67–86.

Strudel, G. (1972). Differénciation d'ébauches chondrogènes d'embryons de poulet cultivées *in vitro* sur différents milieux. *C. R. Acad. Sci. Paris* **274**, 112–115

Telser, A., H. C. Robinson, and A. Dorfman (1965). The biosynthesis of chondroitin-sulfate protein complex. *Proc. Nat. Acad. Sci. U.S.A.* **54**, 912–919.

Thorp, F. K., and A. Dorfman (1967). Differentiation of connective tissues. In *Current Topics in Developmental Biology*, Vol. 2 (A. A. Moscona and A. Monroy, eds.). Academic Press, New York, pp. 151–190.

Trelstad, R. L., A. H. Kang, S. Igarashi, and J. Gross (1970). Isolation of two distinct collagens from chick cartilage. *Biochemistry* **9**, 4993–4998.

Weiss, P. A. (1973). Differentiation and its three facets: Facts, terms and meaning. *Differentiation* **1**, 3–10.

CHAPTER SEVEN

Lectins as Probes for Changes in Membrane Dynamics in Malignancy and Cell Differentiation

LEO SACHS

Molecules that bind specifically to carbohydrate-containing sites on the surface membrane can be used to elucidate changes in the surface membrane associated with changes in the regulation of cell growth (Inbar and Sachs, 1969a; Burger, 1969; Sela et al., 1970). Using as a probe the carbohydrate-binding protein con-canavalin A (Con A) Sumner and Howell, 1936; Edelman et al., 1972), differences between normal and malignant transformed cells have been shown in Con A-induced cell agglutinability (Inbar and Sachs, 1969a; Inbar et al., 1971a, 1972a, 1973a; Ben-Bassat et al., 1970), number and distribution of Con A binding sites (Inbar and Sachs, 1969b, 1973; Ben-Bassat et al., 1971; Shoham and Sachs, 1972; Inbar et al., 1972a; Nicolson, 1972), location of amino acid and carbohydrate transport sites (Inbar et al., 1971), Con A-induced cell toxicity (Shoham et al., 1970; Inbar et al., 1972b; Wollman and Sachs, 1972), and membrane stability and level of cellular ATP (Vlodavsky et al., 1973). Changes in the distribution of Con A binding sites (Ben-Bassat et al., 1971; Inbar and Sachs, 1973) and the movement of antigens on the cell surface (Taylor et al., 1971; Loor et al., 1972; Yahara and Edelman, 1972; Edidin and Weiss, 1972) have indicated that receptors can be mobile in a fluid surface membrane (Singer and Nicolson, 1972).

In the present paper, I will summarize our studies (Inbar and Sachs, 1973; Inbar et al., 1973b, c, d, e, f; Shinitzky et al., 1973; Sachs, 1973a, b; Sachs et al., 1973) to determine the mobility of Con A binding sites on the surface membrane as a probe for membrane fluidity of specific sites in relation to the regulation of cell growth in (*a*) normal and transformed fibroblasts, as examples of cells that form a solid tissue, and (*b*) normal lymphocytes and lymphoma cells, as examples of cells that are in suspension *in vivo,* as well as in relation to the normal differentiation of myeloid leukemic cells to macrophages and granulocytes. I will also present data on the mobility on normal lymphocytes of the sites for the lectins from wheat germ and soybean.

METHODS

Cells and Cell Cultures

The transformed fibroblasts used were a line derived from a Simian virus 40-induced golden hamster tumor. The normal fibroblasts were from tertiary cultures of golden hamster embryos. Normal and transformed fibroblasts were cultured in Eagle's medium with a fourfold concentration of amino acids and vitamins and 10% fetal calf serum. For the experiments, normal and transformed fibroblasts at 3–4 days after seeding were dissociated with a 0.02% EDTA solution (Inbar and Sachs, 1969a), by incubation for 15–30 minutes at 37°C, and the dissociated cells were then washed 3 times with phosphate-buffered saline (PBS) (pH 7.2). Normal lymphocytes were obtained from lymph nodes of 6–8 week old male CR/RAR rats. Lymphocytes were collected by teasing the tissue apart and allowing the pieces to sediment. The lymphoma cells were from an ascites form of a Moloney virus-induced lymphoma grown in A strain mice; 10^5 cells were inoculated intraperitoneally into adult mice, and the cells were used 10 days after inoculation. The normal lymphocytes and the lymphoma cells were collected from animals in PBS and used in the experiments after washing 3 times with PBS. The experiments with the myeloid leukemic cells were carried out with a tissue culture line that consists of two types of clones. One type contains cells (D^+) that can be induced to undergo normal differentiation to mature macrophages and granulocytes. The other type of clone contains cells (D^-) that could not be induced to differentiate (Fibach et al., 1973). The myeloid leukemic cells were cultured in Eagle's medium with 10% inactivated horse serum (56°C for 30 minutes). For the experiments, cells at 3–4 days after seeding were washed 3 times with PBS.

Assay for Agglutination

Concanavalin A (Con A) (Miles-Yeda) at a concentration of 30 mg ml^{-1} was kept as a solution in PBS containing 1 *M* NaCl at $-20°C$. To test for agglutination, 0.5 ml of Con A, diluted at different concentrations in PBS, was mixed with 0.5 ml of cell suspension in a 35-mm petri dish. The density and size of aggregates was scored on a scale from $-$ to $+ + + +$ after 30 minutes of incubation. The agglutination was specific, since it was completely inhibited when Con A was preincubated with 0.1 *M* α-methyl-ᴇ-mannopyranoside (α-MM) as a hapten inhibitor.

Assay for Binding of Radioactive Con A

Concanavalin A (Miles-Yeda) was labeled with ^{3}H-acetic anhydride by the method of Miller and Great (1972). The labeled Con A was purified by affinity chromatography on a Sephadex G-100 column and kept as a solution in PBS containing 1 *M* NaCl at $-20°C$. For the binding of ^{3}H-Con A to cells, 0.5 ml of ^{3}H-Con A, diluted at different concentrations in either PBS or PBS containing 0.1 M α-MM, was mixed with 0.5 ml of cell suspension in a centrifuge tube and incubated for 30 minutes. The cells were then washed 3 times with 5 ml of PBS, the pellet dissolved in O.1 N NaOH, and the radioactivity counted in Triton scintillation fluid. To calculate the amount of ^{3}H-Con A bound specifically, the amount bound in the presence of α-MM was subtracted from the amount bound in the absence of α-MM. The results on Con A binding are given as specific binding. Total cell protein was measured by the method of Lowry et al. (1951).

Assay for Binding of Fluorescent Con A

Fluorescein isothiocyanate-conjugated Con A (F-Con A) (Miles-Yeda) at a fluorescein protein ratio of 1.86 was kept as a solution in PBS containing 1 M NaCl at $-20°$C. For the experiments, cells were incubated with F-Con A for 30 minutes, the cells were washed with PBS, and the fluorescence was determined with a Leitz Ortholux microscope, using transmitted ultraviolet light. With all cell types tested, 95–100% of the cells were stained at a concentration of 100 μg F-Con, A ml^{-1}. The binding of F-Con A to the membrane was specific, since it was inhibited when F-Con A was preincubated with 0.1 M α-MM as a hapten inhibitor. Rotational diffusion analysis of fluorescent Con A and the other fluorescent lectins was carried out as described (Shinitzky et al., 1973).

RESULTS

Final Distribution of Con A Binding Sites on
the Surface Membrane of Normal and Malignant Transformed Fibroblasts

In order to determine the final distribution of Con A binding sites on the surface membrane, the interaction of F-Con A with normal and malignant transformed fibroblasts was examined. The experiments indicated that in 99% of the transformed fibroblasts the surface binding of F-Con A was in clusters of fluorescence over the cell surface, as shown in Fig. 1 *C*. Preincubation of the transformed cells with NaN$_3$ or DNP did not inhibit the cluster formation. However, most of the normal fibroblasts gave a diffuse, semirandom fluorescence covering the surface membrane (Fig. 1B.). Binding of F-Con, A after fixation of the fluid surface membranes of normal and transformed fibroblasts with aldehyde or LaCl$_3$ resulted in an apparently completely random distribution of Con A binding sites (Fig. 1A).

Final Distribution of Con A Binding Sites on the Surface
Membrane of Normal Lymphocytes and Malignant Lymphoma Cells

Experiments on the binding of F-Con A to the surface membranes of normal lymphocytes and lymphoma cells have indicated that in 99% of the lymphoma cells the F-Con A surface binding was in small or large clusters of fluorescence that formed an incomplete ring on the cell periphery (Fig. 1C.), as in the transformed fibroblasts. However, about 30% of the normal lymphocytes gave a polar fluorescence cap, covering about half of the cell surface area (Fig. 1D). Preincubation of cells with NaN$_3$ or DNP inhibited cap formation in normal lymphocytes, but did not inhibit cluster formation in the lymphoma cells (Table 1). Addition of NaN$_3$

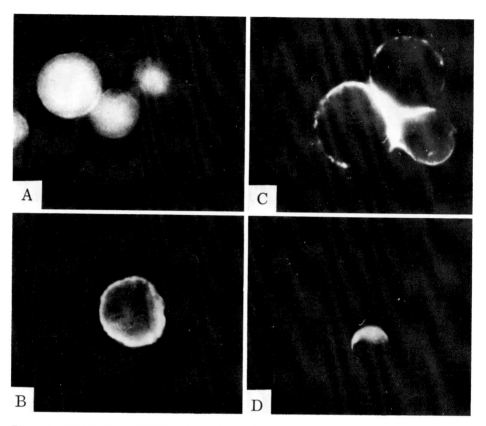

Figure 1. Distribution of F-Con A on the surface membrane of normal and transformed cells. (A) Lymphoma cells after fixation with 2.5% glutaraldehyde. Similar results were obtained with normal lymphocytes, normal fibroblasts, and transformed fibroblasts. The binding of F-Con A shows an apparently completely random distribution. (B) Distribution of F-Con A of the type seen with normal fibroblasts, the formation of a semirandom distribution. (C) Distribution of F-Con A of the type seen with transformed fibroblasts and lymphoma cells, the formation of clusters. (D) Cap formation in normal lymphocytes.

or DNP to cells after the binding of F-Con A resulted in dissociation of the caps, but not the clusters. Cap, but not cluster, formation, therefore, requires energy. Formation of caps was also inhibited by low temperature. The binding of F-Con A after fixation of the surface membranes of normal lymphocytes and lymphoma cells with aldehyde or $LaCl_3$ resulted in an apparently completely random distribution of Con A-binding sites (Fig. 1A).

Fluorescent wheat germ and soybean agglutinins did not produce caps on either normal lymphocytes or malignant lymphoma cells. In normal lymphocytes that showed caps with Con A, the rest of the cell still stained with fluorescent wheat germ agglutinin.

Redistribution of Con A Binding Sites Induced by Con A

Differences in the final distribution of F-Con A on the surface membrane could be due to movement of Con A molecules, movement of membrane sites in the absence of Con A, or movement of Con A-binding sites to form a new distribution only after interaction with Con A molecules. The experiments were carried out at saturation conditions in which all the membrane sites were occupied by Con A, excluding the possibility that the differences were due to movement of Con A molecules. The second possibility (movement of membrane site without Con A) was excluded by the following experiment. Incubation of normal lymphocytes with F-Con A at 0 or 37°C resulted in binding of F-Con A, but the cells formed caps only at 37°C. To determine whether the formation of caps at 37°C is a result of movement of membrane sites without Con A, normal lymphocytes were incubated at 0 and 37°C for 30 minutes, followed by aldehyde fixation at the two temperatures, and F-Con A was added after fixation. The results showed that cap formation was completely abolished when Con A was added after fixation, indicating that caps were not formed in the absence of Con A.

The binding of F-Con A after fixation of the cell membrane (Fig. 1A) suggests that the binding sites for Con A are floating in a fluid membrane in a random distribution in normal and transformed cells. This random distribution can be changed by interaction with Con A molecules, and the final distribution of sites was different in the various cell types studied (Fig. 1.). Cells from a mouse myeloid leukemia cell line (D⁻) in which about 5% of the cells formed caps with F-Con A were used to obtain further evidence on the redistribution of Con A membrane sites by Con A molecules. The results showed that, when a higher cross-linking was induced by adding anti-Con A antibodies or glycogen after the binding of Con A, the percentage of cells with caps increased from about 5 to 25%. Trypsinization of the cells increased cap formation to about 45%, and addition of anti-Con A antibodies or glycogen raised the percentage of the trypsinized cells with caps to about 75% (Table 2).

These data support the conclusion that the redistribution of membrane sites is induced by Con A, and also show that trypsinization of the cells increased the mobility of these sites.

Mobility of Con A Binding Sites and the Differential Agglutination of Normal and Transformed Cells

The experiments on the binding of F-Con A to normal and transformed cells indicate that the degree of site mobility increased from no or almost no change in the random distribution in normal fibroblasts (Fig. 1B) to the formation of clusters in the transformed fibroblasts and lymphoma cells (Fig. 1C) to the formation of caps in normal lymphocytes (Fig. 1D). Agglutination experiments with Con A indicate

TABLE 1. INHIBITION^a OF MOVEMENT OF CON A BINDING SITES

Treatment	Inhibition of Cluster Formation	Inhibition of Cap Formation
Formaldehyde, 10%	+	+
Glutaraldehyde, 2.5%	+	+
LaCl$_3$, 10^{-2}M	+	+
NaN$_3$, 10^{-2}M	—	+
DNP, 10^{-3}M	—	+

^{a} + = Inhibition; — = not inhibited.

TABLE 2. INDUCTION OF CAP FORMATION BY ANTI-CON A ANTIBODIES OR GLYCOGEN IN NONTRYPSINIZED AND TRYPSINIZED MYELOID LEUKEMIC CELLS (D⁻)

Cells	Treatment	Cells with Cap (%)
Nontrypsinized	F-Con A	5 ± 0.5
	F-Con A + anti-Con A	22 ± 2
	F-Con A + glycogen	25 ± 2
Trypsinized	F-Con A	45 ± 2
	F-Con A + anti-Con A	73 ± 5
	F-Con A + glycogen	75 ± 5

Cells were treated with 1 μg purified trypsin for 15 min at 37°C. Anti-Con A antibodies and glycogen (100 μg) were added to cells after binding of F-Con A.

that only the cells with an intermediate degree of site mobility that form clusters are highly agglutinated cells. Normal fibroblasts with a low degree of site mobility and normal lymphocytes with a high degree of mobility have low agglutinability (Table 3). In each cell system, a similar number of radioactively labeled Con A molecules were bound to normal and transformed cells per unit protein (Table 4). The results indicate that a threshold amount of cluster formation of Con A binding sites is associated with agglutination. Further evidence for this assumption was obtained by producing the three degrees of site mobility in the same cell type. Normal lymphocytes were treated with $LaCl_3$ or NaN_3, and the untreated and treated cells tested for Con A agglutinability. The untreated normal lymphocytes with caps and the $LaCl_3$-treated lymphocytes with random distribution showed a

TABLE 3. FINAL DISTRIBUTION OF CON A BINDING SITES AND CELL AGGLUTINATION

Cell Type	Final Distribution of Con A Binding Sites	Agglutination by Con A
Normal fibroblasts	Semirandom	−
Transformed fibroblasts and lymphoma cells	Many clusters	+ + + +
Normal lymphocytes	Caps	±

TABLE 4. SPECIFIC BINDING OF ³H-CON A TO SURFACE MEMBRANES OF NORMAL LYMPHOCYTES AND LYMPHOMA CELLS[a]

Concentration of ³H-Con A (μgml⁻¹)	Input (CPM/10⁷ Cells)	Specific Binding (CPM/100 μg Cell Protein)	
		Normal Lymphocytes	Lymphoma Cells
1	1,215	1235	985
2.5	2,782	2650	2180
5	5,800	4200	4020
10	11,717	5170	5010
50	56,385	6580	7030
100	116,615	7150	7570

[a] Similar results were obtained with normal and transformed fibroblasts.

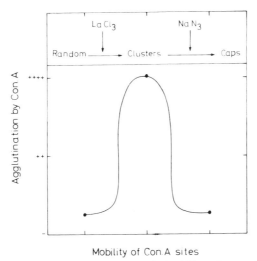

Figure 2. Mobility of Con A binding sites and agglutinability of normal lymphocytes by Con A. The untreated normal lymphocytes with caps and the LaCl₃-treated lymphocytes with random distribution showed a low degree of agglutination (±). However, the NaN₃-treated lymphocytes with clustered distribution showed a high degree of agglutination (+ + + +).

low (±) degree of agglutination by Con A. However, the NaN₃-treated lymphocytes, which had a clustered distribution, also had a high degree of agglutinability (Fig. 2).

The formation of clusters of Con A binding sites on the surface membrane of transformed cells (Fig. 1C) was inhibited by fixation of the fluid state of the membrane (Fig. 1A). Fixation also inhibited cell agglutination by Con A (Fig. 3), although the fixed and the unfixed cells bound similar numbers of radioactively labeled Con A molecules (Fig. 4).

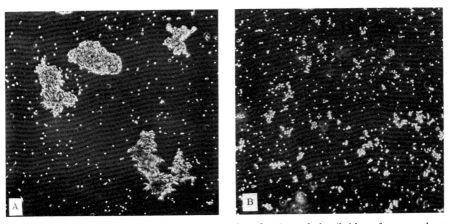

Figure 3. Agglutinability of transformed cells after fixation of the fluid surface membrane. (A) Unfixed lymphoma cells. (B) Lymphoma cells after fixation with 2.5% glutaraldehyde. Agglutination was also inhibited by cell fixation with 10% formaldehyde or 10⁻² M LaCl₃. Similar results were obtained with transformed fibroblasts.

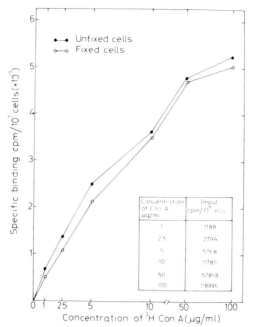

Figure 4. Specific binding of ³H-Con A molecules to the surface membrane after fixation with 10% formaldehyde. Unfixed cells, closed symbols; fixed cells, open symbols.

Mobility of Con A Binding Sites and Normal Differentiation of Myeloid Leukemic Cells to Macrophages and Granulocytes

The experiments with myeloid leukemia were carried out with a tissue culture line of myeloblastic leukemia cells that consists of two types of clones. One type contains cells (D⁺) that can be induced to undergo normal differentiation to mature macrophages and granulocytes. The other type of clone contains cells (D⁻) that could not be induced to differentiate. In soft agar, the D⁻ cells form compact colonies (Fig. 5A), whereas the D⁺ cells form diffuse colonies because of migration of the differentiated cells (Fig. 5B).

In order to determine possible differences in the location and mobility of Con A binding sites on the surface membrane, we examined the interaction of F-Con A with D⁻ and D⁺ cells 3–4 days after seeding. The results indicate that in D⁻ clones about 95% of the cells showed surface binding of F-Con A in the form of a ring of clusters on the cell periphery (Fig. 1C). However, about 50% of the cells in D⁺ clones and about 5% of the cells in D⁻ clones showed a polar fluorescence cap (Fig. 1D). Similar results were obtained with 4D⁻ and 4D⁺ clones (Fig. 6). Measurement of the cell protein and surface area indicate that D⁻ and D⁺ cells have similar contents of cellular protein and similar surface areas per cell. The binding of radioactively labeled Con A molecules indicates that D⁻ and D⁺ cells bind similar numbers of Con A molecules per unit of cell protein and appear to have similar affinities for Con A (Fig. 7). The results show a difference in the mobility of Con A binding sites in these two types of cells and suggest that a gain in the ability of myeloid leukemic cells to undergo normal differentiation is associ-

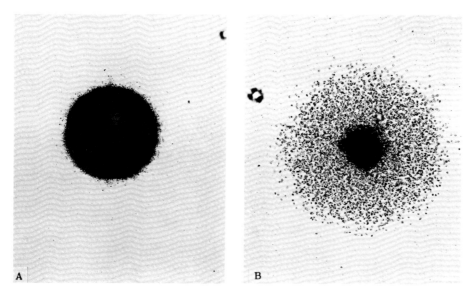

Figure 5. D⁻ and D⁺ colonies grown in soft agar in the presence of the inducer for cell differentiation. (*A*) Compact colony of D⁻ cells. (*B*) Dispersed colony of D⁺ cells.

ated with an increase in the mobility of Con A binding sites on the surface membrane.

Rotational Diffusion of Lectins Bound to the Cell Surface Membrane

The thermal motion of a specific site on a biomembrane can be analyzed by determining the fluorescence polarization characteristics of a fluorophone specifically attached to it. This method is especially adequate for studies on the mobility of receptor sites utilizing specific fluorescent hormones, antigens, or lectins (Shinitzky et al., 1973; Inbar et al., 1973e). It can also be used to study membrane lipids (Inbar et al., 1974).

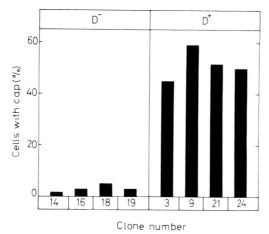

Figure 6. Cap formation after binding of F-Con A in four D⁻ and four D⁺ clones.

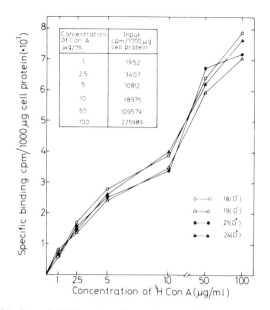

Concentration of Con A µg/ml	Input cpm/1000 µg cell protein
1	1952
2.5	3407
5	10812
10	18975
50	109579
100	225989

Figure 7. Specific binding of ^{3}H-Con A to the surface membrane of D$^-$ and D$^+$ cells.

In order to determine directly the degree of mobility of Con A binding sites on the surface membrane, we have defined a quantitative scale that extends from 0, the immobilized state, to 1, the fully mobile state. The data obtained with this method (Inbar et al., 1973e) support the conclusion that the binding sites for Con A on the surface membrane are mobile and that different cell types have different degrees of mobility (Table 5). Comparison of normal lymphocytes and lymphoma cells have also shown that lymphoma cells with a decreased mobility of Con A binding sites have an increased fluidity of the lipids in the surface membrane. This increased fluidity of the lipids was also found in human chronic lymphatic leukemia (Inbar et al., 1974). The lower fluidity of the surface membrane lipids of normal lymphocytes is due to a higher cholesterol to phospholipid ratio (Vlodavsky and Sachs, 1974).

The mobilities of F-Con A, fluorescent wheat germ, and fluorescent soybean agglutinins bound to normal lymphocytes were 0.82, 0.37, and 0.43, respectively (Table 6). The lower degree of mobility of the wheat germ and soybean agglutinins was associated with the lack of cap formation by these two lectins and their failure to activate DNA synthesis in normal lymphocytes (Inbar et al., 1973a; Shinitzky et al., 1973).

TABLE 5. DEGREE OF MOBILITY OF CON A BINDING SITES ON SURFACE MEMBRANES OF NORMAL AND MALIGNANT TRANSFORMED CELLS, DETERMINED BY FLUORESCENCE POLARIZATION ANALYSIS

Cells	Degree of Mobility[a]
Normal fibroblasts	0.48
Transformed fibroblasts	0.79
Normal lymphocytes	0.82
Lymphoma cells	0.36

[a] 0 = Immobilized state; 1 = fully mobile state.

TABLE 6. DEGREES OF MOBILITY OF CON A, WHEAT GERM, AND SOYBEAN LECTIN BINDING SITES ON THE SURFACE MEMBRANE OF NORMAL LYMPHOCYTES DETERMINED BY FLUORESCENCE POLARIZATION ANALYSIS

Lectin	Degree of Mobility
F-Con A	0.83
F-WGA[a]	0.37
F-SBA[b]	0.43

[a] Fluorescein isothiocyanate-conjugated wheat germ lectin (Miles Yeda).
[b] Fluorescein isothiocyanate-conjugated soybean lectin (Miles Yeda).

CONCLUSIONS

Our results (Inbar and Sachs, 1973; Inbar et al., 1973, b, c, d, e; Shinitzky et al., 1973; Sachs, 1973a, b; Sachs et al., 1973) indicate that carbohydrate-containing structures which are associated with binding sites for Con A can be mobile on the cell surface, and that in the cells studied there was a difference in the mobility of these structures in normal and malignant transformed cells. With the mobility of Con A binding sites used as a probe for fluidity of these sites, the results show that with fibroblasts (cells which form a solid tissue *in vivo*) the transformation of normal into malignant cells is associated with an increase in membrane fluidity of the carbohydrate-containing structures where the Con A sites are located. However, with lymphocytes (cells that are in suspension *in vivo*) the malignant transformation is associated with a decreased fluidity of these structures on the surface membrane. The increased fluidity of specific sites on the membrane in the transformed fibroblasts can explain their lack of contact inhibition, their ability to grow in soft agar, and their malignancy in solid tissue. The studies with normal lymphocytes show that in these cells there is a higher degree of mobility of the Con A sites than of the sites for the lectins from wheat germ and soybean. These two lectins, in contrast to Con A, also do not activate normal lymphocytes to undergo DNA synthesis (Inbar et al., 1973a).

The results also show that there is a difference in the fluid states of the Con A binding structures on the surface membrane in myeloid leukemic cells that can be induced to differentiate normally (D[+] cells) and myeloid leukemic cells that could not be induced to differentiate (D[-] cells). A higher degree of fluidity was associated with a gain in the ability of myeloid leukemic cells to undergo normal differentiation to mature macrophages and granulocytes. The increased fluidity in D[+] cells was associated with an increased migration of the differentiated cells in soft agar. Differences in the membrane fluidity of specific sites may explain differences in the ability of cells to respond to other differentiation-inducing stimuli, in the cellular response to hormones, and in cell migration in embryonic development and carcinogenesis. The differences in fluidity of specific membrane sites may result in differences in the transport and internal concentration of specific chemicals that regulate cell growth and differentiation (Sachs, 1973a).

ACKNOWLEDGMENT

This study was supported by Contract No. NO1-CP-43241 with the Virus Cancer Program of the National Cancer Institute, National Institutes of Health.

REFERENCES

Ben Bassat, H., M. Inbar, and L. Sachs (1970). Requirement for cell replication after SV40 infection for a structural change of the cell surface membrane. *Virology* **40**, 854–859.

Ben-Bassat, H., M. Inbar, and L. Sachs (1971). Changes in the structural organization of the surface membrane in malignant cell transformation. *J. Membrane Biol.* **6**, 183–194.

Burger, M. M. (1969). A difference in the architecture of the surface membrane of normal and virally transformed cells. *Proc. Nat. Acad. Sci. U.S.A.* **62**, 994–1001.

Edelman, G. M., B. A. Cunningham, G. N. Reeke, J. W. Becker, M. J. Waxdal, and J. L. Wang (1972). The covalent and three-dimensional structure of concanavalin A. *Proc. Nat. Acad. Sci. U.S.A.* **69**, 2580–2584.

Edidin, M., and A. Weiss (1972). Antigen cap formation in cultured fibroblasts: A reflection of membrane fluidity and cell motility. *Proc. Nat. Acad. Sci. U.S.A.* **69**, 2456–2459.

Fibach, E., M. Hayashi, and L. Sachs (1973). Control of normal differentiation of myeloid leukemic cells to macrophages and granulocytes. *Proc. Nat. Acad. Sci. U.S.A.* **70**, 343–346.

Inbar, M., and L. Sachs (1969a). Interaction of the carbohydrate binding protein concanavalin A with normal and transformed cells. *Proc. Nat. Acad. Sci. U.S.A.* **63**, 1418–1425.

Inbar, M., and L. Sachs (1969b). Structural differences in sites on the surface membrane of normal and transformed cells. *Nature* **223**, 710–712.

Inbar, M., and L. Sachs (1973). Mobility of carbohydrate containing sites on the surface membrane in relation to the control of cell growth. *FEBS Lett.* **32**, 124–128.

Inbar, M., Z. Rabinowitz, and L. Sachs (1969). The formation of variants with a reversion of properties of transformed cells. III. Reversion of the structure of the cell surface membrane. *Int. J. Cancer* **4**, 690–696.

Inbar, M., H. Ben-Bassat, and L. Sachs (1971a). A specific metabolic activity on the surface membrane in malignant cell-transformation. *Proc. Nat. Acad. Sci. U.S.A.* **68**, 2748–2751.

Inbar, M., H. Ben-Bassat, and L. Sachs (1971b). Location of amino acid and carbohydrate transport sites in the surface membrane of normal and transformed mammalian cells. *J. Membrane Biol.* **6**, 195–209.

Inbar, M., H. Ben-Bassat, and L. Sachs (1972a). Membrane changes associated with malignancy. *Nature New Biol.* **236**, 3–4, 16.

Inbar, M., H. Ben-Bassat, and L. Sachs (1972b). Inhibition of ascites tumor development by concanavalin A. *Int. J. Cancer* **9**, 143–149.

Inbar, M., H. Ben-Bassat, L. Sachs (1973a). Temperature sensitive activity on the surface membrane in the activation of lymphocytes by lectins. *Exp. Cell. Res.* **76**, 143–151.

Inbar, M., H. Ben-Bassat, and L. Sachs (1973b). Difference in the mobility of lectin sites on the surface membrane of normal lymphocytes and malignant lymphoma cells. *Int. J. Cancer* **12**, 93–99.

Inbar, M., H. Ben-Bassat, E. Fibach, and L. Sachs (1973c). Mobility of carbohydrate-containing structures on the surface membrane and the normal differentiation of myeloid leukemic cells to macrophages and granulocytes. *Proc. Nat. Acad. Sci. U.S.A.* **70**, 2577–2581.

Inbar, M., C. Huet, A. R. Oseroff, H. Ben-Bassat and L. Sachs (1973d). Inhibition of lectin agglutinability by fixation of the cell surface membrane. *Biochim. Biophys. Acta* **311**, 594–599.

Inbar, M., M. Shinitzky and L. Sachs (1973e). Rotational relaxation time of concanavalin A bound to the surface membrane of normal and malignant transformed cells. *J. Mol. Biol.* **81**, 245.

Inbar, M., M. Shinitzky, and L. Sachs (1974). Microviscosity in the surface membrane lipid layer of intact normal lymphocytes and leukemic cells. *FEBS Lett.* **38**, 268.

Loor, F., L. Forni, and B. Pernis (1972). The dynamic state of the lymphocyte membrane: factors affecting the distribution and turnover of surface immunoglobulins. *Eur. J. Immunol.* **2**, 203–212.

Lowry, O. H., N. J. Rosenbrough, A. L. Farr, and R. L. Randall (1951). Protein measurement by the folin phenol reagent. *J. Biol. Chem.* **193**, 265–276.

Miller, I. R., and H. Great (1972). Protein labelling by acetylation. *Biopolymers* **11**, 2533–2536.

Nicolson, G. L. (1972). Topography of membrane concanavalin A sites modified by proteolysis. *Nature New Biol.* **239**, 193–197.

Sachs, L. (1973a). Regulation of membrane changes, differentiation and malignancy in carcinogenesis. *Harvey Lectures*, in press.

Sachs, L. (1973b). Control of growth and differentiation in normal hematopoietic and leukemic cells. In *Control of Proliferation in Animal Cells.* Cold Spring Harbor Laboratories, New York (in press).

Sachs, L., M. Inbar, and M. Shinitzky (1973). Mobility of lectin sites on the surface membrane and the control of cell growth and differentiation. In *Control of Proliferation in Animal Cells.* Cold Spring Harbor Laboratories, New York (in press).

Sela, B., H. Lis, N. Sharon, and L. Sachs (1970). Different location of carbohydrate containing sites in the surface membrane of normal and transformed mammalian cells. *J. Membrane Biol.* **3**, 267–279.

Shinitzky, M., M. Inbar, and L. Sachs (1973). Rotational diffusion of lectins bound to the surface membrane of normal lymphocytes. *FEBS Lett.* **34**, 247–250.

Shoham, J., and L. Sachs (1972). Differences in the binding of fluorescent concanavalin A to the surface membrane of normal and transformed cells. *Proc. Nat. Acad. Sci. U.S.A.* **69**, 2479–2482.

Shoham, J., M. Inbar, and L. Sachs (1970). Differential toxicity on normal and transformed cells *in vitro* and inhibition of tumor development *in vivo* by concanavalin A. *Nature* **227**, 1244–1246.

Singer, S. H., and G. L. Nicolson (1972). The fluid mosaic model of the structure of cell membranes. *Science* **175**, 720–731.

Sumner, J. B., and S. F. Howell (1936). The identification of the hemagglutinin of the Jack bean with concanavalin A. *J. Bacteriol.* **32**, 227–238.

Taylor, R. B., W. P. H. Duffus, M. C. Raff, and S. De Petris (1971). Redistribution and pinocytosis of lymphocyte surface immunoglobulin molecules induced by anti-immunoglobulin antibody. *Nature New Biol.* **233**, 225–229.

Vlodavsky, I., and Sachs, L. (1974). Difference in the cellular cholesterol to phospholipid ratio in normal lymphocytes and lymphocytic leukemic cells. *Nature*, in press.

Vlodavsky, I., M. Inbar, and L. Sachs (1973). Membrane changes and adenosine triphosphate content in normal and malignant transformed cells. *Proc. Nat. Acad. Sci. U.S.A.* **70**, 1780–1784.

Wollman, Y., and L. Sachs (1972). Mapping of sites on the surface membrane of mammalian cells. II. Relationship of sites for concanavalin A and an ornithine-leucine copolymer. *J. Membrane Biol.* **10**, 1–10.

Yahara, I., and G. M. Edelman (1972). Restriction of the mobility of lymphocyte immunoglobulin receptors by concanavalin A. *Proc. Nat. Acad. Sci. U.S.A.* **69**, 608–612.

CHAPTER EIGHT

Receptor Specificity and Mitogenesis in Lymphocyte Populations

GERALD M. EDELMAN, PATRICIA G. SPEAR, URS RUTISHAUSER,
AND
ICHIRO YAHARA

The system of lymphoid cells and antibodies that gives rise to immune responses promises to be a valuable tool for analyzing problems in developmental biology. Moreover, the immune response is itself a useful model for analyzing growth control and other associated phenomena. Lymphocytes are particularly suitable for studying cell surface interactions and growth control because they are readily available as dissociated cells, because they produce, in addition to specific surface markers, gene products of known structure such as immunoglobulins, and because they can be stimulated from the resting state by various mitogenic agents that bind to the cell surface.

Most of the available evidence (Cairns, 1967) suggests that the immune response is selective, that is, the information for generating antibodies of various specificities already exists before contact with an antigen. Interactions with an antigen of antigen-binding cells having appropriately complementary antibodies provokes clonal expansion, which consists of cellular maturation, cell division, and increased protein synthesis. In the present paper, we shall consider some experiments designed to analyze the origin of antigen-binding cells during development, to determine their receptor specificities and ranges of binding, and to study the effects of molecular interactions with lymphocyte surface receptors.

The results of these experiments suggest that antigen-binding cells appear rapidly at the beginning of organogenesis of the spleen. They also suggest that the specificity of lymphocyte responses depends not only on their capacity to bind antigen but also on their threshold for the triggering of clonal expansion. Additional experiments prompt the hypothesis that lymphocyte receptors interact with a protein capable of modulating their mobility in the plane of the cell membrane. It is possible that such mechanisms play a role in the early steps of mitogenic stimulation.

ONTOGENY OF SPECIFIC ANTIGEN-BINDING CELLS

The production of humoral antibodies in response to antigen depends on the interaction of two kinds of antigen-specific lymphocytes that differentiate along two separate pathways, one mediated by the thymus, producing T cells, and the other mediated by the bursa in chickens or its equivalent in mammals, producing B cells capable of secreting immunoglobulins (Mitchell and Miller, 1968). The development of immune competence and specificity depends both on the appearance of a sufficient number and variety of antigen-binding cells (both T cells and B cells) and on their maturation to a state capable of responding to antigens.

An understanding of the maturation of the immune system requires some knowledge concerning the times of appearance of T cells, B cells, and specific antigen-binding cells in relation to the onset of immune responses. Recent experiments (Spear et al., 1973) on the fetal immunology of the mouse have provided the beginnings of a quantitative analysis of the generation of various lymphoid cell types and antigen-binding specificites *in utero* and after birth.

In Fig. 1 are shown the total numbers of nucleated cells, B cells, and T cells in the spleens of Swiss-L mice at various ages. As early as 15 days of gestation, T cells were detected by their θ antigen and B cells by their immunoglobulin receptors. By day 16 of gestation, 1% of nucleated cells were Ig-positive and almost 8% were θ positive. The B cells increased in number at a faster rate than the T cells and became more numerous than T cells within 24 hours after birth (Fig. 1). During this interval, however, most of the cells in the spleen were neither T nor B cells, a situation consistent with evidence that the spleen of the fetal mouse is predominantly erythropoietic and granulocytic. During the first week after birth the rate of increase in the numbers of all cell types in the spleen decreased, reaching a plateau during the second week. Moreover, the ratio of T cells to B cells remained relatively constant during this interval. After 14 days, there was a significant increase in the number of spleen cells, due primarily to increases in the B-cell population; this change was reflected in a decrease in the T cell/B cell ratio.

In the mouse, the splenic anlage is first detectable on day 13 of gestation. It is clear from the experiments summarized in Fig. 1 that lymphocytic infiltration of the spleen occurs earlier than was previously suggested on the basis of conventional histologic techniques (Metcalf and Moore, 1971). Presumably, the T cells detected in the spleen migrate from the thymus. If so, there is a substantial flow of cells from the thymus to the spleen at a time when the thymus is undergoing many changes during its own development.

The site at which B-cell precursors differentiate to become Ig-positive B cells remains to be determined for mammalian species. By transferring fetal tissues to

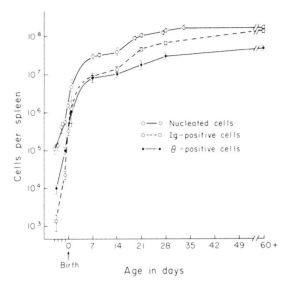

Figure 1. Numbers of nucleated cells, Ig-positive cells or B cells, and θ-positive cells or T cells per spleen in Swiss-L mice as a function of age. Numbers of nucleated cells were determined directly by counting spleen cell suspensions in hemocytometers. The numbers of animals at the same age used to prepare each spleen cell suspension ranged from 3 for adults to as many as 40 (4 litters) for fetuses. Each value shown is the arithmetic mean of 4–7 independent determinations ± the standard error of the mean. The fraction of nucleated cells bearing Ig receptors or the θ antigen at each age was determined by immunofluorescence. Multiplication of the fractions obtained by the mean number of nucleated cells per spleen at the appropriate age yielded the numbers of Ig-positive and θ-positive cells per spleen. These calculated values are shown with vertical bars representing the propagated standard error of the product of the means. (From Spear et al., 1973.)

irradiated adults, Tyan and Herzenberg (1968) have demonstrated that the liver is the major source of B-cell precursors from day 10 of gestation until birth. By 2 or 3 days before birth, these cells can also be found in other organs, including the spleen, bone marrow, and blood. Nossal and Pike (1972) found that, during the development of CBA mice, Ig-positive cells appeared in the bone marrow and liver later than in the spleen and blood; these experiments appear to rule out the fetal liver and bone marrow as sites for the earliest detectable expression of Ig receptors. These workers found also that the percentage of B cells among nucleated cells before birth was slightly higher in the blood than in the spleen. Because the *total* number of B cells was probably considerably larger in the spleen than in the blood, it is still possible that the commitment of B-cell precursors to the production and expression of one kind of immunoglobulin receptor occurs first in the spleen.

The role of endogenous antigens in the proliferation of various types of lymphocytes and their migration to the spleen remains to be assessed. The appearance of lymphocytic cells in the spleen and the rapid increases in their numbers before and immediately after birth are probably not influenced by foreign antigens, however, for the fetus is largely protected from foreign substances. Furthermore, the exposure to antigens from the environment at birth apparently does not stimulate faster rates of increase for B cells, T cells, or antigen-binding cells (Figs. 1 and 2).

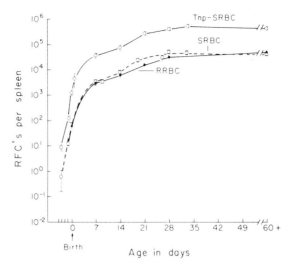

Figure 2. Numbers of rosette-forming cells (RFC) specific for Tnp-derivatized sheep red blood cells, sheep red blood cells, and rabbit red blood cells in the spleens of Swiss-L mice as a function of age. To detect antigen-binding cells by the rosette assay, 0.5-ml aliquots of medium containing 1–2 × 10^6 nucleated spleen cells and 10^7 red blood cells were centrifuged at 200g for 10 minutes. The pellets were gently resuspended, and aliquots of known volume were placed between two microscope slides and scanned microscopically for rosettes (lymphocytes with at least 7 adherent red blood cells). Multiplication of the number of RFC per 10^6 nucleated cells by the mean number of nucleated cells per spleen (Fig. 1) yielded the number of RFC per spleen for each antigen at each age tested. Vertical bars represent the propagated standard error of the product of the means. (From Spear et al., 1973.)

What can we say about the appearance of antigen-binding cells of different specificities? Cells capable of binding to each of three different exogenous antigens [trinitrophenylated (Tnp) sheep red blood cells, sheep red blood cells, and rabbit red blood cells] were first detected by rosette assay between days 15 and 16 of gestation in the spleens of animals pooled according to age (Fig. 2). Their numbers increased rapidly and in parallel until about 1 week after birth. At all ages tested, there were approximately 10 times as many Tnp-specific cells as cells specific for either erythrocyte type. The cells that bound to the two kinds of underivatized erythrocytes were probably partially overlapping populations. Nevertheless, a significant proportion of the lymphocytes that bound to sheep red blood cells did not bind to rabbit red blood cells and vice versa, inasmuch as rosette assays performed with a mixture of the two kinds of erythrocytes yielded 50% more rosette-forming cells (RFC) than were obtained with either type of erythrocyte alone.

In addition to these studies on rates of appearance, an attempt was made to compare the range of avidities for the Tnp hapten in adult and fetal antigen-binding cells. The relative avidity with which a lymphocyte binds to a soluble Tnp-protein conjugate can be estimated from the concentration of conjugate required to inhibit rosette formation with Tnp-erythrocytes. The formation of rosettes by the cells of higher avidity will be inhibited by the lower concentrations of the soluble competitor, and differences in the avidity distributions of two cell populations can readily be demonstrated by using several concentrations of soluble competitor to inhibit rosette formation.

Spleen cells pooled from several unimmunized adults, as well as from a number of 18-day fetuses, were mixed with various concentrations of Tnp-hemocyanin before centrifugation in the rosette assay. The results shown in Fig. 3 indicate that the degree of inhibition of fetal and adult Tnp-RFC at several Tnp-hemocyanin concentrations was not significantly different. Within the error of this test, therefore, the fetal antigen-binding cells exhibited the same range of avidities for the antigen as the adult cells. In view of the time of appearance of cells specific for the three antigens tested and the observation that the proportions of cells with different specificities remained constant with age, it appears that, for the antigens used, the expression of one specificity does not lag more than 24 hours behind the expression of another. Because the range of avidities is great, it seems likely that the variety of different specificities that can be expressed in the fetus will also be found to be large.

According to certain theories of the origin of antibody diversity, the various specificities are generated by a process of somatic mutation of a few germ line genes and subsequent selection of suitable mutations after the expression of antigen-binding receptors on cell surfaces. If this were correct, one would expect the variety of specificities expressed *early* in the fetus to be quite limited. For the few antigens tested so far, however, no restriction was found in the range of specificities that can be expressed in the fetus. If the results obtained with the three antigens used can be extended to other antigens, severe constraints would be placed on somatic mutation theories of antibody diversity. It would become necessary, for example, to postulate that many suitable mutations occur before day 16 of gestation, at a time when the total number of lymphoid cells is quite small. Furthermore, because

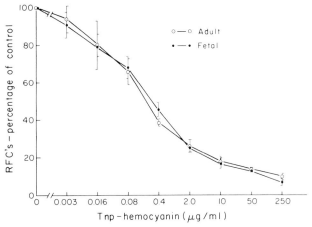

Figure 3. Number of rosettes by fetal or adult spleen cells with Tnp-derivatized sheep red blood cells in the presence of Tnp-hemocyanin at several concentrations, expressed as the percentage of uninhibited control values. The spleen cell suspensions were prepared from 50 fetuses (4 litters) at 18 days of gestation and from 3 unimmunized adults. These suspensions were assayed for Tnp-specific RFC (see legend to Fig. 2) in the presence or absence of Tnp-hemocyanin as indicated. Each value shown is the mean of 3 independent determinations ± the standard error of the mean. (From Spear et al., 1973.)

immunoglobulin receptors are not detected before 15 days of gestation, an indirect mechanism would have to operate to afford selective advantage of the mutated genes over the germ line genes. Our results are more compatible with germ line or somatic recombination theories for the generation of antibody diversity. According to these theories, the different antigen-binding specificities arise during evolution or are generated by recombinational events among evolutionarily selected genes. The expression of the receptor specificities could occur at random within a short time interval in accord with the observed kinetics of appearance of antigen-binding cells.

Is the presence of antigen-binding cells sufficient for an immune response? In mice, the capacity to respond to antigen by the production of humoral antibody appears after birth, in contrast to some species in which this capacity arises *in utero*. Our experiments indicate that Swiss-L mice do not respond to immunization until the second week after birth, although antigen-binding cells are present earlier. The data shown in Fig. 4 indicate that the mice were able to produce antibody-

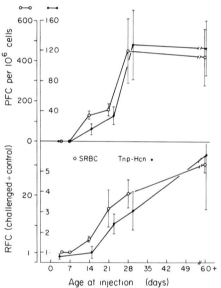

Figure 4. Numbers of antibody-secreting cells or plaque-forming cells (PFC) and RFC produced in response to sheep red blood cells and to Tnp-hemocyanin as a function of age in Swiss-L mice. Several mice of the same age were injected with sheep red blood cells, Tnp-hemocyanin absorbed onto bentonite, saline, or bentonite alone at the ages indicated. The spleens from the mice in each experimental group (numbers in each group ranged from 2 for adults to 5 for the youngest mice tested) were pooled before the assays were performed 5 days after injection. Mice injected with saline had less than 1 PFC per 10^6 spleen cells in assays with sheep red blood cells. Animals injected with bentonite had about 5 PFC per 10^6 spleen cells at every age tested when assayed against Tnp-derivatized sheep red blood cells. These background levels of PFC were subtracted from the values obtained for the numbers of PFC per 10^6 spleen cells in mice injected with the antigens. Each value shown in the top half of the figure represents the arithmetic mean of 3 independent determinations ± the standard error of the mean. The numbers of RFC per 10^6 spleen cells were determined for the experimental and control groups of animals in 3 separate series of experiments. The arithmetic means of the values obtained were used to calculate the ratios presented in the bottom half of this figure. The vertical bars indicate the propagated standard error of the quotient of the means. (From Spear et al., 1973.)

secreting cells in response to either of two antigens within 5 days when injected at 14 days of age or later, but not when injected at 7 days or earlier. A response to antigen consisting of elevated numbers of RFC was detected in animals injected with sheep red blood cells at 14 days of age or later, and in animals injected with Tnp-hemocyanin at 21 days or later.

Several studies, predominantly with lamb fetuses (Silverstein et al., 1963) but also with neonatal mice (Playfair, 1968), have revealed that the onset of response to different antigens occurs at different times. Furthermore, all individuals of the same strain follow the same temporal sequence in responding to a series of antigens. It has been suggested that this process may reflect the appearance or maturation of antigen-reactive clonal precursors at different times and rates. Because the mouse has a much shorter developmental period than the lamb, it is more difficult to demonstrate temporal sequences in the response to antigens and in the appearance of antigen-binding cells. In fact, our results show an almost simultaneous onset of response to at least two different antigens (Fig. 4), as well as the appearance of cells specific for these two antigens in the same 24-hour interval during gestation (Fig. 2). The additional finding that the ratio of RFC to B cells remains constant with age for three different antigens (Spear et al., 1973) suggests that other antigen-binding cells will also be found to arise at more or less the same time. It remains to be determined whether mice can respond at different times to two different antigens for which antigen-binding cells arise almost simultaneously.

Inasmuch as the response to antigen consists of the expansion of clones of antigen-binding cells as well as the production of antibodies, the proliferation of cells observed in Swiss-L mice between 2 and 3 weeks of age is probably a result rather than a cause of immunologic competence. Moreover, there was very little change in the number of B cells, T cells, or antigen-binding cells between 1 and 2 weeks after birth (Fig. 1). It therefore seems likely that the capacity of a Swiss-L mouse to respond to antigen at 2 weeks but not at 1 week of age is due to a qualitative rather than a quantitative change in the spleen cell population during this interval. Preliminary evidence from studies on the antigenic stimulation of spleen cells *in vitro* suggests that the maturation of T-cell helper function may be the limiting factor in the development of immunologic competence.

THE RELATIONSHIP BETWEEN SPECIFICITY AND CELL TRIGGERING

What is the mechanism by which a particular antigen induces specific clonal proliferation or immune tolerance in certain populations of lymphoid cells? Although many means are being used to study this question, two approaches seem to be particularly suitable for its analysis at the molecular level. The first and most direct approach is to fractionate lymphocytes according to the specificity of their receptor antibodies for subsequent studies of their responses to antigens of known molecular geometry. The second approach is to analyze the structure and activity of molecules such as lectins that can stimulate lymphocytes regardless of their antigen-binding specificities.

We have been attempting to approach the problem of the specific fractionation of lymphocytes by using nylon fibers to which antigens have been covalently coupled (Edelman et al., 1971; Rutishauser et al., 1972; Rutishauser and Edelman, 1972). The derivatized fibers are strung tautly in a tissue culture dish so that cells in suspension may be shaken in order to collide with them. Some of the cells colliding with the fibers are bound to the covalently coupled antigens by means of their surface receptors. Bound cells may be counted microscopically *in situ* by focusing on the edge of the fiber. After unbound cells have been washed away, the bound cells may be removed by plucking the fibers and shearing the cells quantitatively from their sites of attachment. The removed cells retain their viability (about 90%) provided that the tissue culture medium contains serum.

The specificity of the binding has been evaluated in a number of ways (Table 1). Binding to an antigen fiber is inhibited by the same antigen in soluble form, but not in general by other antigens. Binding is also prevented by preincubation of the cells with antibodies against immunoglobulin receptors on the cell surface. As expected, specific immunization increases the number of cells bound to the fiber. Finally, we have recently demonstrated that the fractionated cells are greatly enhanced in their ability to rebind to fibers of the same specificity (Table 2), but do not bind to fibers derivatized with a different antigen. Moreover, the magnitude of the enrichment obtained with cell populations from both immunized and unimmunized animals (Table 2) indicates that 80–100% of the fractionated cells are specific for the antigen used to isolate them.

Derivatized nylon fibers have the ability to bind both thymus-derived lympho-

TABLE 1. THE SPECIFICITY OF BINDING OF MOUSE SPLEEN CELLS TO ANTIGEN DERIVATIZED FIBERS

Antigen on fiber:	Dnp.	Dnp	Tosyl	Tosyl
Immunization:	None	Dnp	None	Tosyl
Cells bound to fiber (per cm):	1200	4000	800	2000
Inhibition of binding (%) by:				
Dnp	90	95	5	10
Tosyl	1	2	75	87
Anti-Ig	85	93	73	90

Dnp and Tosyl represent, respectively, 2,4-dinitrophenyl and *p*-toluene sulfonyl groups conjugated to protein before their use in the derivatization of the fibers, immunization, or inhibition of fiber binding. Anti-Ig is the γ-globulin fraction of a rabbit antiserum to mouse immunoglobulin.

TABLE 2. REBINDING OF CELLS FRACTIONIZED WITH DINITROPHENYLATED FIBER TO FIBERS OF THE SAME SPECIFICITY

Number of Immunizations	Specific ABC in Unfractionated Population (%)	Predicted Enrichment	Observed Enrichment
None	1–2	50–100X	44–78X
One	3–5	20–33X	23X
Two	10–17	6–10X	6–9X

Animals were immunized with dinitrophenylated proteins where indicated. The observed enrichment was calculated by determining the concentrations of fractionated and unfractionated cells required to yield the same number of fiber-bound cells during a 1-hour incubation.

cytes (T cells) and bone marrow-derived lymphocytes (B cells) according to the specificity of their receptors for a given antigen (Rutishauser and Edelman, 1972) (Fig. 5). About 60% of the spleen cells specifically isolated are B cells and the remainder are T cells. By the use of appropriate antisera to cell surface receptors (Takahashi et al., 1971; Raff, 1969), the cells of each type can be counted on the fibers and most of the cells of one type or the other can then be destroyed by the subsequent addition of serum complement. In this way, one can obtain populations of either T or B cells that are highly enriched in their capacity to bind a given antigen.

Cells of either kind may be further fractionated according to the relative affinities of their receptors. This can be accomplished by prior addition of a chosen amount of the free antigen, which, by binding to their receptors, serves to inhibit specific attachment of subpopulations of cells to the antigen-derivatized fibers. As defined by this technique, cells capable of binding specifically to a particular antigen con-

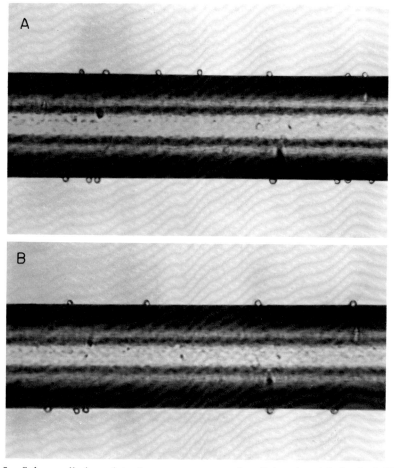

Figure 5. Spleen cells bound to the same segment of a dinitrophenyl-derivatized fiber (A) before and (B) after removal of thymus-derived cells by treatment with anti-θ serum and complement.

stitute as much as 1–2% of a mouse spleen cell population. Very few of these original antigen-binding cells appear to proliferate in response to immunization, however (Rutishauser et al., 1972), and the cells that do respond are those having receptors of higher relative affinities (Fig. 6). This is in agreement with the conclusions of Siskind and Benacerraf (1969).

By use of this method, together with serological means of distinguishing T and B cells, the cells may be compared for their range of antigen-binding specificities and their avidities for antigens. Our studies suggest that T and B cells do not differ greatly in the range of their antigen-binding specificities, at least for several hapten and protein antigens (Rutishauser and Edelman, 1972). On comparison, the avidities of T and B cells were the same for the monovalent ε-DNP-lysine. In contrast, T cells showed a consistently higher apparent avidity for multivalent DNP-bovine serum albumin containing an average of 10 DNP groups per molecule. This suggests the possibility that the receptors on the surface of T cells may be arranged or are arrangeable in clusters differing from those of B cells.

It should be noted that plaque-forming cells (Jerne and Nordin, 1963) do not bind to antigen-derivatized fibers (Rutishauser et al., 1972), and therefore antigen-binding cells can be fractionated from cells that are already actively secreting antibodies. Our recent experiments indicate that the antigen-binding cells isolated from immunized animals by this method may be transferred to irradiated animals to reconstitute a response to the antigen used to isolate them (Table 3). The response obtained with the fractionated cells was equivalent to that obtained with about 10 times as many unfractionated cells. Rebinding studies with these cells indicated that the fractionation had increased the frequency of specific antigen-binding cells also about 10 times, from 10% to over 90%. Since the enrichment of antigen-binding cells is directly reflected in the response, it can be concluded

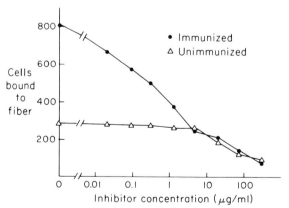

Figure 6. Inhibition by soluble dinitrophenylated-bovine serum albumin of spleen cell binding to dinitrophenyl-derivatized fibers. Cell numbers are the fiber edge counts for a 2.5-cm fiber segment, and each point represents a separate binding experiment in the presence of the indicated amount of inhibitor. Note that only a small proportion of the nonimmune cell binding is inhibited at low concentrations of inhibitor, whereas over 60% of immune cell binding is prevented by concentrations of less than 4 μg ml^{-1}. Below 4 μg ml^{-1} the two cell populations behave essentially the same. This indicates that immunization results in a selective increase in the number of cells with a high affinity for the soluble inhibitor.

TABLE 3. RESPONSE OF CELLS ISOLATED WITH DINITROPHENYL-FIBERS AFTER TRANSFER TO IRRADIATED RECIPIENTS

Transferred Cells			Response	
Fractionated Dnp-FBC	Unfractionated Dnp-Hcy cells	Unfractionated Hcy cells	IgM PFC per Spleen	IgG PFC per Spleen
0	0	10^7	4163 ± 290	6,503 ± 752
0	7×10^4	10^7	5077 ± 1583	9,537 ± 2158
0	7×10^5	10^7	4873 ± 246	25,533 ± 2801
7×10^4	0	10^7	4487 ± 354	30,733 ± 3177

Donor animals were immunized twice with dinitrophenylated hemocyanin (Dnp-Hcy) or hemocyanin (Hcy). Recipient mice were irradiated (700 R) to suppress their immune response. Note that 7×10^4 Dnp-fractionated cells (Dnp-FBC) give an IgG response over the control value (Hcy cells only) which is equivalent to that of 10 times as many unfractionated Dnp-Hcy cells. The limitation of the enhancement to antibodies of the IgG class is characteristic of cells from the Dnp-immunized animals.

that the fractionated cells include precursors of plaque-forming cells and that the function of these cells is not impaired by the isolation procedure.

The comparison of the distribution of antigen-binding cells before and after immunization (Fig. 6) suggests that the concept of immunological specificity based solely on the specificity of binding antibodies must be re-examined. To a certain extent the diversification of receptor antibodies may have come about by selective pressures resulting from exposure to certain classes of chemical substances during evolution. According to the theory of clonal selection, however, it is *impossible* for *each* cell receptor to have been selected for or against during evolution. Instead, a great number of antibody variants have to be generated (by whatever process, germ line or somatic), many of which will never be selected during the lifetime of the organism. Under these circumstances how specific can such a system be? This question can be posed in the following terms: Is the probability of cross-reactivity with various antigens the same in the antigen-binding cell population and in the population of humoral antibodies?

The evidence obtained by studying antigen-binding and antigen-reactive cells suggests that the degree of cross-reactivity may be much greater among receptor antibodies in the antigen-binding cell population (D'Eustachio et al., 1973). If this is the case, selection for specificity cannot merely be the result of antigen binding, but must depend also on a second factor. The most likely candidate is the triggering threshold for stimulation of the cell carrying the antibody receptor. Thus, if a cell population contains cells that can bind two different antigens, there is a chance that specificity could be lost. Specificity would be preserved, however, if a particular cell capable of binding both antigens is much more likely to be *triggered* by only one of the two. The trigger threshold might depend on the state of the cell but probably would depend on steric factors (reflected in the free-energy of binding for each particular antigen) and on the *cell surface density* of the antigen, which would increase the *avidity* of the binding.

Whatever the detailed mechanism of triggering, the implication of this two-factor hypothesis is that variation at the level of antigen-binding cells leads to a relatively nonspecific set of immunoglobulin molecules many of which are capable of binding various antigens with relatively low specificity. Some of these molecules can also bind certain antigens with higher specificities. According to this idea, the selective forces that yield specificity are a product of both the probability of binding and the probability of lymphocyte stimulation above a certain triggering threshold. Such a system behaves as an amplifier with a high pass filter or "high free-energy filter" on its input. The behavior of such a system is a result of complex cell-cell interactions, as well as of specific mechanisms of triggering with particular thresholds. It is therefore necessary to study these mechanisms in some detail, particularly their initial conditions and the state of cell surface receptors.

RECEPTOR DYNAMICS AND THEIR POTENTIAL RELATION TO MITOGENESIS

The foregoing analysis suggests that, although as many as 1 in 10^2 cells of a spleen cell population may bind a particular antigen, as few as only 1 in 10^5 cells are antigen reactive. For this reason, it is difficult to study directly the mitotic process that occurs in response to an antigen. Fortunately, this process can be studied by other means, for the capacity to be triggered is to some extent specificity independent, and a large number of different agents in addition to antigens are mitogenic for lymphocytes. These agents include mitogenic lectins (plant proteins with specificity for various carbohydrates) (Sharon and Lis, 1972), lipopolysaccharides (Andersson et al., 1972a), and chemical reagents such as periodate (Novogrodsky and Katchalski, 1972). The availability of a variety of mitogens provides an opportunity to study the stimulation of lymphocytes without extensively purifying them according to the specificity of their Ig receptors. The detailed structure of the surface receptors for these various mitogenic agents is unknown, but they are in general different from each other, at least in their carbohydrate portions. Because the events of lymphocyte stimulation by each mitogen are similar, however, it is probable that mitogenesis is mediated by a final common pathway.

There are two requirements for mitogenic stimulation:

1. The antigen or mitogenic agent must bind to the appropriate receptors in the proper conformation or arrangement.

2. After this binding, coupling to the chain of metabolic alterations preceding transformation and mitosis must take place. These alterations include immediate changes in lipid turnover (Fisher and Mueller, 1971), in cyclic GMP levels (Hadden et al., 1972), and in transport of ions and small molecules (Quastel and Kaplan, 1970; Peters and Hausen, 1971). Somewhat later, there are changes in RNA and DNA metabolism (Cooper, 1971).

It should be noted that binding events occur in seconds or minutes, whereas metabolic events occur in minutes to hours. Moreover, irreversible stimulation does not in general occur until 20 hours after initial binding (Novogrodsky and Katchalski, 1971). Thus, stimulation involves a complex series of events, the detailed kinetics of which remain to be worked out.

Even before the kinetics are known, however, it is of definite value to ask whether the lectin must act first at the cell surface, and, if so, to consider the effects it can induce at that surface. It is known that only some of the glycoprotein receptors on the cell surface are responsive to stimulation, for there exist several lectins that can bind to cell surface glycoproteins but are not mitogenic. Of a given population of receptors responsive to a mitogenic lectin such as Con A, as few as 6% have to be bound to induce transformation (Inbar et al., 1973). Some evidence exists to suggest that lectins act directly at the cell surface, for covalent coupling of Con A and PHA (Andersson et al., 1972b; Greaves and Bauminger, 1972) at solid surfaces does not abolish their ability to stimulate cells. Nevertheless, the possibility remains that lectins induce both surface effects and effects within the cell after endocytosis of a few molecules.

In determining how the surface interactions might be transmitted to the interior of the cell, the grouping, mobility, and attachment of the lectin receptors must be considered. Recent findings indicate that the lipid portion of the cell membrane is fluid (Hubbell and McConnell, 1969), that cell receptors are mobile (Frye and Edidin, 1970), and that immunoglobulins and other receptors on the lymphocyte may be cross linked by specific divalent antibodies to form patches and ultimately caps at one pole of the cell (Taylor et al., 1971; de Petris and Raff, 1972; Yahara and Edelman, 1972). It has been suggested that cap formation from patches may be correlated with cell mobility, while patch formation depends on diffusion of receptor molecules on the plane of the cell surface (de Petris and Raff, 1972; Edidin and Weiss, 1972). Receptors of an individual type can be "patched" and "capped" independently of other receptors (Preud'homme et al., 1972), and thus, with one exception to be discussed in detail here, they appear to be independent of each other (i.e., they are not physically associated in terms of their mobility).

Although there is no direct structural knowledge about the detailed mode of anchorage of any of the receptors to the membrane, certain studies indicate that membrane proteins may go through the membrane (Bretscher, 1971) or be connected to structures within the membrane. Despite the absence of details about the connection of surface receptors to structures in the cytoplasm, valuable inferences can be drawn by using lectins to agglutinate cells or otherwise alter their function. We have recently found a quite striking effect of Con A on the mobility of receptors that sheds light on the nature of receptor anchorage and receptor interactions. Native concanavalin A strongly affects the ability of cell receptors to form patches and caps (Yahara and Edelman, 1972; Yahara and Edelman, 1973). If Con A is added in doses greater than 5 μg ml^{-1} to lymphocytes at 21° or 37°C before treatment with anti-Ig, both patch formation and cap formation are inhibited (Fig. 7). This effect, which is dose dependent, is reversed by addition of α-methyl-D-mannoside, a competitive inhibitor of Con A. Therefore the system responsible for inhibition of receptor mobility can recover from the effects of Con A binding, and the lymphoid cell is not killed by the interaction under the conditions of these experiments.

As shown in Fig. 8, the effect of Con A is seen at doses as low as 2 μg ml^{-1} in PBS at 21°C and it is virtually complete at 20 μg ml^{-1}. Concanavalin A also inhibits the capping of antibody receptors induced by forming a sandwich of antigen and antibody. Similar effects are seen in the prevention of patching and capping by antibodies to the θ antigen of thymocytes. Concanavalin A also inhibits patch and cap formation by its own receptors at 21° and 37°C (Yahara and Edelman, 1973).

In contrast to these findings, if Con A is added to the cells at 4°C, the excess Con A is washed away, and the cells are then brought to 37°, Con A forms patches and caps with its own receptors (Unanue et al., 1972; Yahara and Edelman, 1973).

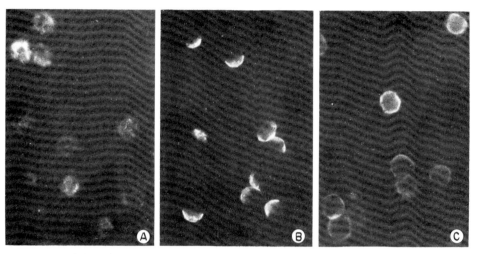

Figure 7. Labeling patterns of cells with fl-anti-Ig. (*A*) Cells incubated with fl-anti-Ig (80 μg ml^{-1}) plus NaN$_3$ (10 mM) at 21°C for 30 minutes, showing patches. (*B*) Cells incubated with fl-anti-Ig (80 μg ml^{-1}) at 20°C for 30 minutes, showing caps. (*C*) Cells incubated with fl-anti-Ig (80 μg ml^{-1}) plus Con A, showing diffuse patterns.

Native tetravalent Con A therefore has two antagonistic actions that depend both on the temperature and on the concentration of the lectin. The results of the bind-ing-washing procedure at 4° show that capping of Con A receptors is optimal at a binding level of 3×10^5 molecules per cell, and the curve resembles a precipitation curve (Fig. 9). At a given concentration of Con A, fewer molecules are bound to splenic lymphocytes at 4° than at 37°C. Nevertheless, except for a small amount of capping at a concentration of 2 μg ml^{-1}, the Con A inhibition at 37° is seen at

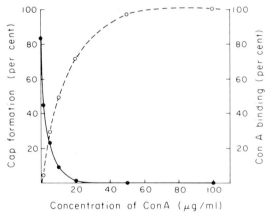

Concentration of ConA (μg/ml)

Figure 8. Effect of Con A on cap formation (——): 2×10^7 cells ml^{-1} were incubated with various concentrations of Con A in PBS-BSA at 21°C for 10 minutes. Fl-anti-Ig was added to the mixture to give a concentration of 80 μg ml^{-1}, and the mixture was incubated for 30 minutes at 21°C. Cap formation was determined after washing cells. Binding [^{125}I]Con A to cells ($-\bigcirc-\bigcirc-$): 2×10^7 cells ml^{-1} were incubated with various concentrations of [^{125}I]Con A (2.4×10^4 cpm μg^{-1}) at 21°C for 30 minutes in PBS-BSA. Cells were washed with PBS-BSA, and the radioactivity was determined.

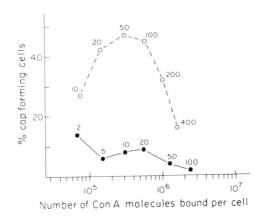

Figure 9.   Relation of cap formation to the number of Con A molecules bound to lymphoid cells at 4°C (○) and at 37°C (●). Numbers above each point refer to the concentrations of Con A used.

levels of binding ranging from 10^5 to 10^6 molecules per cell. These experiments show that the antagonistic actions of Con A depend on the temperature, the concentration, and the state of the lymphoid cell. The results cannot be explained, however, by the fact that only about one third as many molecules of Con A are bound per cell at 4° as at 37°, for even when equal numbers of molecules are bound at the two different temperatures the inhibition is still seen at 37° (Fig. 9).

What properties of the lectin are responsible for these effects, and what structure in the cell mediates them? Several recent experiments on the structure of Con A and on the effects of drugs have yielded some useful working hypotheses and provisional answers to these questions.

Studies (Edelman et al., 1972; Gunther et al., 1973) on the subunit and three-dimensional structure of Con A indicate that, at pH 7 and above, it is a tetramer with four saccharide binding sites. This suggests that receptor cross-linkage by the tetravalent Con A molecule may play an important role in its action on cells. Because it is multivalent, Con A may also function to cross-link receptors between two different cells and agglutinate them. For these reasons, it is of particular interest to determine the effects of a change in valence on the inhibition of the mobility of cell surface receptors.

Recent experiments (Gunther et al., 1973) indicate that the valence of Con A may be changed by suitable chemical treatment. Succinylation of Con A can be achieved by treatment with succinic anhydride. A variety of findings (Table 4) indicate that the major effect of succinylation is to alter the association properties of the Con A subunit so that it forms dimers rather than tetramers. Unlike the native molecule, dimeric succinyl-Con A lacks both the capacity to induce patches and caps on splenic lymphocytes and the capacity to inhibit patch and cap formation by its own and other receptors. More pertinent to the question of the mechanism of inhibition of patch formation is the finding that treatment with anti-Con A can restore the ability of bound succinyl-Con A to restrict the movement of immunoglobulin receptors. Anti-Con A itself has no effect on cap formation induced by anti-Ig; conversely, the presence of anti-Ig neither enhances nor inhibits cap

TABLE 4. COMPARISON OF THE BIOLOGICAL ACTIVITIES OF
CON A AND SUCCINYL-CON A

Property	Con A	Succinyl-Con A
1. Number of binding sites per cell		
(a) Sheep erythrocytes	1.1×10^6	2.8×10^6
(b) Mouse spleen cells	1.4×10^6	4.4×10^6
2. Agglutination (μg ml^{-1})		
(a) Sheep erythrocytes	1	>500
(b) Sheep erythrocytes +		
succinyl-Con A (330 μg ml^{-1})	8	—
(c) Mouse spleen cells	4.5	40
3. Percentage of cells forming lectin-receptor caps		
(a) Lectin (5 μg ml^{-1}, 37°)	0.2–2	0
(b) Lectin (100 μg ml^{-1}, 37°)	≤0.2	0
(c) Lectin (170 μg ml^{-1})		
preincubated in ice bath, washed,		
then brought to 37°	62	—
(d) Lectin (20 μg ml^{-1}) + anti-Con A		
(100 μg ml^{-1})	18	82
4. Percentage of inhibition of anti-Ig capping		
(a) Lectin (100 μg ml^{-1})	100	0
(b) Lectin (50 μg ml^{-1}) + anti-Con A		
(100 μg ml^{-1})	100	40
5. Mitogenesis		
(a) Lectin (5 μg ml^{-1})	Positive	Positive
(b) Lectin (50 μg ml^{-1})	Negative	Positive

formation by succinyl-Con A plus fl-anti-Con A. These results are summarized in Fig. 10.

The failure of divalent Con A to influence receptor mobility may be the result of a change in valence or a change in the surface charge of the molecule. At the concentration necessary for inhibition of patch formation, Con A covers only about 1% of the cell surface. This is probably not enough for the altered charge to affect the agglutination process. Most convincing, however, is the observation that the addition of anti-Con A to cells *after* succinyl-Con A has been bound can bring about all three phenomena shown in Fig. 10. Fab fragments of antibodies to Con A do not restore these phenomena (Yahara, Edelman, and Wang, unpublished observations), and anti-Con A alone has no effect. It seems likely, therefore, that treatment with anti-Con A is equivalent to an increase in the effective valence of the bound succinyl-Con A molecules, and that the main effects do not result from changes in net charge.

All of these experiments suggest that the cross-linkage of lectin receptors plays a large role in inhibition of the mobility of cell surface receptors as well as in other activities of mitogenic lectins. They leave unanswered, however, the question concerning the nature of the cellular structures that modulate receptor mobility. Our recent observations on certain drugs that reverse the inhibition of mobility provide a clue to the nature of these structures. It has been found (Edelman et al., 1973) that colchicine, colcemid, vinblastine, and vincristine will partially reverse the effect of Con A on receptor mobility and thus permit the formation again of both Con A caps and anti-Ig caps (Table 5). Colchicine does not bind to Con A, nor

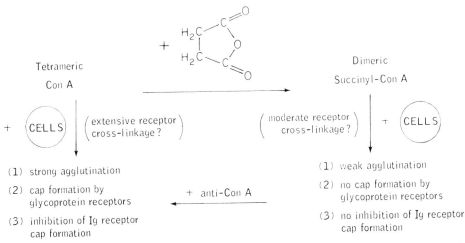

Figure 10. Schematic comparison of the biological properties of native Con A and succinyl-Con A. The activities listed include the immediate cellular reactions mediated by the lectin.

does it cause disaggregation of its subunits. Furthermore, colchicine does not inhibit either Con A-saccharide interactions or the cell-binding activity of Con A. Although it remains to be seen whether colchicine alters the fluidity of the lipid portion of the membrane, these observations suggest an alternative interpretation: Con A binding and the cross-linking of its receptors, may alter the association-dissociation equilibrium of cytoplasmic structures interacting with the cell surface receptors, and this alteration may in turn affect the mobility of the receptors. Microtubular proteins (Weisenberg et al., 1968) appear to be good candidates for direct or indirect interaction with receptors because they have been shown to be sensitive to all four of the drugs listed in Table 5 (Wilson, 1970), because they have singular temperature- and drug-dependent association-dissociation properties (Borisy and Taylor, 1967; Porter, 1966), and because they are ubiquitous in the cell.

It should be noted that no extensive microtubular structures have been observed directly under the plasma membrane except perhaps in the cases of platelets (White, 1971; Hoak, 1972) and of lens epithelial cells in culture (Piatigorsky et al., 1973). We have found that cytochalasin B, which partially affects cap formation, but not patch formation, probably by altering microfilaments (Taylor et al., 1971), has no effect on inhibition of receptor mobility by Con A. We can tentatively conclude, therefore, that although it is possible that a filament structure other than microtubules may mediate the inhibition of receptor mobility, this structure is not likely to be the same as the structure mediating cap formation. The observations so far implicate a protein that can be altered by colchicine and *Vinca* alkaloids; although its identity is unknown, this colchicine-binding protein (CBP) may be related to certain of the actin-like proteins (Berl et al., 1973) or microtubules (Borisy and Taylor, 1967). One possible candidate is a protein similar to spectrin (Marchesi et al., 1970), found in association with red cell membranes.

A working hypothesis to explain the observations on modulation of receptor mobility requires that some of the cell surface receptors be anchored on a common assembly which is on, in, or under the lymphocyte plasma membrane. Our detailed

TABLE 5. EFFECTS OF VARIOUS DRUGS ON THE INHIBITION BY
CON A OF CAP FORMATION IN SPLENIC LYMPHOCYTES

Treatment	Percentage of Cap-Forming Cells		
	With fl-anti-Ig (100 μg ml^{-1})[a]	With fl-anti-Ig (100 μg ml^{-1}) + Con A (100 μg ml^{-1})[b]	With fl-Con A (100 μg ml^{-1})[c]
Control	85	2	2
Colchicine[d]			
(10^{-4}M)	87	22	31
Colcemid			
(10^{-4}M)	88	25	24
Vinblastine[d]			
(10^{-4}M)	91	55	42
Vincristine			
(10^{-4}M)	83	15	19
Low temperature			
(4°C)[e]	88	30	45
Cytochalasin B	62	—	1

[a] In order to test for cap formation by immunoglobulin receptors, the percentage of cap-forming cells obtained with fluorescein-labeled anti-immunoglobulin (fl-anti-Ig) was measured.

[b] In order to test for the inhibition by Con A of immunoglobulin receptor cap formation, the percentage of cap-forming cells obtained with fl-anti-Ig was measured in the presence of Con A.

[c] In order to test for cap formation by Con A receptors, the percentage of cap-forming cells obtained with fluorescein-labeled Con A (fl-Con A) was measured.

[d] Colchicine (10^{-4}M) and vinblastine (10^{-4}M) did not affect the amount of [^{125}I] Con A bound to splenic lymphocytes.

[e] The amount of [^{125}I] Con A bound to splenic lymphocytes at 4°C was 30% of the value obtained at 37°.

hypothesis (Fig. 11) incorporates the following assumptions (Edelman et al., 1973):

1. Certain surface receptors interact reversibly with colchicine-binding proteins, possibly the microtubular assemblies of the cytoplasm. If we call A the anchored state of the receptors (attached to the CBP) and F the state of the receptors that are free from the CBP, these two states are assumed to exist in an equilibrium, $A \rightleftharpoons F$. Through this anchorage, the distribution of the receptors on the cell surface is affected by the state of the CBP. A similar suggestion has been made by Berlin and Ukena (1972), who found that colchicine and vinblastine inhibited the agglutination of polymorphonuclear leukocytes by Con A.

2. Not only is the distribution of the cell receptors affected by the state of the CBP, but, conversely, the mobility and state of this assembly are affected by cross-linking interactions and aggregations of particular receptors. This provides a means by which receptor states can be communicated to the interior of the cell. The valence of external ligands can therefore by a critical factor in cell surface-cytoplasmic interactions.

3. The mobility of the membrane or its receptors is affected by the state of the CBP, and therefore alteration of the CBP by one set of cell surface receptors may affect the movement of the other receptors.

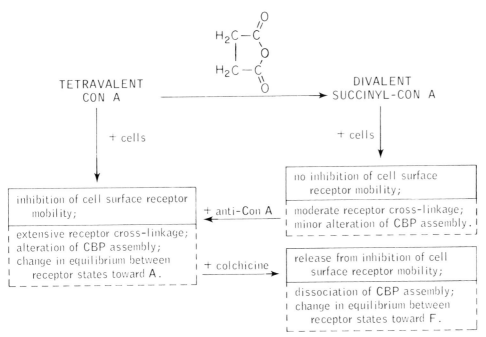

Figure 11. Summary of the effects of binding Con A or succinyl-Con A on mouse splenic lymphocytes. The experimentally observed effect on restriction of receptor mobility is boxed by solid lines. The hypothesized changes in the colchicine-binding protein assembly (CBP) and the shift in equilibrium of receptor states between A (anchored to the CBP) and F (free from the CBP) are boxed by dotted lines. The addition of anti-Con A to succinyl-Con A bound on cells mimics the effects of Con A.

4. Finally, the equilibrium between the two states (A and F) of the receptors is affected by colchicine and related agents. Alteration of the equilibrium may occur either because structures such as microtubules are dissociated by these agents (Borisy and Taylor, 1968) or because receptors are released from attachment with the CBP assembly or both.

The proposed hypothesis assumes a means of coupling between certain receptors (such as Con A receptors) and an ordered state of the CBP assembly, and therefore it is important to consider how the cell surface receptors might be linked to this common network. The observations suggest two possible means of coupling, direct and indirect (Fig. 12). Direct coupling can occur either through intramembranous particles (Mandel, 1972) or independently of them by extension of the lectin receptors through the membrane to interact noncovalently with proteins such as those of the CBP. Recent freeze-fracture studies show no alteration of intramembranous particles by various doses of Con A or colchicine and tend to favor a direct mode (Yahara et al., unpublished observations). An additional possibility is that lectin binding and receptor aggregation lead to a change in membrane transport or of enzymatic activity, which, in turn, may lead to alterations in the mobility of the CBP or its interaction with receptors.

It is obviously premature to construct a detailed hypothesis on the relationship between the surface alterations induced by Con A and the mechanism of mitogenesis. Mitogenesis is a complex series of events with a long delay between the initial

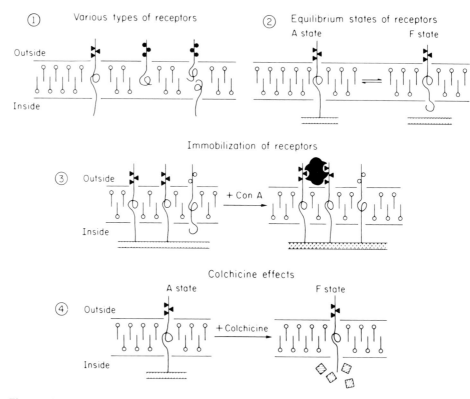

Figure 12. Diagram illustrating some hypothesized interactions between cell surface receptors and colchicine-binding proteins (CBP) and the effects of concanavalin A and colchicine on these interactions. (*1*) Types of receptors: Some receptors may penetrate the lipid bilayer and interact directly or indirectly with cytoplasmic structures, while other receptors may terminate in or on the membrane. (*2*) Some of the first type of receptors are assumed to interact with CBP in an equilibrium consisting of two states, anchored (A) and free (F). (*3*) Cross-linkage of Con A receptors by Con A may lead to structural changes in the CBP assembly and affect the equilibria, A⇌F, of other receptors. (*4*) Colchicine dissociates CBP into subunits resulting in a shift in the equilibrium, A⇌F.

signal from the mitogen and the final metabolic responses of the cell leading to replication and cytokinesis. Moreover, the effects of mitogens are complex, showing very strong dose dependence as well as possible requirements for factors from other cells. The roles of transport alteration, endocytosis of lectins, and comitogenic factors remain to be assessed in detail. Nevertheless, there is some justification for suggesting that the surface events described here may be connected with the early events of mitogenesis.

Recent experiments have shown that the mitogenic activity of Con A is inhibited by drugs such as colchicine, vinblastine, and vincristine (Table 6) at concentrations as low as 10^{-5} *M*. This effect is not attributable to inhibition of DNA synthesis, for we have found that thymidine incorporation into stimulated cells can take place in the presence of the drugs. In addition, there appears to be a correlation of the mitogenic activities of some lectins with their capacities to immobilize the cell surface receptors. Preliminary experiments have shown that several mitogenic lectins (Con A, PHA, lentil, and pea lectins, and extracts from fava and black turtle

TABLE 6. EFFECTS OF VARIOUS DRUGS ON THE MITOGENIC ACTIVITY OF CON A

Treatment	Percentage of Optimal Mitogenic Response[a]
Control	100
Colchicine[b] (10^{-4} M)	14
Vinblastine[b] (10^{-4} M)	2
Vincristine (10^{-4} M)	15

[a] Optimal mitogenic response was obtained by culturing mouse spleen cells with 3 μg ml^{-1} Con A as described by Gunther et al. (1973).

[b] Colchicine (10^{-4} M) and vinblastine (10^{-4} M) did not affect the amount of [^{125}I] Con A bound to splenic lymphocytes.

beans) inhibited cap formation. In contrast, nonmitogenic lectins (wheat germ agglutinin, extracts of Idaho red, small California white and pink beans) did not inhibit cap formation. This suggests the possibility that nonmitogenic lectins may attach to receptors that are not directly connected wih the CBP system.

A simple interpretation consistent with all of the current observations is that mitogenic stimulation involves the formation of micropatches containing relatively mobile receptors in reversible equilibrium with the CBP assembly. The formation of these micropatches may lead to alterations in the CBP and also initiate the various metabolic events in stimulation. At higher concentrations, Con A may inhibit the mitogenic response by extensive alteration of the CBP, instantaneously "freezing" the cell surface receptors and preventing the formation of micropatches; eventually, this state would lead to cell death. A divalent lectin such as succinyl-Con A cannot immobilize the receptors but may be able to form micropatches and is therefore mitogenic at high concentrations without being toxic (Table 4). Colchicine and *Vinca* alkaloids may act to inhibit mitogenesis by affecting the postulated CBP assembly in such a way that, even if micropatches are formed, they cannot alter cytoplasmic function via this assembly.

Obviously, colchicine and *Vinca* aklaloids may act at different sites to produce their effects, for it is known that these drugs have multiple actions (Creasey and Markin, 1965; Erbe et al., 1966; Madoc-Jones and Mauro, 1968). In the case of their effects on mitogenesis, the multiple actions of these drugs must be much more extensively explored and isolated from each other before firm conclusions can be drawn.

The hypotheses discussed here have a number of implications for cell-ligand and cell-cell interactions. They may, for example, provide a reasonable basis for understanding how certain receptors might remain relatively fixed on the membrane while others move. In addition, they explain avidity effects, in which clusters of receptors participate in specific interactions that dispersed receptors cannot carry out. Finally, because of the possible relationship of microtubules to morphogenesis and of microfilaments to cell movement, the present working hypotheses may help to provide a connection between specific surface signals and alterations of cellular form. Indeed, recent observations have revealed distinct effects of *Vinca* alkaloids on the shape of fiber-bound lymphocytes (Yahara et al., unpublished observations).

It will be of particular significance if these findings can be extended to a variety of eukaryotic cells.

ACKNOWLEDGMENTS

This work was supported by grants from the U.S. Public Health Service and the National Science Foundation.

REFERENCES

Andersson, J., G. M. Edelman, G. Möller, and O. Sjöberg (1972a). Activation of B lymphocytes by locally concentrated concanavalin A. *Eur. J. Immunol.* **2**, 233.

Andersson, J., O. Sjöberg, and G. Möller (1972b). Induction of immunoglobulin and antibody synthesis *in vitro* by lipopolysaccharides. *Eur. J. Immunol.* **2**, 349.

Berl, S., S. Puszkin, and W. J. Nicklas (1973). Actomyosin-like protein in brain. *Science* **179**, 441.

Berlin, R. D., and T. E. Ukena (1972). Effect of colchicine and vinblastine on the agglutination of polymorphonuclear leucocytes by concanavalin A. *Nature New Biol.* **238**, 120.

Borisy, G. G., and E. W. Taylor (1967). The mechanism of action of colchicine; binding of colchicine-³H to cellular protein. *J. Cell Biol.* **34**, 525.

Bretscher, M. (1971). Major human erythrocyte glycoprotein spans the cell membrane. *Nature New Biol.* **231**, 229.

Cairns, J. (Ed.) (1967). Antibodies. *Cold Spring Harbor Symp. Quant. Biol.* **32**.

Cooper, H. L. (1971) Biochemical alterations accompanying initiation of growth in resting cells. In *The Cell Cycle and Cancer* (R. Baserga, ed.). M. Dekker, New York, p. 191.

Creasey, W. A., and M. E. Markin (1965). Biochemical effects of the *Vinca* alkaloids. III. The synthesis of ribonucleic acid and the incorporation of amino acids in Ehrlich ascites cells *in vitro*. *Biochim. Biophys. Acta* **103**, 635.

De Petris, S., and M. C. Raff (1972). Distribution of immunoglobulins on the surface of mouse lymphoid cells as determined by immunoferritin electron microscopy: Antibody-induced, temperature-dependent redistribution and its implications for membrane structure. *Eur. J. Immunol.* **2**, 523.

D'Eustachio, P. G., U. Rutishauser, and G. M. Edelman (1973). Unpublished results.

Edelman, G. M., U. Rutishauser, and C. F. Millette (1971). Cell fractionation and arrangement on fibers, beads, and surfaces. *Proc. Nat. Acad. Sci. U.S.A.* **68**, 2153.

Edelman, G. M., B. A. Cunningham, G. N. Reeke, Jr., J. W. Becker, M. J. Waxdal, and J. L. Wang (1972). The covalent and three-dimensional structure of concanavalin A. *Proc. Nat. Acad. Sci. U.S.A.* **69**, 2580.

Edelman, G. M., I. Yahara, and J. L. Wang (1973). Receptor mobility and receptor-cytoplasmic interactions in lymphocytes. *Proc. Nat. Acad. Sci. U.S.A.* **70**, 1442.

Edidin, M., and A. Weiss (1972). Antigen cap formation in cultured fibroblasts: A reflection of membrane fluidity and of cell motility. *Proc. Nat. Sci. U.S.A.* **69**, 2456.

Erbe, W., J. Preiss, R. Seifert, and H. Helz (1966). Increase in RNase and DNase activities in ascites tumor cells induced by various cytostatic agents. *Biochem. Biophys. Res. Commun.* **23**, 392.

Fisher, D. B., and G. C. Mueller (1971). Studies on the mechanism by which phytohemagglutinin rapidly stimulates phospholipid metabolism of human lymphocytes. *Biochim. Biophys. Acta,* **248**, 434.

Frye, C. D., and M. Edidin (1970). The rapid intermixing of cell surface antigens after formation of mouse-human heterokaryons. *J. Cell Sci.* **7**, 319.

Greaves, M. F., and S. Bauminger (1972). Activation of T and B lymphocytes by insoluble phytomitogens. *Nature New Biol.* **235**, 67.

Gunther, G. R., J. L. Wang, I. Yahara, B. A. Cunningham, and G. M. Edelman (1973). Concanavalin A derivatives with altered biological activities. *Proc. Nat. Acad. Sci. U.S.A.* **70**, 1012.

Hadden, J. W., E. M. Hadden, M. K. Haddox, and N. O. Goldberg (1972). Guanosine 3′:5′-cyclic monophosphate: A possible intracellular mediator of mitogenic influences in lymphocytes. *Proc. Nat. Acad. Sci. U.S.A.* **69**, 3024.

Hoak, J. C. (1972). Freeze-etching studies of human platelets. *Blood* **40**, 514.

Hubbell, W. L., and H. M. McConnell (1969). Motion of steroid spin labels in membranes. *Proc. Nat. Acad. Sci. U.S.A.* **63**, 16.

Inbar, M., H. Ben-Bassat, and L. Sachs (1973). Temperature-sensitive activity on the surface membrane in the activation of lymphocytes by lectins. *Exp. Cell Res.*, **76**, 143.

Jerne, N. K., and A. A. Nordin (1963). Plaque formation in agar by single antibody producing cells. *Science* **140**, 405.

Madoc-Jones, H., and F. Mauro (1968). Interphase action of vinblastine and vincristine: Differences in their lethal action through the mitotic cycle of cultured mammalian cells. *J. Cell Physiol.* **72**, 185.

Mandel, T. E. (1972). Intramembraneous marker in T-lymphocytes. *Nature New Biol.* **239**, 112.

Marchesi, S. L., E. Steens, V. T. Marchesi and T. W. Tillack (1970). Physical and chemical properties of protein isolated from red cell membranes. *Biochemistry* **9**, 50.

Metcalf, D., and M. A. S. Moore (1971). *Haemopoietic Cells*. American Elsevier Publishing Co., New York, p. 214.

Mitchell, G. F., and J. F. A. P. Miller (1968). Cell to cell interaction in the immune response. II. The source of hemolysin-forming cells in irradiated mice given bone marrow and thymus or thoracic duct lymphocytes. *J. Exp. Med.* **128**, 821.

Nossal, G. J. V., and B. L. Pike (1972). Differentiation of B lymphocytes from stem cell precursor. In *Microenvironmental Aspects of Immunity* (B. D. Jankovic and K. Isakovic, eds.). Plenum Publishing Co., New York, pp. 11–18.

Novogrodsky, A., and E. Katchalski (1971). Lymphocyte transformation induced by concanavalin A and its reversion by methyl-α-D-mannopyranoside. *Biochim. Biophys. Acta* **228**, 579.

Novogrodsky, A., and E. Katchalski (1972). Membrane site modified on induction of the transformation of lymphocytes by periodate. *Proc. Nat. Acad. Sci. U.S.A.* **69**, 3207.

Peters, J. H., and P. Hausen (1971). Effect of phytohemagglutinin on lymphocyte membrane transport. 1. Stimulation of uridine uptake. *Eur. J. Biochem.* **19**, 502.

Piatigorsky, J., S. S. Rothschild, and M. Wollberg (1973). Stimulation by insulin of cell elongation and microtubule assembly in embryonic chick-lens epithelia. *Proc. Nat. Acad. Sci. U.S.A.* **70**, 1195.

Playfair, J. H. L. (1968) Strain differences in the immune response of mice. I. The neonatal response to sheep red cells. *Immunology* **15**, 35.

Porter, K. R. (1966). *Cytoplasmic Microtubules and Their Function* (O. Wolstenholme and M. O'Conner, eds.). CIBA Foundation Symposium, London, p. 308.

Preud'homme, J. L., C. Neauport-Sautes, S. Piat, D. Silvestre and F. M. Kourilsky (1972). Independence of HL-A antigens and immunoglobulin determinants on the surface of human lymphoid cells. *Eur. J. Immunol.* **2**, 297.

Quastel, M. R., and J. G. Kaplan (1970). Early stimulation of potassium uptake in lymphocytes treated with PHA. *Exp. Cell Res.* **63**, 230.

Rutishauser, U., and G. M. Edelman (1972). Binding of thymus- and bone marrow-derived lymphoid cells to antigen-derivatized fibers. *Proc. Nat. Acad. Sci. U.S.A.* **69**, 3774.

Rutishauser, U., C. F. Millette, and G. M. Edelman (1972). Specific fractionation of immune cell populations. *Proc. Nat. Acad. Sci. U.S.A.* **69**, 1596.

Rutishauser, U., P. D. D'Eustachio, and G. M. Edelman (1973). Immunological function of lymphocytes fractionated with antigen-derivatized fibers. *Proc. Nat. Acad. Sci, U.S.A.* **70**, 3894.

Sharon, N., and H. Lis (1972). Lectins: cell-agglutinating and sugar specific proteins. *Science* **177**, 949.

Silverstein, A. M., J. W. Uhr, K. L. Kraner, and R. J. Lukes (1963). Fetal response to antigenic stimulus. II. Antibody production by the fetal lamb. *J. Exp. Med.* **117**, 799.

Siskind, G. P., and B. Benacerraf (1969). Cell selection by antigen in the immune response. *Adv. Immunol.* **10**, 1.

Spear, P. G., A.-L. Wang, U. Rutishauser, and G. M. Edelman (1973). Characterization of splenic lymphoid cells in fetal and newborn mice. *J. Exp. Med.* (in press).

Takahashi, T., L. J. Old, K. R. McIntire, and E. A. Boyse (1971). Immunoglobulin and other surface antigens of cells of the immune system. *J. Exp. Med.* **134**, 815.

Taylor, R. B., W. P. H. Duffus, M. C. Raff, and S. de Petris (1971). Redistribution and pinocytosis of lymphocyte surface immunoglobulin molecules induced by anti-immunoglobulin antibody. *Nature New Biol.* **233**, 225.

Tyan, M. L., and L. A. Herzenberg (1968). Studies on the ontogeny of the mouse immune system. II. Immunoglobulin-producing cells. *J. Immunol.* **101**, 446.

Unanue, E. R., W. D. Perkins, and M. J. Karnovsky (1972). Ligand-induced movement of lymphocyte membrane macromolecules. I. Analysis by immunofluorescence and ultrastructural radioautography. *J. Exp. Med.* **136**, 885.

Weisenberg, R. C., G. G. Borisy, and E. W. Taylor (1968). The colchicine-binding protein of mammalian brain and its relation to microtubules. *Biochemistry* **7**, 4466.

White, J. G. (1971). In *The Circulating Platelet* (S. A. Johnson, ed.). Academic Press, New York, p. 59.

Wilson, L. (1970). Properties of colchicine binding protein from chick embryo brain: Interactions with *Vinca* alkaloids and podophyllotoxin. *Biochemistry* **9**, 4499.

Yahara, I., and G. M. Edelman (1972). Restriction of the mobility of lymphocyte immunoglobulin receptors by concanavalin A. *Proc. Nat. Acad. Sci. U.S.A.* **69**, 608.

Yahara, I., and G. M. Edelman (1973). The effect of concanavalin A on the mobility of lymphocyte surface receptors. *Exp. Cell Res.* (in press).

CHAPTER NINE

Surface Antigens and Differentiation of Thymus-Dependent Lymphocytes

IRVING GOLDSCHNEIDER

The prevailing view of lymphopoiesis (Fig. 1) is that the lymphoid system is composed of two ontogenetically and functionally distinct populations of lymphocytes: thymus-derived lymphocytes (T cells) and bone marrow-derived lymphocytes (B cells) (Roitt et al., 1969; Raff, 1973). These designations are somewhat misleading, inasmuch as T cells, as well as B cells, originate from precursors (perhaps a common precursor) in bone marrow in late fetal and postnatal life (Goldschneider and McGregor, 1966; McGregor, 1968; Tyan, 1968; Tyan and Herzenberg, 1968). What is important is that T cells are dependent on the thymus for their maturation, whereas B cells are not (Miller and Osoba, 1967). Indeed, in birds a distinct gut-associated generative organ for B cells, the bursa of Fabricius, has been identified (Cooper et al., 1966; Warner et al., 1962). An analog of the bursa has not been found in mammals, although there is some evidence to suggest that other gut-associated lymphoid tissues may subserve this function (Cooper et al., 1968).

The thymus and the bursa of Fabricius direct the proliferation and differentiation of an immigrant population of stem cells (Moore and Owen, 1966, 1967; Owen and Ritter, 1969) at least in part through a hormonal mechanism (Luckey, 1973; Pierre and Ackerman, 1965; Jankovic and Leskowitz, 1965). A major function of these "central" lymphoid organs appears to be generation of immunological diversity among the progeny of the stem cells, although the genetic mechanisms involved are unclear (Burnet, 1962; Jerne, 1971). This diversity is expressed phenotypically by the appearance on the lymphocyte plasma membrane of re-ceptors for specific antigens (Wigzell, 1970; Paul, 1970; Basten, 1971; Roelants, 1972; Goldstein et al., 1973). Such antigen-recognition cells (ARC) emigrate from the central lymphoid organs and join the pools of immunologically competent lymphocytes which populate the lymph nodes, spleen, and other "peripheral" lymphoid tissues (Nossal, 1964; Murray and Woods, 1964; Weissman, 1967; Linna, 1968; Toivanen et al., 1972; Durkin et al., 1972).

After antigenic stimulation, ARCs of both the T- and the B-cell systems undergo blast transformation, clonal proliferation, and further differentiation (Gowans and McGregor, 1965; Kennedy et al., 1966; Shearer and Cudkowicz, 1969). In the B-cell system, antibody-producing plasma cells are formed; in the T-cell system, lymphocytes which can initiate so-called cell-mediated immunological responses (e.g., homograft rejection, delayed-type hypersensitivity reactions, protection against infection with intracellular parasites) are formed. Despite the fact that T and B cells are able to function independently, cooperation between T and B cells and perhaps macrophages is necessary in most instances for the most efficient production of antibodies by B cells (Claman and Chaperon, 1969; Davies, 1969; Miller and Mitchell, 1969; Rajewsky et al., 1969; Mosier and Coppleson, 1968). Also, cooperation between two subpopulations of T cells (T_1 and T_2 cells) has been demonstrated in certain cell-mediated immunological responses (Cantor and Asofsky, 1972; Stobo et al, 1973a).

In addition to these functional differences, T and B cells differ with respect to a number of other parameters, including buoyant density (Howard et al., 1972), electrophoretic mobility (Zeiller et al., 1972), surface topography (Polliack, 1973), RNA metabolism (Rieke, 1966), responsiveness to phytomitogens (Greaves and Bauminger, 1972; Andersson et al., 1972), formation of nonimmune rosettes (Jondal et al., 1972), anatomical localization (Waksman et al., 1962; Parrot et al.,

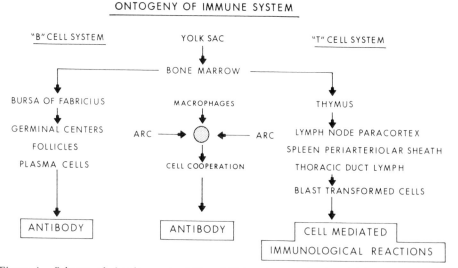

Figure 1. Scheme of development of T and B lymphocytes in the chicken. ARC, antigen-recognition cells from T- and B-cell systems. See text for discussion.

1966; Goldschneider and McGregor, 1968a), and migratory behavior (Howard et al., 1972; Sprent, 1973).

One of the major advances in the study of the development and functions of the lymphoid system in birds and mammals has been the discovery of antigens that are specific for the surfaces of subpopulations of lymphocytes. Some of these antigens occur in multiple allelic forms (alloantigens) and can be detected by antisera prepared in animals of the same species (alloantisera). Other antigens, while organ specific, are not known to occur in allelic forms and must be detected by antisera prepared in animals of different species (heteroantisera). The latter species-specific antisera require extensive absorption to render them organ specific. As examples of selective gene action between different cell types in a single individual, both alloantigens and heteroantigens can be properly referred to as "differentiation antigens" (Boyse and Old, 1969).

With the exception of the cell surface receptors for antigens, nothing is known about the functions of lymphocyte-specific antigens. However, they have been intensively studied by investigators interested in the genetic control of phenotypic variation (Boyse and Old, 1969) and in the dynamic aspects of cell membrane structure (Taylor et al., 1971; Unanue et al., 1973). Inasmuch as these topics have been considered in detail elsewhere, the following discussion will be restricted to the use of lymphocyte-specific cell surface antigens as phenotypic markers of lymphocyte differentiation. The discussion will be further restricted to a consideration of the T-cell system, for which a reasonably comprehensive picture of the major stages of differentiation has begun to emerge.

THE PRECURSOR OF THYMOCYTES

Although earlier evidence suggested that thymocytes were derived from thymic epithelial cells in the embryo (Ackerman and Knouff, 1965; Auerbach, 1961), it subsequently has been demonstrated by experiments in parabiotic chickens and in mice that thymocytes are derived from blood-borne stem cells which invade the epithelial anlage of the thymus (Moore and Owen, 1967; Owen and Ritter, 1969). Moreover, it is clear that migration of stem cells into the thymus continues throughout adult life (Ford and Micklem, 1963; Metcalf and Wakonig-Vaartaja, 1964; Harris et al., 1964). The sources of stem cells are the placenta, yolk sac, and liver in the embryo, and almost exclusively the bone marrow in the adult (Taylor, 1964; Tyan, 1968; Goldschneider and McGregor, 1966; McGregor, 1968, 1969; Moore and Metcalf, 1970).

The suspected precursors of thymocytes in the embryo are large basophilic cells which localize between the epithelial cells in the presumptive cortex of the thymus (Moore and Owen, 1967). In the adult, the first cells that have been observed to migrate into regenerating thymus grafts are small lymphocytes which have prominent nucleoli, abundant polyribosomes, and short segments of endoplasmic reticulum (Blackburn and Miller, 1967). Although no direct proof exists that small lymphocyte-like cells are the precursors of thymocytes in the adult, there is reason to think that this may be the case. Thus, Cudkowicz et al. (1964) and Blomgren (1969) have shown that the hemopoietic and lymphopoietic colony-forming potential of bone marrow cell suspensions is proportional to the lymphocyte content of the cell suspensions. Haskill and Moore (1970) have characterized the hemopoietic colony-forming units (CFUs) in adult mouse bone marrow as small, high-density cells; whereas CFUs in fetal mouse liver are large, low-density cells.

Zeiller et al. (1972) have described a population of bone marrow lymphocytes in the adult mouse which shares two properties with thymocytes: low electrophoretic mobility, and a specific cell surface antigen (RLTA = rat low-electrophoretic-mobility-thymocyte antigen). These properties are not shared by mature T cells in peripheral lymphoid tissues or in bone marrow. Indeed, with this one possible exception, the bone marrow precursor of thymocytes has been shown not to resemble thymocytes antigenically (Schlesinger, 1972).

We have recently identified a population of small lymphocytes in adult rat bone marrow which lack antigens specific for the surfaces of mature T or B cells (Goldschneider and McGregor, 1973). Unlike the antigenically identifiable T cells in bone marrow, these antigen-negative cells, which comprise approximixately 41% of bone marrow lymphocytes, are unable to initiate a graft-versus-host reaction and are not depleted by prolonged drainage of the thoracic duct. A similar antigenically "null" population of lymphocytes, which lack the θ antigen and immunoglobulin molecules characteristically found on T and B cells, respectively, has been identified in the mouse spleen (Stobo et al., 1973b). It remains to be seen whether purified suspensions of these unique lymphocytes from bone marrow and spleen can serve as precursors of thymocytes.

DIFFERENTIATION OF THYMOCYTES

Under the influence of the thymic microenvironment, the precursors of thymocytes are committed to a pathway of differentiation that eventuates in the production of immunologically competent T lymphocytes. It has been calculated that precursor cells in the thymus undergo six to eight divisions (mean generation time approximately 9 hours), resulting in the production of populations of progressively smaller thymocytes (Sainte-Marie and Leblond, 1964; Borum, 1968; Bryant, 1972). The mitotic wave appears to progress from the subcapsular region of the thymus cortex to the deeper parts of the cortex. In the embryonic thymus, as many as 10,000 thymocytes may be produced from a single precursor cell (Metcalf and Moore, 1971). Inasmuch as many times more thymocytes are generated each day than is necessary to replace effete T cells in the peripheral lymphoid tissues, it has been suggested (Metcalf, 1967) that most newly formed thymocytes die within the cortex of the thymus, perhaps as part of the process by which lymphocytes with diverse immunological specificities to nonself antigens are selected (Burnet, 1962; Jerne, 1971).

In any event, the thymocytes which are destined to join the pool of peripheral T cells appear to enter the medullary region of the thymus. Here they cease their proliferative activities and undergo further differentiation. By the end of their stay in the medulla, thymocytes become immunologically competent (Blomgren and Andersson, 1971) and acquire the capacity to migrate specifically to lymph nodes, spleen, and other peripheral lymphoid tissues (Goldschneider and McGregor, 1968a, b; Colley et al., 1970b; Lance et al., 1971). Another interpretation of these observations is that cortical and medullary thymocytes are ontogenetically distinct populations of cells, each arising from its own set of precursor cells (Schlesinger, 1972). However, most evidence favors the derivation of medullary thymocytes from cortical thymocytes, a view which will be advanced in the ensuing discussion.

The differentiation of thymocytes is characterized not only by changes in cell size, density, electrophoretic mobility, immunological competence, and susceptibility to cortisone and irradiation (*vide infra*) but also by the appearance and disappearance of certain antigens on the cell surface. Some of these antigens are specific for the plasma membranes of thymocytes or subpopulations of thymocytes, whereas others are shared by thymocytes and peripheral T cells but not by B cells. The distribution of these antigens among the cells which constitute the T lymphocyte system is listed in Tables 1 and 2.

Differentiation Antigens of Cortical Thymocytes

Alloantigens

Only one system of cell surface alloantigens has been described thus far which is specific for cortical thymocytes, the TL (thymus-leukemia) system of the mouse (Boyse et al., 1969). The TL antigens, of which there are four known alleles, are coded by a single locus which is adjacent to the D end of the major histocompatibility locus (*H-2*) in the IXth linkage groups. The TL antigens have been studied intensively as models of selective gene action, since leukemia cells from TL-negative mice are frequently TL positive; and as models of antigenic modulation, since

TABLE 1. T-CELL ALLOANTIGENS IN THE MOUSE

Stage of Differentiation of T Cells	Relative concentration of Alloantigens on Cell Surface[a]			
	H-2	TL	θ	Ly-A, B, C
Stem cell	+ + +	−	−	−
Cortical thymocyte	+	+ + +	+ + +	+ + +
Medullary thymocyte	+ +	−	+ +	+ +
Mature T cell	+ + +	−	+	+

[a] Expressed in arbitrary units: + + +, high concentration; + +, intermediate concentration; +, low concentration; −, undetectable. See text for description of antigens and for references.

TABLE 2. T-CELL HETEROANTIGENS

Stage of Differentiation of T Cells	Representation of Heteroantigens on Cell Surface[a]						
	Mouse		Rat				Chicken
	MSLA	MPLA	RSLA	RLTA	RTA	RMTA	CTLA
Stem cell	−	−	−	+	−	−	?
Cortical thymocyte	+	−	+	+	+	−	+
Medullary thymocyte	+	+	+	+	+	(Masked)	+
Mature T cell	+	+	+	−	−	+	+

[a] Plus (+) or minus (−) sign denotes presence or absence of detectable antigen, respectively. See text for description of antigens and for references.

cells from TL-positive mice become TL negative and express increased quantities of H-2 antigens on their surface when incubated in the presence of anti-TL serum. The TL antigens appear to be expressed under the inductive influence of the thymus cortex microenvironment. Neither the precursors of cortical thymocytes in the bone marrow nor the progeny of cortical thymocytes in the thymus medulla or peripheral lymphoid tissues bear these antigens (Schlesinger and Hurvitz, 1968b; Owen and Raff, 1970; Leckband and Boyse, 1971). It may be significant that another lymphocyte-specific cell surface alloantigen, G_{1X}, which shares many properties with TL antigens, is associated with murine leukemia virus infection (Stockert et al., 1971).

Four other systems of lymphocyte alloantigens are represented in high concentrations on cortical thymocytes in the mouse; Ly-A, Ly-B, Ly-C (Boyse et al., 1968, 1971), and θ (Thy-1) (Reif and Allen, 1964). These antigens are not present on the bone marrow precursor of thymocytes (Schlesinger and Hurvitz, 1968b; Boyse and Old, 1969; Owen and Raff, 1970). However, unlike the TL antigens, they continue to be expressed, albeit in markedly decreased concentrations, on medullary thymocytes and peripheral T cells (Aoki et al., 1969; Konda et al., 1973; Schlesinger et al., 1973). The Ly-A, Ly-B, and Ly-C alloantigens are strictly lymphocyte specific, whereas the θ alloantigens are also present in high concentration in adult brain tissue. There is no evidence of linkage between any of the four systems of alloantigens, each of which is composed of two alleles. None of these antigens is present on cells of the B lymphocyte system. An alloantigen system having a tissue distribution similar to that of the θ system has recently been described in the rat (Douglas, 1972; Peter et al., 1973; Lubaroff, 1973).

Precursors of thymocytes appear in the mouse thymus rudiment between the 10th and 11th days of gestation, and the thymus becomes frankly lymphoid in character by the 13th day. Yet it is not until the 16th day of gestation that some thymocytes begin to express TL, Ly, and θ alloantigens. By the 18th day virtually all thymocytes possess these antigens (Schlesinger and Hurvitz, 1968a; Owen and Raff, 1970). It is of great interest therefore that the acquisition of these antigens coincides with another major event in the fetal thymus, the abrupt transition of rapidly dividing large and medium lymphoid cells to typical small thymocytes (Ball, 1963). It would appear that the expression of these cell surface alloantigens is an integral part of the differentiation of thymocytes from precursor cells, and that the thymus (presumably the epithelial cell component) acquires the capacity to direct such differentiation on or about the 16th day of gestation in the mouse.

Further evidence that the thymus directs both the expression of T-cell-specific antigens and the production of cells that bear these antigens is derived from studies in congenitally thymusless (*nu/nu*) mice, neonatally thymectomized mice, and adult mice that have been thymectomized, lethally irradiated, and reconstituted with bone marrow cells. These animals are markedly lymphopenic, and almost all the remaining lymphocytes lack the Ly and θ alloantigens (Raff and Wortis, 1970; Schlesinger and Yron, 1970). Recently subpopulations of lymphocytes from the spleen and bone marrow of adult normal and *nu/nu* mice and the livers of 14-day-old normal mouse embryos were induced to express TL, Ly and θ antigens by exposure to a thymus extract for approximately 2 hours *in vitro* (Komuro and Boyse, 1973a, b). This presumably is analogous to the inductive process which normally occurs within the microenvironment of the intact thymus.

Histocompatibility alloantigens are normally found on the surfaces of almost all cells of the body. Nevertheless, they are of especial interest in considerations of T-cell development for two reasons. First, mature T cells are exceptionally reactive immunologically to foreign histocompatibility antigens (Nisbet et al., 1969; Wilson and Nowell, 1970). Second, histocompatibility antigens are present in high concentrations on peripheral T cells and probably also on the precursors of thymocytes, but they are present in significantly lower concentrations on cortical thymocytes (Aoki et al., 1969; Hamerling et al., 1969). These two observations may be coincidental, but it is possible that they are related in some manner to the process by which so-called forbidden clones of thymocytes are eliminated and immunological diversity is generated (Burnet, 1962; Jerne, 1971).

Heteroantigens

A number of heteroantigens that are specific for thymocytes have been described, although none is unique for cortical thymocytes. Colley et al. (1970a) and Williams et al. (1971) injected rabbits with rat thymocytes and calf thymocytes, respectively, and absorbed the resultant antisera with lymph node and spleen cells. The adsorbed antisera detected antigens (RTA = rat thymus antigen; BTA = bovine thymus antigen) on the surface of essentially all thymocytes, but not on the precursors of thymocytes or on peripheral lymphocytes of adult animals. However, numerous lymphocytes in lymph node and spleen of fetal rats were RTA positive. This latter finding suggests that thymocytes are released to the periphery in a less mature state in the fetus than in the adult animal, a hypothesis that has received support from other quarters (Weissman, 1967). Bachvaroff et al. (1969) have isolated and partially purified an antigen from the plasma membranes of rat thymocytes which has the same tissue distribution as RTA. The thymus-specific antigen was contained in a protein fraction that moved as a single homogeneous band in polyacrylamide disk electrophoresis.

Potworowski and Nairn (1967) prepared antisera in mice against a microsome fraction from rat thymus cells. After absorption with homogenates of rat liver and peripheral lymphoid tissues, the antisera were specific for thymocytes, reacting at least in part with their plasma membranes. However, unlike RTA, this thymus-specific antigen was present on thymocytes only during fetal life and early postnatal life. The antigen was absent from thymocytes in adult rats and did not reappear during thymus regeneration in irradiated rats. These findings suggest either that the microenvironment of the thymus differs in prenatal and postnatal life, or that thymocytes originate from different populations of stem cells during these two periods. The studies of Haskill and Moore (1970) lend credence to the latter possibility. These authors found that colony-forming units (CFUs) from mouse fetal liver were large, low-density cells and that CFUs from adult bone marrow were small, high-density cells. Moreover, CFUs from adult bone marrow were incapable of reverting to the embryonic profile even under conditions of rapid proliferation.

Another heteroantigen, RLTA (Zeiller and Dolan, 1972), was discussed earlier (see p. 168) in relation to the precursor of thymocytes. This antigen is present in high concentration on thymocytes of low electrophoretic mobility (cortical thymocytes), and in lower concentration on thymocytes of high electrophoretic mobility (medullary thymocytes). Although RLTA is present on a subpopulation of bone marrow lymphcytes of low electrophoretic mobility, it is absent from mature T cells.

In addition to the above antigens in the rat, heteroantigens that are specific for thymocytes or for both thymocytes and peripheral T cells have been described in the mouse, chicken, guinea pig, and human being (Shigeno et al., 1968; Forget et al., 1970; McArthur et al., 1971; Malchow et al., 1972; Shevach et al., 1972; Aiuti and Wigzell, 1973; Smith et al., 1973).

Differentiation Antigens of Medullary Thymocytes

The population of thymocytes that is present in the medulla of the thymus is generally considered to contain the immediate precursors of periperal T cells. Like T cells, and unlike cortical thymocytes, medullary thymocytes are (1) resistant to cortisone and irradiation; (2) immunologically competent; (3) responsive to phytomitogens; (4) long lived; (5) able to migrate to thymus-dependent zones of lymph node and spleen when injected intravenously; (6) highly mobile electrophoretically; (7) TL-antigen negative; (8) θ and Ly-antigen poor; and (9) H-2 antigen rich (Ishidate and Metcalf, 1963; Cerottini and Brunner, 1967; Everett and Tyler, 1967; Blomgren and Andersson, 1971; Lance et al., 1971; Stobo and Paul, 1972; Zeiller and Dolan, 1972).

It could be argued that medullary thymocytes are in fact mature T cells which have migrated from the peripheral lymphoid tissues to the thymus. The fact that very few peripheral T cells have the ability to migrate to the thymus in the adult animal (Gowans and Knight, 1964; Goldschneider and McGregor, 1968b) does not refute this argument, inasmuch as medullary thymocytes comprise only about 5% of lymphoid cells in the thymus. Nonetheless, several telling differential points exist between medullary thymocytes and peripheral T cells. First, the appearance of medullary thymocytes precedes that of peripheral T cells during ontogeny; medullary thymocytes are the first cells to become immunologically competent; and the degree of immunological competence of thymocytes is equivalent in adults and newborn animals (Sosin et al., 1966; Schwarz, 1967; MacGillivray et al., 1970). Second, immunologically competent cells develop in the embryonic thymus after several days of culture in vitro or in diffusion chambers in vivo (Metcalf and Moore, 1971). Third, medullary thymocytes have a lower buoyant density than peripheral T cells (Colley et al., 1970b; Konda et al., 1973; Schlesinger et al., 1973). And fourth, medullary thymocytes bear thymus-specific antigens (e.g., RTA, BTA, RLTA) that are absent from mature T cells ((Colley et al., 1970b; Williams et al., 1971; Zeiller and Dolan, 1972).

Recently, we described an antigen which is fully expressed on the surface of rat peripheral T cells, but is present in a masked form on medullary thymocytes and apparently not at all on cortical thymocytes (Goldschneider, 1973). Antibodies to the antigen (RMTA = rat masked thymocyte antigen) were present in certain antisera that were raised in rabbits against rat thoracic duct lymphocytes. After extensive absorption with erythrocytes, peritoneal exudate cells, B lymphocytes and thymocytes, the resultant antisera reacted strongly with peripheral T cells but not detectably with viable thymocytes as determined by indirect immunofluorescence. Thus RMTA differs from MPLA, a heteroantigen that is normally expressed on both medullary thymocytes and peripheral T cells in the mouse (Raff and Cantor, 1971).

However, the antisera *did* react with the plasma membranes of medullary thymocytes in frozen sections of whole thymus, suggesting that the act of sectioning had mechanically unmasked an antigen. This antigen could also be unmasked enzymatically, by treating suspensions of viable thymocytes with neuraminidase. Between 5 and 10% of neuraminidase-treated thymocytes from normal rats were RMTA positive. Moreover, the percentage of RMTA-positive thymocytes could be progressively increased to 50% by pretreating the donor of the thymocytes with cortisone-acetate for 1–3 days, a procedure known to markedly enrich the proportion of medullary thymocytes present in the thymus.

As proof that RMTA was the same antigen as that normally expressed on the surface of peripheral T cells, antisera were prepared against neuraminidase-treated thymocytes and were absorbed with untreated thymocytes. Such antisera contained antibodies that reacted with and could be absorbed by peripheral T cells and neuraminidase-treated medullary thymocytes. Antisera raised against untreated thymocytes, on the other hand, did not contain antibodies to RMTA.

It is tempting to speculate, as a result of these experiments, that the unmasking of RMTA *in vivo* may be an event of functional consequence in the terminal differentiation of medullary thymocytes to mature T cells. At the least, the existence of RMTA adds additional weight to the hypothesis that medullary thymocytes are the immediate precursors of peripheral T cells.

MATURE T CELLS

The study of T- and B-cell-specific antigens not only has helped to establish the existence of these two developmentally distinct populations of lymphocytes, but also has provided information about other properties of T and B cells, such as their distribution in peripheral lymphoid tissues and the nature of the antigen-recognition sites on their surfaces. The ability to migrate to lymphoid tissues and to recirculate from blood to lymph is another property of T cells that is probably related to the presence of specific cell surface receptors. These topics are briefly discussed below.

Anatomical Distribution of T Cells

T and B lymphocytes occupy anatomically segregated compartments within peripheral lymphoid tissues. This conclusion has emerged from studies involving thymic ablation, thoracic duct drainage, and lymphocyte reconstitution (Gowans and Knight, 1964; Waksman et al., 1962; Cooper et al., 1966; Parrott et al., 1966; Goldschneider and McGregor, 1968a; Howard et al., 1972), and from the study of lymphocyte-specific cell surface antigens. An example of our findings in the rat will illustrate the value of the latter approach (Goldschneider and McGregor, 1973).

Antisera raised in rabbits against rat thoracic duct lymphocytes were made specific for T-cell antigens by absorption with rat erythrocytes, peritoneal exudate cells, and lymphocytes from T-cell-depleted donors. Antibodies to nuclear and cytoplasmic constituents were removed by washing after reacting the antisera with viable TDL. The T-cell-specific antibodies were recovered from the cell surfaces by eluting with O.1 M citric acid. Antisera specific for surface antigens of peripheral B lymphocytes were prepared in a similar manner, except that thymocytes were used in the absorption procedure rather than B lymphocytes. The anatomical distributions of T and B cells in the rat were studied by indirect immunofluorescence. The results are similar to those reported for the mouse, where different cell surface markers were used (Dukor et al., 1970; Gutman and Weissman, 1972).

Rabbit antirat T-cell serum (ALS$_T$) reacted selectively with the surfaces of lymphocytes in the paracortex of lymph node and in the periarteriolar sheath of white pulp of spleen (Fig. 2). These are the same anatomical regions that are depleted of lymphocytes after neonatal thymectomy or chronic drainage of lymphocytes from a thoracic duct fistula. Rabbit anti-B-cell serum (ALS$_B$) reacted selectively with the surfaces of lymphocytes in the follicular cortex of lymph node and in the lymphoid follicles and marginal zones of white pulp of spleen, the so-called thymus-independent zones of peripheral lymphoid tissues (Fig. 2).

Essentially all lymphocytes in cell suspensions from lymphoid tissues and body fluids reacted with ALS$_T$ or ALS$_B$, with the exception of bone marrow (Table 3). Approximately 41% of lymphocytes in bone marrow lacked detectable T- or B-cell surface antigens. The role, if any, of this "null" population of bone marrow lymphocytes in the scheme of development of T and B lymphocytes is unknown (see "The Precursor of Thymocytes," p. 167, for further discussion).

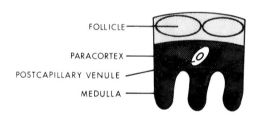

LYMPH NODE

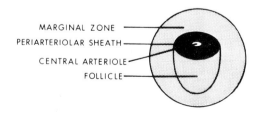

SPLEEN

Figure 2.   Anatomical distribution of T and B lymphocytes in rat lymphoid tissues. ▨ , T cells; ▢ , B cells. B lymphocytes, in the form of plasma cells, are also present in the lymph node medulla. See text for discussion.

TABLE 3. DISTRIBUTION OF T AND B LYMPHOCYTES IN THE RAT

Class of Lymphocyte	Percentage of Lymphocytes in Each Class[a]					
	Thymus	Thoracic Duct Lymph	Lymph Node	Blood	Spleen	Bone Marrow
T cell	> 99	87	81	61	∾ 50	13
B cell	—	11	18	40	∾ 50	38
Other	—	—	—	—	?	41

[a] Determined by indirect immunofluorescence, using differentially absorbed lymphocyte-specific antisera (see text).

The Nature of the Antigen-Recognition Sites on T Cells

Both T and B cells have receptors on their surfaces which enable them to bind antigens and to initiate immunological responses. The antigen-recognition molecules on B cells are membrane-bound immunoglobulins, mainly monomeric IgM (Vitetta et al., 1971). It has been estimated that B lymphocytes have approximately 100,000 surface Ig molecules per cell on the average (Rabellino et al., 1971). This presumably accounts for the preponderance of antibodies against rat Ig molecules that are present in rabbit antisera raised against rat B cells (Goldschneider and McGregor, 1973).

There is considerable controversy, however, about the nature of the antigen-recognition molecules on the surface of T cells. First, T cells bind much less antigen to their surfaces than do B cells (Roelants, 1972). Second, attempts to unequivocally demonstrate Ig molecules on the surface of T cells by immuno-fluorescence, cytotoxic, or autoradiographic techniques have been unsuccessful (Raff, 1971; Takahashi et al., 1971; Unanue et al., 1971; Nossal et al., 1972). And, third, in the hands of some investigators chemical extraction techniques that have been successful in isolating Ig molecules from the surfaces of B cells have failed to isolate Ig molecules from the surfaces of T cells (Vitetta et al., 1972). Thus, it has been postulated that the antigen receptors on T cells either belong to an unidentified class of Ig molecules ("IgX") (Mitchinson, 1969) or are unidenti-fied products of the immune response (*Ir*) genes of the major histocompatibility locus, that is, a new type of antigen-recognition molecule (McDevitt et al., 1972).

Despite these arguments, a number of authors have found that antibodies to Ig molecules can interefere with the ability of T cells to initiate delayed-type hyper-sensitivity and graft-rejection reactions (Greaves, 1970; Mason and Warner, 1970; Basten et al., 1971), and to cooperate with B cells (Lesley et al., 1971). On the basis of experiments with anti-Ig sera, Greaves and Hogg (1971) hypothesized that Ig molecules are present on T cells in a masked form. In support of this hypothesis, Marchalonis et al. (1972), using an enzymatic radioiodination technique to label cell surface proteins, claim to have isolated monomeric IgM molecules in equal quantities from the surfaces of resting and of antigen-stimulated mouse thymocytes, as well as from B cells. Significantly, these authors found that longer segments of the Ig heavy chains from antigen-stimulated thymocytes and from B cells were labeled with ^{125}I than from nonstimulated thymocytes. This suggested that surface Ig molecules were normally exposed on resting B cells, partially masked on resting T cells, and unmasked on activated T cells. The studies of Feldman (1972) have indicated that Ig molecules are released from the surface of activated T cells and, in the presence of antigen, are able to act through the mediation of macrophages to cooperate with B cells in humoral antibody responses.

Using immunofluorescence techniques, we too have found evidence that Ig molecules are present in increased quantity or in a more readily detectable form on the surface of antigen-stimulated rat T cells (Goldschneider and Cogen, 1973). From 60 to 92% of T cells from thoracic duct lymph that underwent blast transformation after stimulation with histocompatibility antigens or tuberculoprotein displayed readily detectable surface Ig determinants. Unstimulated or mitogen-stimulated T cells were Ig negative. Recent experiments using cortisone-resistant thymocytes as a pure source of antigen-reactive T cells, and highly purified rabbit F(ab')$_2$ fragments specific for rat Ig light chains, have confirmed these results (Goldschneider, unpublished observations). It is likely, therefore, that the Ig molecules on the surface of antigen-stimulated rat T cells are produced by T cells.

From a teleological viewpoint, the unmasking of Ig molecules during T-cell activation is consistent with the increased efficiency with which antigen-stimulated ("educated") T cells engage in immunological reactions. For example, antigen-stimulated T cells are more efficient than nonstimulated T cells at binding antigens (Roelants, 1972), implementing contact-mediated target cell destruction (effector cells) (Cerrotini et al., 1970), cooperating with B cells (helper cells) (Miller et al., 1971), and initiating accelerated cell-mediated immunological responses (memory cells) (Wilson and Nowell, 1971). It must be emphasized, however, that the demonstration of Ig molecules on the surfaces of activated T cells does not exclude the possibility that another species of antigen-recognition molecule, such as the *Ir* gene product, is also present.

Migration Pathways of T Cells

The ability to migrate from the blood to paracortex of lymph node and to the periarteriolar sheath of spleen is a unique property of peripheral T cells and some medullary thymocytes (Fig. 3). Gowans and Knight (1964) demonstrated that a population of small lymphocytes in the rat continually recirculated from blood to lymph via specialized venules in the paracortex of lymph nodes. It was originally thought that these lymphocytes were transported by emperipolesis through the cytoplasm of specialized endothelial cells which line segments of the postcapillary venues (Marchesi and Gowans, 1964); but Schoefl (1972) has recently suggested that T cells pass between these endothelial cells en route to lymph node cortex. In any event, the physiological state of these endothelial cells appears to depend on the traffic of lymphocytes across the vessel wall (Goldschneider and McGregor, 1968a). Once in the paracortex of lymph node, T lymphocytes migrate into the medullary cords, where they gain entrance to the lymph sinuses and thence to the thoracic duct. The cycle is completed when the thoracic duct lymph drains into the venous circulation.

The migration pathway of T cells into the white pulp of spleen appears to involve the directional movement of lymphocytes between endothelial cells lining blood sinusoids in the marginal zone and across the intervening tissues to the periarteriolar lymphoid sheath (PALS) (Goldschneider and McGregor, 1968a). This presumably is a reversible process, inasmuch as cells which have migrated to the PALS can be recovered a few hours later in the splenic vein (Ford, 1969).

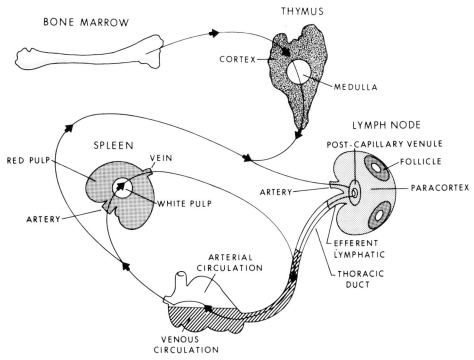

Figure 3.   Migration pathways of T cells. See text for discussion.

The mechanism(s) by which T cells are attracted to the paracortex of lymph node and to the periarteriolar lymphoid sheath of spleen is not known. However, it appears from the studies of Gesner et al. (1969) that recognition molecules on the surface of T cells are probably important. Thus, treatment of thoracic duct lymphocytes with neuraminidase prevents them from "homing" to lymph node and spleen, properties which the lymphocytes recover concomitantly with regeneration of the sialic acid residues on their surfaces. Moreover, treatment of thoracic duct lymphocytes with trypsin interferes with the ability of T cells to migrate to lymph node but not to spleen. The latter observation suggests that at least two different mechanisms and perhaps two different cell surface recognition sites are involved in the migration of T cells to lymph node and spleen. It is not yet known whether any of the antigens that have been detected on the surfaces of T cells are involved in determining the migratory behavior of this class of lymphocytes.

SUMMARY

The peripheral lymphoid tissues of higher vertebrates are populated by two developmentally and functionally distinct populations of lymphocytes. Thymus-processed lymphocytes (T cells) initiate cell-mediated immunological reactions, whereas bursa-processed lymphocytes (B cells) in chickens, and their equivalents in mammals, initiate humoral antibody-mediated immunological reactions. The study of alloantigens and heteroantigens that are specific for the surfaces of subpopulations of lymphocytes ("differentiation antigens") has provided valuable insights into the differentiative pathways of lymphocytes, especially T cells.

In our laboratory, antisera raised in rabbits against rat thoracic duct lymphocytes were made specific for T and B cells by differential absorption. The distributions and anatomical localizations of T and B cells in normal lymphoid tissues were determined by indirect immunofluorescence. Approximately 100% thymus, 90% thoracic duct, 80% lymph node, 60% blood, 50% spleen, and 10% bone marrow lymphocytes were identified as T cells. The rest of the lymphocytes were B cells, except in bone marrow, where approximately 41% of lymphocytes did not bear either T- or B-cell surface antigens. In frozen sections of lymph node and spleen, T cells were found to occupy the paracortical and periarteriolar sheath regions exclusively, whereas B cells occupied the lymphoid follicles and marginal zones. Antisera to rat γ-globulins reacted strongly with rat B cells but not with T cells. However, after stimulation with antigen, rat T cells underwent blast transformation and expressed readily detectable surface immunoglobulin molecules. In addition, an antigen that normally is expressed on the surface of functionally mature T cells was found to exist in a masked form on the surface of medullary thymocytes.

These observations and many others in the literature suggest that cells in the T lymphocyte system undergo three major differentiative events during their development, each of which is heralded by qualitative and quantitative changes in the representation of lymphocyte-specific cell surface antigens. The first occurs when blood-borne stem cells are committed to a pathway of lymphocytic differentiation within the microenvironment of the thymus cortex. The second occurs when thymocytes acquire immunological competence within the microenvironment of the thymus medulla. And the third occurs when peripheral T cells are stimulated by antigen to engage in immunological reactions.

ACKNOWLEDGMENT

The work described in this paper was supported by U.S. Public Health Service Grant AI-09649 from the National Institute of Allergy and Infectious Diseases.

REFERENCES

Ackerman, G. A., and R. A. Knouff (1965). The epithelial origin of the lymphocytes in the thymus of the embryonic hamster. *Anat. Rec.* **152**, 35–41.

Aiuti, F., and H. Wigzell (1973). Function and distribution pattern of human T lymphocytes. I. Production of anti-T lymphocyte specific sera as estimated by cytotoxicity and elimination of function of lymphocytes. *Clin. Exp. Immunol.* **13**, 171–181.

Andersson, J., G. M. Edelman, G. Moller, and O. Sjoberg (1972). Activation of B lymphocytes by locally concentrated concanavalin A. *Eur. J. Immunol.* **2**, 233–235.

Aoki, T., U. Hammerling, E. de Harven, E. A. Boyse, and L. J. Old (1969). Antigenic structure of cell surfaces: An immunoferritin study of the occurrence and topography of H-2, θ, and TL alloantigens on mouse cells. *J. Exp. Med.* **130**, 979–1001.

Auerbach, R. (1961). Experimental analysis of the origin of cell types in the development of the mouse thymus. *Develop. Biol.* **3**, 336–354.

Bachvaroff, R., F. Galdiero, and P. Grabar (1969). Anti-thymus cytotoxicity: Identification and isolation of rat thymus-specific membrane antigens and purification of the corresponding antibodies. *J. Immunol.* **103**, 953–961.

Ball, W. D. (1963). A quantitative assessment of mouse thymus differentiation. *Exp. Cell Res.* **31**, 82–88.

Basten, A., J. F. A. P. Miller, N. L. Warner, and J. Pye (1971). Specific inactivation of thymus-derived (T) and non-thymus-derived (B) lymphocytes by ^{125}I-labelled antigen. *Nature New Biol.* **231**, 104–106.

Blackburn, W. R., and J. F. A. P. Miller (1967). Electron microscopic studies of thymus graft regeneration and rejection. II. Syngeneic-irradiated grafts. *Lab. Invest.* **16**, 833–846.

Blomgren, H. (1969). The influence of the bone marrow on the repopulation of the thymus in X-irradiated mice. *Exp. Cell Res.* **58**, 353–364.

Blomgren, H., and B. Andersson (1971). Characteristics of the immunocompetent cells in the mouse thymus: Cell population changes during cortisone-induced atrophy and subsequent regeneration. *Cell. Immunol.* **1**, 545–560.

Borum, K. (1968). Pattern of cell production and cell migration in mouse thymus studied by autoradiography. *Scand. J. Haematol.* **5**, 339–352.

Boyse, E. A., and L. J. Old (1969). Some aspects of normal and abnormal cell surface genetics. *Ann. Rev. Genet.* **3**, 269–291.

Boyse, E. A., M. Miyazawa, T. Aoki, and L. J. Old (1968). Two systems of lymphocyte isoantigens in the mouse. *Proc. Roy. Soc.* **B170**, 175–193.

Boyse, E. A., E. Stockert, and L. J. Old (1969). Properties of four antigens specified by the TLa locus; similarities and differences. In *International Convocation on Immunology* (N. Rose and F. Milgrom, eds.). S. Karger, Basel, pp. 353–357.

Boyse, E. A., K. Itakura, E. Stockert, C. A. Iritani, and M. Miura (1971). A third locus specifying alloantigens expressed only on thymocytes and lymphocytes. *Transplantation* **11**, 351–353.

Bryant, B. J. (1972). Renewal and fate in the mammalian thymus: mechanisms and inferences of thymocytokinetics. *Eur. J. Immunol.* **2**, 38–45.

Burnet, M. (1962). The immunological significance of the thymus: An extension of the clonal selection theory of immunity. *Australas. Ann. Med.* **11**, 79–91.

Cantor, H., and R. Asofsky (1972). Synergy among lymphoid cells mediating the graft-versus-host response. *J. Exp. Med.* **135**, 764–779.

Cerottini, J. C., and K. T. Brunner (1967). Localization of mouse isoantigens on the cell surface as revealed by immunofluorescence. *Immunology* **13**, 395–403.

Cerottini, J. C., A. A. Nordin, and K. T. Brunner (1970). *In vitro* cytotoxic activity of thymus cells sensitized to alloantigens. *Nature* **227**, 72–73.

Claman, H. N., and E. A. Chaperon (1969). Immunologic complementation between thymus and marrow cells—a model for the two-cell theory of immunocompetence. *Transplant. Rev.* **1**, 92–113.

Colley, D. G., A. Malakian, and B. H. Waksman (1970a). Cellular differentiation in the thymus. II. Thymus-specific antigens in rat thymus and peripheral lymphoid cells. *J. Immunol.* **104**, 585–592.

Colley, D. G., A. Y. Shih, and B. H. Waksman (1970b). Cellular differentiation in the thymus. III. Surface properties of rat thymus and lymph node cells separated on density gradients. *J. Exp. Med.* **132**, 1107–1121.

Cooper, M. D., R. D. A. Peterson, M. A. South, and R. A. Good (1966). The function of the thymus system and the bursa system in the chicken. *J. Exp. Med.* **123**, 75–102.

Cooper, M. D., D. Y. Perey, A. E. Gabrielsen, D. E. R. Sutherland, M. F. McKneally, and R. A. Good (1968). Production of an antibody deficiency syndrome in rabbits by neonatal removal of organized intestinal lymphoid tissues. *Int. Arch. Allergy* **33**, 65–88.

Cudkowicz, G., M. Bennett, and G. M. Shearer (1964). Pluripotent stem cell function of the mouse marrow "lymphocyte." *Science* **144**, 866–868.

Davies, A. J. S. (1969). The thymus and the cellular basis of immunity. *Transplant. Rev.* **1**, 43–91.

Douglas, T. C. (1972). Occurrence of a theta-like antigen in rats. *J. Exp. Med.* **136**, 1054–1062.

Dukor, P., C. Bianco, and V. Nussenzweig (1970). Tissue localization of lymphocytes bearing a membrane receptor for antigen-antibody-complement complexes. *Proc. Nat. Acad. Sci. U.S.A.* **67**, 991–997.

Durkin, H. G., G. A. Theis, and G. J. Thorbecke (1972). Bursa of Fabricius as site of origin of germinal centre cells. *Nature New Biol.* **235**, 118–119.

Everett, N. B., and R. W. Tyler (1967). Lymphopoiesis in the thymus and other tissues: Functional implications. *Int. Rev. Cytol.* **22**, 205–237.

Feldman, M. (1972). Cell interactions in the immune response in vitro. V. Specific collaboration via complexes of antigen and thymus-derived cell immunoglobulin. *J. Exp. Med.* **136**, 737–760.

Ford, C. E., and H. S. Micklem (1963). The thymus and lymph nodes in radiation chimearas. *Lancet* **i**, 359–362.

Ford, W. L. (1969). The kinetics of lymphocyte recirculation within the rat spleen. *Cell Tissue Kinet.* **2**, 171–191.

Forget, A., E. F. Potworowski, G. Richer, and A. G. Borduas (1970). Antigenic specificities of bursal and thymic lymphocytes in the chicken. *Immunology* **19**, 465–468.

Gesner, B. M., J. J. Woodruff, and R. T. McCluskey (1969). An autoradiographic study of the effect of neuraminidase or trypsin on transfused lymphocytes. *Am. J. Pathol.* **57**, 215–230.

Goldschneider, I. (1973). Antigenic relationship between medullary thymocytes and peripheral T cells in the rat: Description of a "masked" antigen (submitted for publication).

Goldschneider, I., and R. B. Cogen (1973). Immunoglobulin molecules on the surface of activated T lymphocytes in the rat. *J. Exp. Med.* **138**, 163–175.

Goldschneider, I., and D. D. McGregor (1966). Development of immunologically competent cells in the rat. *Nature* **212**, 1433–1435.

Goldschneider, I., and D. D. McGregor (1968a). Migration of lymphocytes and thymocytes in the rat. I. The route of migration from blood to spleen and lymph nodes. *J. Exp. Med.,* **127**, 155–168.

Goldschneider, I., and D. D. McGregor (1968b). Migration of lymphocytes and thymocytes in the rat. II. Circulation of lymphocytes and thymocytes from blood to lymph. *Lab. Invest.* **18**, 397–406.

Goldschneider, I., and D. D. McGregor (1973). Anatomical distribution of T and B lymphocytes in the rat: Development of lymphocyte-specific antisera. *J. Exp. Med.* (in press).

Goldschneider, I., and A. A. Moscona (1972). Tissue-specific cell-surface antigens in embryonic cells. *J. Cell. Biol.* **53**, 435–449.

Goldstein, P., H. Blomgren, and E. A. Svedmyr (1973). The extent of specific adsorption of cytotoxic educated thymus cells: evolution with time and number of injected cells. *Cell. Immunol.* **7**, 213–221.

Gowans, J. L., and E. J. Knight (1964). The route of recirculation of lymphocytes in the rat. *Proc. Roy. Soc.* **B159**, 257–282.

Gowans, J. L., and D. D. McGregor (1965). The immunological activities of lymphocytes. *Prog. Allergy* **9**, 1–78.

Greaves, M. F. (1970). Biological effects of anti-immunoglobulins: Evidence for immunoglobulin receptors on "T" and "B" lymphocytes. *Transplant. Rev.* **5**, 45–75.

Greaves, M. F., and S. Bauminger (1972). Activation of T and B lymphocytes by insoluble phytomitogens. *Nature New Biol.* **235**, 67–70.

Greaves, M. F., and N. W. Hogg (1971). Antigen binding sites on mouse lymphoid cells. In *Cell Interactions and Receptor Antibodies in Immune Responses* (O. Makela, A. Cross, and T. U. Kosunen, eds). Academic Press, New York, pp. 145–155.

Gutman, G. A., and I. L. Weissman (1972). Lymphoid tissue architecture: Experimental analysis of the origin and distribution of T-cells and B-cells. *Immunology* **23**, 465–479.

Hammerling, U., N. Shigeno, L. J. Old, and E. A. Boyse (1969). Labelling of mouse alloantibody with tritiated DL-alanine. *Immunology* **17**, 999–1006.

Harris, J. E., D. W. Barnes, C. E. Ford, and E. P. Evans (1964). Cellular traffic of the thymus: experiments with chromosome markers. Evidence from parabiosis for an afferent stream of cells. *Nature* **201**, 884–887.

Haskill, J. S., and M. A. S. Moore (1970). Two dimensional cell separation: comparison of embryonic and adult haemopoietic stem cells. *Nature* **226**, 853–854.

Howard, J. C., S. V. Hunt, and J. L. Gowans (1972). Identification of marrow-derived and thymus-derived small lymphocytes in the lymphoid tissue and thoracic duct of normal rats. *J. Exp. Med.* **135**, 200–219.

Ishidate, M., and D. Metcalf (1963). The pattern of lymphopoiesis in the mouse thymus after cortisone administration or adrenalectomy. *Aust. J. Exp. Biol.* **41**, 637–649.

Jankovic, B. D., and S. Leskowitz (1965). Restoration of antibody producing capacity in bursectomized chickens by bursal grafts in Millipore chambers. *Proc. Soc. Exp. Biol. Med.* **118**, 1164–1166.

Jerne, N. K. (1971). The somatic generation of immune recognition. *Eur. J. Immunol.* **1**, 1–9.

Jondal, M., G. Holm, and H. Wigzell (1972). Surface markers on human T and B lymphocytes. I. A large population of lymphocytes forming nonimmune rosettes with sheep red blood cells. *J. Exp. Med.* **136**, 207–215.

Kennedy, J. C., J. E. Till, L. Siminovitch, and E. A. McCulloch (1966). The proliferative capacity of antigen-sensitive precursors of hemolytic plaque-forming cells. *J. Immunol.* **96**, 973–980.

Komuro, K., and E. A. Boyse (1973a). *In-vitro* demonstration of thymic hormone in the mouse by conversion of precursor cells into lymphocytes. *Lancet* **i**, 740–743.

Komuro, K., and E. A. Boyse (1973b). Induction of T lymphocytes from precusor cells *in vitro* by a product of the thymus. *J. Exp. Med.* **138**, 479–482.

Konda, S., E. Stockert, and R. T. Smith (1973). Immunologic properties of mouse thymus cells: membrane antigen patterns associated with various cell subpopulations. *Cell. Immunol.* **7**, 275–289.

Lance, E. M., S. Cooper, and E. A. Boyse (1971). Antigenic change and cell maturation in murine thymocytes. *Cell. Immunol.* **1**, 536–544.

Leckband, E., and E. A. Boyse (1971). Immunocompetent cells among mouse thymocytes: A minor population. *Science* **172**, 1258–1260.

Lesley, J. F., J. R. Kettman, and R. W. Dutton (1971). Immunoglobulins on the surface of thymus-derived cells engaged in the initiation of a humoral immune response. *J. Exp. Med.* **134**, 618–629.

Linna, T. J. (1968). Cell migration from the thymus to other lymphoid organs in hamsters of different ages. *Blood* **31**, 727–746.

Lubaroff, D. M. (1973). An alloantigenic marker on rat thymus and thymus-derived cells. *Transplant. Proc.* **5**, 115–118.

Luckey, T. D., ed. (1973). *Thymic Hormones.* University Park Press, Baltimore, Maryland.

MacGillivray, M. H., B. Mayhew, and N. R. Rose (1970). A comparison of the immunologic function of thymus cells of varying stages of maturation. *Proc. Soc. Exp. Biol. Med.* **133**, 688–692.

Malchow, D., W. Droege, and J. L. Strominger (1972). Solubilization and partial purification of lymphocyte specific antigens in the chicken. *Eur. J. Immunol.* **2**, 30–35.

Marchalonis, J. J., R. E. Cone, and J. L. Atwell (1972). Isolation and partial characterization of lymphocyte surface immunoglobulins. *J. Exp. Med.* **135**, 956–971.

Marchesi, V. T., and J. L. Gowans (1964). The migration of lymphocytes through the endothelium of venules in lymph nodes: an electron microscope study. *Proc. Roy. Soc.* **B159**, 283–290.

Mason, S., and N. L. Warner (1970). The immunoglobulin nature of the antigen recognition site on cells mediating transplantation immunity and delayed hypersensitivity. *J. Immunol.* **104**, 762–765.

McArthur, W. P., J. Chapman, and G. J. Thorbecke (1971). Immunocompetent cells of the chicken. I. Specific surface antigenic markers on bursa and thymus cells. *J. Exp. Med.* **134**, 1036–1045.

McDevitt, H. O., B. D. Deak, D. C. Shreffler, J. Klein, J. H. Stimpfling, and G. D. Snell (1972). Genetic control of the immune response. Mapping of the *Ir-1* locus. *J. Exp. Med.* **135**, 1259–1278.

McGregor, D. D. (1968). Bone marrow origin of immunologically competent lymphocytes in the rat. *J. Exp. Med.* **127**, 953–966.

McGregor, D. D. (1969). Effect of tritiated thymidine and 5-bromodeoxyuridine on development of immunologically competent lymphocytes. *Immunology* **16**, 83–90.

Metcalf, D. (1967). Relation of the thymus to the formation of immunologically reactive cells. *Cold Spring Harbor Symp. Quant. Biol.* **32**, 583–590.

Metcalf, D., and M. A. S. Moore, eds. (1971). *Haemopoietic Cells.* North-Holland Publishing Co., Amsterdam, p. 242.

Metcalf, D., and Wakonig-Vaartaja (1964). Stem cell replacement in normal thymus grafts. *Proc. Soc. Exp. Biol. Med.* **115**, 731–735.

Miller, J. F. A. P., and G. F. Mitchell (1969). Thymus and antigen-reactive cells. *Transplant. Rev.* **1**, 3–42.

Miller, J. F. A. P., and D. Osoba (1967). Current concepts of the immunological function of the thymus. *Physiol. Rev.* **47**, 437–520.

Miller, J. F. A. P., A. Basten, J. Sprent, and C. Cheers (1971). Interaction between lymphocytes in immune responses. *Cell. Immunol.* **2**, 469–495.

Mitchison, N. A. (1969). Cell populations involved in immune responses. In *Immunological Tolerance* (M. Landy and W. Braun, eds.). Academic Press, New York, p. 113.

Moore, M. A. S., and D. Metcalf (1970). Ontogeny of the haemopoietic system: Yolk sac origin of *in vivo* and *in vitro* colony forming cells in the developing mouse embryo. *Brit. J. Haemat.* **18**, 279–296.

Moore, M. A. S., and J. J. T. Owen (1966). Experimental studies on the development of the bursa of Fabricius. *Develop. Biol.* **14**, 40–51.

Moore, M. A. S., and J. J. T. Owen (1967). Experimental studies on the development of the thymus. *J. Exp. Med.* **126**, 715–725.

Mosier, D. E., and L. W. Coppleson (1968). A three-cell interaction required for the induction of the primary immune response in vitro. *Proc. Nat. Acad. Sci. U.S.A.* **61**, 542–547.

Murray, R. G., and P. A. Woods (1964). Studies on the fate of lymphocytes. III. The migration and metamorphosis of *in situ* labelled thymic lymphocytes. *Anat. Rec.* **150**, 113–128.

Nisbet, N. W., M. Simonsen, and M. Zaleski (1969). The frequency of antigen-sensitive cells in tissue transplantation: A commentary on clonal selection. *J. Exp. Med.* **129**, 459–467.

Nossal, G. J. V. (1964). Studies on the rate of seeding of lymphocytes from the intact guinea pig thymus. *Ann. N.Y. Acad. Sci.* **120**, 171–181.

Nossal, G. J. V., N. L. Warner, H. Lewis, and J. Sprent (1972). Quantitative features of a sandwich radioimmunolabeling technique for lymphocyte surface receptors. *J. Exp. Med.* **135**, 405–428.

Order, S. E., and B. H. Waksman (1969). Cellular differentiation in the thymus: Changes in size, antigenic character, and stem cell function of thymocytes during thymus repopulation following irradiation. *Transplantation* **8**, 783–800.

Owen, J. J. T., and M. C. Raff (1970). Studies on the differentiation of thymus-derived lymphocytes. *J. Exp. Med.* **132**, 1216–1232.

Owen, J. J. T., and M. A. Ritter (1969). Tissue interaction in the development of thymus lymphocytes. *J. Exp. Med.* **129**, 431–437.

Parrott, D. M. V., M. A. B. de Sousa, and J. East (1966). Thymus dependent areas in the lymphoid organs of neonatally thymectomized mice. *J. Exp. Med.* **123**, 191.

Paul, W. E. (1970). Functional specificity of antigen-binding receptors of lymphocytes. *Transplant. Rev.* **5**, 130–166.

Peter, H-H., J. Clagett, J. D. Feldman, and W. O. Weigle (1973). Rabbit antiserum to brain-associated thymus antigens of mouse and rat. I. Demonstration of antibodies cross-reacting to T cells of both species. *J. Immunol.* **110**, 1077–1084.

Pierre, St. R. L., and G. A. Ackerman (1965). Bursa of Fabricius in chickens: Possible humoral factor. *Science* **147**, 1307–1308.

Polliack, A., N. Lampen, B. D. Clarkson, E. de Harven, Z. Bentwich, F. P. Siegal, and H. G.

Kunkel (1973). Identification of human B and T lymphocytes by scanning electron microscopy. *J. Exp. Med.* **138**, 607–624.

Potworowski, E. F., and R. C. Nairn (1967). Origin and fate of a thymocyte-specific antigen. *Immunology* **13**, 597–602.

Rabellino, E., S. Colon, H. M. Grey, and E. R. Unanue (1971). Immunoglobulins on the surface of lymphocytes. I. Distribution and quantitation. *J. Exp. Med.* **133**, 156–167.

Raff, M. C. (1969). Theta isoantigen as a marker of thymus-derived lymphocytes in mice. *Nature* **224**, 378–379.

Raff, M. C. (1971). Surface antigenic markers for distinguishing T and B lymphocytes in mice. *Transplant. Rev.* **6**, 52–80.

Raff, M. C. (1973). T and B lymphocytes and immune responses. *Nature* **242**, 19–23.

Raff, M. C., and H. Cantor (1971). Subpopulations of thymus cells and thymus-derived lymphocytes. In *Progress in Immunology* (B. Amos, ed.). Academic Press, New York, pp. 83–93.

Raff, M. C., and H. H. Wortis (1970). Thymus dependence of θ-bearing cells in the peripheral lymphoid tissues of mice. *Immunology* **18**, 931–942.

Rajewsky, K., V. Schirrmacher, S. Nase, and N. K. Jerne (1969). The requirement of more than one antigenic determinant for immunogenicity. *J. Exp. Med.* **129**, 1131–1143.

Reif, A. E., and J. M. Allen (1964). The AKR thymic antigen and its distribution in leukemias and nervous tissues. *J. Exp. Med.* **120**, 413–433.

Rieke, W. O. (1966). Lymphocytes from thymectomized rats: Immunologic, proliferative, and metabolic properties. *Science* **152**, 535–538.

Roelants, G. (1972). Quantification of antigen specific T and B lymphocytes in mouse spleens. *Nature New Biol.* **236**, 252–254.

Roitt, I. M., G. Torrigiani, M. F. Greaves, and J. Brostoff (1969). The cellular basis of immunological responses. *Lancet* **ii**, 367–371.

Sainte-Marie, G., and C. P. Leblond (1964). Cytologic features and cellular migration in the cortex and medulla of thymus in the young adult rat. *Blood J. Hematol.* **23**, 275–299.

Schlesinger, M. (1972). Antigens of the thymus. *Prog. Allergy* **16**, 214–299.

Schlesinger, M., and D. Hurvitz (1968a). Differentiation of the thymus-leukemia (TL) antigen in the thymus of mouse embryos. *Israel J. Med. Sci.* **4**, 1210–1215.

Schlesinger, M., and D. Hurvitz (1968b). Serological analysis of thymus and spleen grafts. *J. Exp. Med.* **127**, 1127–1137.

Schlesinger, M., and I. Yran (1970). Serologic demonstration of a thymus-dependent population of lymph node cells. *J. Immunol.* **104**, 798–804.

Schlesinger, M., S. Gottesfeld, and Z. Korzash (1973). Thymus cell subpopulations separated on discontinuous BSA gradients: Antigenic properties and circulation capacity. *Cell. Immunol.* **6**, 49–58.

Schoefl, G. I. (1972). The migration of lymphocytes across the vascular endothelium in lymphoid tissue. *J. Exp. Med.* **136**, 568–588.

Schwarz, R. M. (1967). Transformation of rat small lymphocytes with allogenic lymphoid cells. *Am. J. Anat.* **121**, 559–570.

Shearer, G. M., and G. Cudkowicz (1969). Distinct events in the immune response elicited by transferred marrow and thymus cells. I. Antigen requirements and proliferation of thymic antigen-reactive cells. *J. Exp. Med.* **130**, 1243–1261.

Shevach, E., I. Green, L. Ellman, and J. Maillard (1972). Heterologous antiserum to thymus-derived cells in the guinea pig. *Nature New Biol.* **235**, 19–21.

Shigeno, N., V. Hammerling, C. Arpels, E. A. Boyse, and L. J. Old (1968). Preparation of lymphocyte-specific antibody from anti-lymphocyte serum. *Lancet* **ii**, 320–323.

Smith, R. W., W. D. Terry, D. N. Buell, and K. W. Sell (1973). An antigenic marker for human thymic lymphocytes. *J. Immunol.* **110**, 884–887.

Sosin, H., H. Hilgard, and C. Martinez (1966). The immunologic competence of mouse thymus cells measured by the graft vs. host spleen assay. *J. Immunol.* **96**, 189–195.

Sprent, J. (1973). Circulating T and B lymphocytes of the mouse. I. Migratory properties. *Cell Immunol.* **7**, 10–39.

Stobo, J. D., and W. E. Paul (1972). Functional heterogeneity of murine lymphoid cells. II. Acquisition of mitogen responsiveness and of antigen during the ontogeny of thymocytes and "T" lymphocytes. *Cell. Immunol.* **4**, 367–380.

Stobo, J. D., W. E. Paul, and C. S. Henney (1973a). Functional heterogeneity of murine lymphoid cells. IV. Allogeneic mixed lymphocyte reactivity and cytolytic activity as functions of distinct T cell subsets. *J. Immunol.* **110**, 652–660.

Stobo, J. D., A. S. Rosenthal, and W. E. Paul (1973b). Functional heterogeneity of murine lymphoid cells. V. Lymphocytes lacking detectable surface θ or immunoglobulin determinants. *J. Exp. Med.* **138**, 71–88.

Stockert, E., L. J. Old, and E. A. Boyse (1971). The G_{IX} system: A cell surface alloantigen associated with murine leukemia virus; implications regarding chromosomal integration of the viral genome. *J. Exp. Med.* **133**, 1334–1355.

Takahashi, T., L. J. Old, K. R. McIntire, and E. A. Boyse (1971). Immunoglobulin and other surface antigens of cells of the immune system. *J. Exp. Med.* **134**, 815–832.

Taylor, R. B. (1964). Pluripotential stem cells in mouse embryo liver. *Brit. J. Exp. Pathol.* **46**, 376–383.

Taylor, R. B., W. P. Duffus, M. C. Raff, and S. de Petris (1971). Redistribution and pinocytosis of lymphocyte surface immunoglobulin molecules induced by anti-immunoglobulin antibody. *Nature New Biol.* **233**, 225–229.

Toivanen, P., A. Toivanen, and R. A. Good (1972). Ontogeny of bursal function in chicken. III. Immunocompetent cells for humoral immunity. *J. Exp. Med.* **136**, 816–831.

Tyan, M. L. (1968). Studies on the ontogeny of the mouse immune system. I. Cell-bound immunity. *J. Immunol.* **100**, 535–542.

Tyan, M. L., and L. A. Herzenberg (1968). Immunoglobulin production by embryonic tissues: Thymus independent. *Proc. Soc. Exp. Biol. Med.* **128**, 952–954.

Unanue, E. R., H. M. Grey, E. Rabellino, P. Campbell, and J. Schmidtke (1971). Immunoglobulins on the surface of lymphocytes. II. The bone marrow as the main source of lymphocytes with detectable surface-bound immunoglobulin. *J. Exp. Med.* **133**, 1188–1198.

Unanue, E. R., M. J. Karnovsky, and H. D. Engers (1973). Ligand-induced movement of lymphocyte membrane macromolecules. III. Relationship between the formation and the fate of anti-Ig–surface Ig complexes and cell metabolism. *J. Exp. Med.* **137**, 675–689.

Vitetta, E. S., S. Baur, and J. W. Uhr (1971). Cell surface immunoglobulin. II. Isolation and characterization of immunoglobulin from mouse splenic lymphocytes. *J. Exp. Med.* **134**, 242–264.

Vitetta, E. S., C. Bianco, V. Nussenzweig, and J. W. Uhr (1972). Cell surface immunoglobulins. IV. Distribution among thymocytes, bone marrow cells, and their derived populations. *J. Exp. Med.* **136**, 81–93.

Waksman, B. H., B. G. Arnason, and B. D. Jankovic (1962). Role of the thymus in immune reactions in rats. III. Changes in the lymphoid organs of thymectomized rats. *J. Exp. Med.* **116**, 187–206

Warner, N. L., A. Szenberg, and F. M. Burnet (1962). The immunological role of different lymphoid organs in the chicken. I. Dissociation of immunological responsiveness. *Aust. J. Exp. Biol. Med. Sci.* **40**, 373–387.

Weissman, I. L. (1967). Thymus cell migration. *J. Exp. Med.* **126**, 291–304.

Wigzell, H. (1970). Specific fractionation of immunocompetent cells. *Transplant. Rev.* **5**, 76–104.

Williams, R. M., A. D. Chanana, E. P. Cronkite, and B. H. Waksman (1971). Antigenic markers on cells leaving calf thymus by the way of the efferent lymph and venous blood. *J. Immunol.* **106**, 1143–1146.

Wilson, D. B., and P. C. Nowell (1970). Quantitative studies on the mixed lymphocyte in rats. IV. Immunologic potentiality of the responding cells. *J. Exp. Med.* **131**, 391–407.

Wilson, D. B., and P. C. Nowell (1971). "Primary" and "secondary" reactivity of lymphocytes to major histocompatibility antigens: a consideration of immunologic memory. In *Cell Interactions and Receptor Antibodies in Immune Responses* (O. Makela, A. Cross, and T. U. Kosunen, eds). Academic Press, New York, pp. 277–289.

Zeiller, K., and L. Dolan (1972). Thymus-specific antigen on electrophoretically separated rat lymphocytes. Tracing of the differentiation pathway of bone marrow-derived thymocytes by use of a surface marker. *Eur. J. Immunol.* **2**, 439–444.

Zeiller, K., E. Holzberg, G. Pascher, and K. Hannig (1972). Free flow electrophoretic separation of T and B lymphocytes. Evidence for various subpopulations of B cells. *Hoppe-Seylers Z. Physiol. Chem.* **353**, 105–110.

CHAPTER TEN

Low-Resistance Junctions: Some Functional Considerations

JUDSON D. SHERIDAN

The importance of cell interactions has long been recognized by developmental biologists. Only recently, however, have we begun to identify the mechanisms underlying these interactions. One putative mechanism receiving much recent attention involves the direct transfer of inorganic ions and other small molecules from cell to cell via specialized intercellular junctions (Furshpan and Potter, 1968; Loewenstein, 1968). Although we have learned much about the ultrastructure, permeability, and distribution of these so-called low-resistance junctions, we still know little about their functions. Two suggested functions have provided most of the impetus for junctional studies. The first function might be called "electrical communication" and involves the role of low-resistance junctions in the transmission of electrical signals between certain excitable cells (Bennett, 1966). This function is clearly important in cardiac (DeHaan and Sachs, 1972) and smooth muscle (M. R. Bennett, 1972) and in cases when the junctions connect certain nerve cells (M. V. L. Bennett, 1966, 1972). The second function might be called "nonelectrical communication" and involves the transfer of metabolically important molecules that help to regulate cellular processes (Loewenstein, 1966, 1973; Furshpan and Potter, 1968). This second function is particularly interesting to developmental biologists but has not been clearly demonstrated in any system, developing or mature. We might anticipate that further insights into the physiology of junctional communication will be gained from studies like those discussed by Dr. Revel (Chapter 4 in this volume) on the formation of junctions, and other studies currently in progress in various laboratories concerning metabolic cooperation (see "Different Genes Expressed," p. 196), junctional alterations in neoplasms

(Loewenstein, 1973; Sheridan and Johnson, 1974), and the biochemistry and biophysics of isolated junctional elements (Goodenough and Stoeckenius, 1972). However, in this discussion I wish to apply another, more deductive approach, that is, to consider some of the ways in which two adjacent cells can acquire different concentrations of small molecules and to explore some of the logical consequences of having low-resistance junctions between these differing cells.

I will begin by discussing some reasonable generalizations about the permeability, structure, and distribution of low-resistance junctions.

FEATURES OF LOW-RESISTANCE JUNCTIONS

Permeability

The most important feature of low-resistance junctions is their ability to allow substances to diffuse directly from the interior of one cell to the interior of the next (Bennett, 1966; Loewenstein, 1966; Furshpan and Potter, 1968). A reasonable generalization is that the junctions pass small molecules (up to about 1000 molecular weight or about 10–20 Å in diameter) but exclude larger molecules. The evidence for this generalization comes from three main sources. Studies with microelectrodes show that electric current carried predominantly by small inorganic ions passes readily between adjacent cells. As discussed below, this movement presumably occurs at gap junctions. The current flow through the junctions is responsible for electrically coupling cells; that is, a potential change induced across the membrane of the one cell is accompanied by a similar though smaller potential change across the membrane of the other cell. An example of this effect is shown in Fig. 1.

In a few selected cases it has been possible to estimate the permeability of the junctions to small ions or, in electrical terms, their resistance. Values ranging from 10^5 to 10^7 Ω have been found in various systems (Furshpan and Potter, 1959; Loewenstein et al., 1967; Asada and Bennett, 1971; Bennett et al., 1972). Attempts

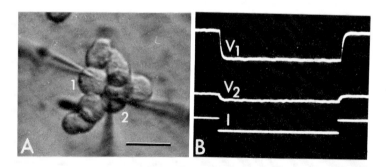

Figure 1. Electrical coupling between adjacent Novikoff hepatoma cells in small clump taken from suspension culture. As shown in the electrical records in (B), a current pulse, I (4.2×10^{-9} A), passed in cell 1 produced a voltage change, V_1 (20 mV), in cell 1 and a voltage change, V_2 (4 mV), in cell 2. Calibration in (A) is 25 μm.

to determine the degree of membrane specialization required for these low resistances have been quite crude because it is difficult to measure the area of the
junctions involved in the ion movements. The best estimates indicate that the junctional membranes have resistances for a unit area many orders of magnitude lower
than do nonjunctional membranes (Spira, 1971; Woodbury and Crill, 1970).

Evidence that molecules larger than small inorganic ions also pass via the low-
resistance junctions comes primarily from experiments in which tracers are injected
into one cell and are observed to move to adjacent, electrically coupled cells. Many
different tracers have been used, although only two fluorescent substances, fluorescein (MW 330) (Lowenstein, 1966; Bennett, 1966; Furshpan and Potter, 1968;
Johnson and Sheridan, 1971; Azarnia and Loewenstein, 1971) and Procion yellow
(MW ca. 650) (Payton et al., 1969; Kaneko, 1971; Johnson and Sheridan, 1971;
Rose, 1971), have been studied in any detail. These two substances have been
shown to move between coupled cells in a variety of adult and tissue culture systems (see Figs. 2 and 3). Attempts to demonstrate movement of molecules larger
than these have met with variable success (Kanno and Loewenstein, 1966; Bennett,
1973; J. D. Pitts—personal communication), and at present the safest conclusion is
that the junctions do not pass molecules with molecular weights over 1000.

In general, intercellular movement of tracers goes hand in hand with electrical
coupling. Most coupled cells that have been tested readily exchange small tracers.
Also, disruption of electrical coupling invariably interrupts tracer movement
(Payton et al., 1969; Johnson and Sheridan, 1971; Oliveira-Castro and Loewen-

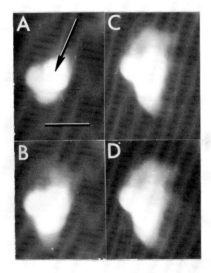

Figure 2. Exchange of fluorescein between reaggregated Novikoff hepatoma cells. Clumps of
cells were dissociated with EDTA and allowed to reaggregate for 4 hours at 37°C. Fluorescein
was injected into one cell [indicated by arrow in (A)], and movement of fluorescein to other
cells in the clump was monitored photographically. Exposures of 20 sec were made 15 sec
(A), 1 minute (B), 2 minutes (C), and 3½ minutes (D) after the beginning of injection.
Calibration in (A) is 25 µm.

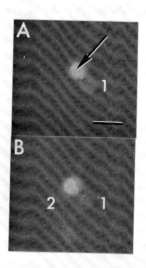

Figure 3.   Exchange of Procion yellow between Novikoff cells, dissociated and reaggregated as in Fig. 2. Procion yellow was injected into one cell [indicated by arrow in (A)], and movement to two adjacent cells (1 and 2) photographed at 11 and 18 minutes, respectively, after injection was begun. Calibration in (A) is 25 μm.

stein, 1971). In a few systems from very early embryos, however, electrical coupling occurs but tracer movement cannot be demonstrated (Slack and Palmer, 1969; Tupper and Saunders, 1972; Bennett et al., 1972). These results are difficult to interpret because of quantitative factors such as the size of the cells and the uncertainty about the threshold of detectability of the tracer (Sheridan, 1971a). Nevertheless, they suggest that some junctions may be permeable only to inorganic ions and not to larger molecules.

Another, more indirect approach has corroborated the tracer experiments and has added a more functional dimension to the description of junctional permeability. Mutant fibroblasts have been isolated that lack the enzyme hypoxanthine-guanine phosphoribosyltransferase (HGPRT, sometimes called inosinic pyrophosphorylase, or IPP). This enzyme converts hypoxanthine into inosine monophosphate and is required for the incorporation of exogenous hypoxanthine into nucleic acids. However, the mutant cells acquire the ability to incorporate labeled hypoxanthine when they are in contact with normal fibroblasts. This phenomenon of "metabolic cooperation," first described by Subak-Sharpe et al. (1966), can best be attributed to the movement of labeled inosine monophosphate (or its derivatives) from the normal cells into the defective cells. Careful experiments have ruled out the movement of enzyme, messenger or labeled polynucleotides (Cox et al., 1970; J. D. Pitts—personal communication). In an important recent study, Gilula et al. (1972) have shown that metabolic cooperation occurs only between cells that are electrically coupled and that have gap junctions (see the next section). Thus, studies of metabolic cooperation strongly indicate that small molecules of functional importance can be transferred via low-resistance junctions and that larger molecules of protein and polynucleotides are excluded.

Ultrastructure

Electron microscopists have identified many types of cell junctions, but only one type, the "gap junction" (Revel and Karnovsky, 1967) or nexus (Dewey and Barr, 1964), has consistently been found between electrically coupled cells (see McNutt and Weinstein, 1973, for a review). Other circumstantial evidence strongly indicates that gap junctions and low-resistance junctions are identical. For example, disruption (Barr et al., 1965) or congenital loss (Gilula et al., 1972) of these junctions uncouples cells electrically and prevents metabolic cooperation (Gilula et al., 1972; Cox et al., 1973).

The fine structure of gap junctions, reviewed by Dr. Revel (Chapter 4 in this volume), has been well characterized, and the essential features are illustrated in Fig. 4. In thin sections viewed with the electron microscope, the junctions appear as small regions of close membrane apposition. Extracellular tracers and freeze-fracture reveal tight aggregates of complementary subunits in the junctional membranes. These subunits appear to cross the extracellular space and may contain channels linking the cytoplasms of the two cells. An important point for our

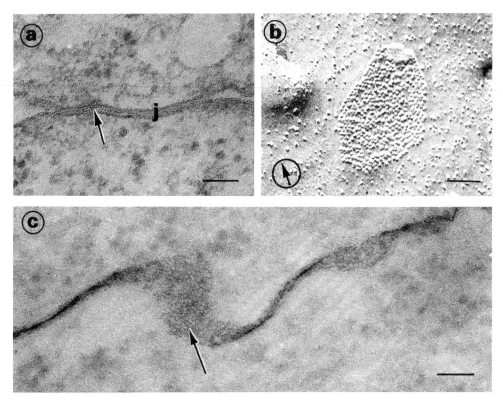

Figure 4. Gap junctions between Novikoff hepatoma cells prepared with thin section (a), freeze-fracture (b), and colloidal lanthanum (c) methods. In (a) the arrow indicates the typical 20–40 Å "gap." In (b) aggregated junctional particles are seen on a membrane A face. (The angle of shadowing is indicated by the circled arrow.) Note that the particles are grouped into "domains," a typical feature of junctions in this system. In (c), the arrow indicates subunits outlined by lanthanum in a region where the junction is viewed *en face*. Calibration marks in each micrograph are 0.1 μm.

present purpose is that the basic structure of gap junctions is similar whether the junctions occur in primitive invertebrates, such as *Hydra* (Hand and Gobel, 1972), or in higher vetebrates, such as mice (Goodenough and Revel, 1970). Arthropods as a group appear to have gap junctions with the greatest divergence from the common pattern (see Satir and Gilula, 1973). Yet it is unlikely that even these variations alter the basic ability of the arthropod junctions to pass small molecules, for both electrical coupling and tracer movement have been extensively documented in these organisms (Loewenstein, 1966; Payton et al., 1969). Therefore, we are probably justified in including arthropod gap junctions in our considerations, but should be cautious in generalizing from data obtained exclusively from arthropods.

Although gap junctions have a constant basic structure, they vary in extent. For example, gap junctions occur with diameters of less than 0.1 to over 4 μ (Revel et al., 1971; Goodenough and Revel, 1970). Furthermore, there are variations in the number of junctions per interface and the percentage of contact area made up of junctions (Revel et al., 1971). As might be expected, the greatest variations are seen when one compares cells of different types, for example, fibroblasts and liver cells. Some evidence suggests that the ratio of junctional area to volume may be more constant than these other parameters (Sheridan, 1973), but more extensive studies are needed for confirmation. The constancy of basic junctional structure and the variability of junctional area suggest that junctions may be controlled primarily by alterations in size (and number) rather than in basic structure or even permeability per unit area (Sheridan and Johnson, 1973).

Distribution

Only a few systems have been studied with both physiological and ultrastructural techniques. Therefore, in order to make the broadest possible statements about the distribution of low-resistance junctions, we must rely on the generalization developed in the last section: that gap junctions can provide for electrical coupling and tracer movement. On the basis of the more rigorous ultrastructural criteria, gap junctions have an extremely wide distribution. They are found throughout the metazoa, from the more primitive invertebrates (Satir and Gilula, 1973) to the more highly evolved veterbrates (McNutt and Weinstein, 1973). They are found in embryos (Revel, Chapter 4 in this volume) as well as adults and occur in most epithelial (Loewenstein et al., 1965; Friend and Gilula, 1972) as well as non-epithelial (Revel et al., 1971; Sheridan, 1971b) tissues. Gap junctions form between cells in tissue culture (Johnson and Sheridan, 1971; Gilula et al., 1972) and even between cells of different species (Michalke and Loewenstein, 1971) and differentiated types (Johnson et al., 1973). In short, nearly all cells studied have low-resistance junctions. There are, however, a few notable exceptions. Gap junctions are found only infrequently between nerve cells and are absent between vertebrate skeletal muscle cells. They are also absent between circulating blood cells, although they may form under special circumstances, for example, when lymphocytes are stimulated by phytohemmaglutinin (Hülser and Peters, 1972).

FUNCTIONAL CONSIDERATIONS

I have made the generalization that low-resistance junctions provide channels for the passive diffusion of small molecules from cell to cell. For net movement of molecules to take place, the concentrations of the substance must be different in the two cells and/or there must be other net forces (e.g., electrical potential differences) acting on the molecules. Since electrical potential differences may reflect differences in ion concentrations, we can focus on differences in concentrations of small molecules and the ways in which these differences might arise in natural biological systems [but see categories 3 (p. 198) and 4 (p. 199)].

Origins of Differences between Coupled Cells

There are four basic categories describing the ways in which two adjacent cells might acquire different concentrations of small molecules:

1. The cells have different genetic information, that is, different genomes.
2. The cells have identical genomes, but express different genes.
3. The cells have identical genomes and express identical genes, but make different quantities of certain gene products or the same quantities at different times.
4. The cells have identical genomes, express identical genes, and make identical quantities of all gene products, but are exposed to different quantities or types of environmental stimuli.

1. Different Genomes

At first glance, we might question the relevance of this category to our discussion, but there are two important examples that we can consider.

The first example involves somatic mutations which occur at a low but significant rate. Although many of these mutations will have no phenotypic effect and others will probably be lethal for the mutant cells, some might ultimately result in altered concentrations of small molecules. For example, a mutation in a gene controlling a membrane transport enzyme might produce a lowered transport rate in the mutant cell relative to its neighbors. Such an altered transport rate would tend to lower the concentration of the poorly transported substance relative to its concentration in neighboring cells. Because most substances transported across the membrane are small molecules, we would expect junctional transfer of the substance from the neighboring, normal cells to the mutant cell. This transfer would have two effects: (*a*) It would raise the concentration of the substance in the mutant cell, and (*b*) it would lower the concentration of the substance in the donor cells. In the specific case that I am considering, however, the mutant cell would have junctions with a number of normal cells (two to eight, depending on tissue organization). Consequently the mutant cell could acquire nearly normal concentration of the substance without a marked lowering of the concentration in the donor cells.

We have here an example of an important principle: each cell in a group of cells connected by junctions will tend to acquire a concentration of small molecules like that in the majority of the cells. This means that low-resistance junctions make it difficult for an individual cell to set up and maintain a higher concentration of small molecules than is found in its immediate neighbors.

The second example in this category is viral-induced transformation, in which the genomes of a few cells in a population are altered by the addition and integration of viral genetic information. Transcription and translation (Todaro and Huebner, 1972) (or "reverse" transcription, transcription, and translation for RNA viruses: Temin, 1972) of part of the viral genome leads to phenotypic transformation by mechanisms not presently understood. It has been proposed (Pardee, 1964; Holley, 1972) that one important feature of transformation is an increase in the ability of the cell to accumulate critical nutrients or other small molecules, for example, as a result of increased membrane transport. This accumulation, it has been suggested, causes increased growth. However, I have noted that a single cell, coupled to many adjacent cells, will find it difficult to accumulate small molecules to a concentration greater than that in adjacent cells. If this principle holds in viral transformation, two alternative conclusions can be envisioned. The transport rates in the transformed cell might be so great that the normal cells could not drain away the critical substance (s) fast enough to prevent accumulation. As a result, the normal cells would also begin to accumulate the critical substance (s) that they receive from the transformed cells. Then not only would the transformed cell be stimulated, but so also would the normal cells. Alternatively, the alteration in transport might be accompanied by a defect in the ability of the junctions to carry the substance (s) away. Such a defect might arise from an inability of the virally transformed cell to make or maintain sufficient numbers or sizes of junctions with its neighbors. Both electron microscopical and electrophysiological evidence suggests that some transformed cells may have such defects in their junctions (Martinez-Palomo, 1971; Loewenstein, 1973; Sheridan and Johnson, 1973).

We can conclude that in this category the possible role of low-resistance junctions is to maintain the normal intracellular concentrations of small molecules in a few genetically abnormal cells. The result may be to ameliorate some of the possible consequences (e.g., increased growth) of the genetic abnormality.

Having considered some important, but rare, cases of genetically aberrant cells, I will turn now to more common sources of cell differences.

2. Different Genes Expressed

This category is of special interest to developmental biologists, for it includes interactions between cells of different phenotypes. It might seem appropriate at this point to discuss some developmentally important example such as embryonic induction, which by definition involves interactions between cells expressing different genes (see "A Development Illustration," p. 201). However, I have chosen instead other examples for which the distribution of gap junctions is more clearly defined.

I will begin with brief descriptions of two systems in which gap junctions are present between cells of different types ("heterocellular" gap junctions). The first example comes from a recent study by Johnson et al. (1973) of the midgut and associated structures in the horseshoe crab, *Limulus polyphemus*. The midget is composed of columnar epithelial cells (E cells), underlain by a dense basement membrane which separates the E cells from a variety of other cell types, including reserve cells (R cells), smooth muscle cells, and vascular elements. Extensive gap junctions are found in three locations: between the E cells, between the R cells, and, surprisingly, between the E and the R cells. The E-R junctions are formed between the basal surface of the E cells and the ends of long, narrow processes of the R cells. The R-cell processes must pass through the basement membrane in order to contact the E cells. The authors suggest that E-R junctions may allow nutrients, absorbed and perhaps processed by the E cells, to pass into the R cells, where they can be stored.

The second example is the enamel organ of the developing mammalian tooth. At the time when the enamel matrix is being secreted, the enamel organ contains three different types of cells, ameloblasts (A), stratum intermedium cells (SI), and outer enamel epithelial cells (OEE) (Garant, 1972). Extensive gap junctions occur not only between cells of the same type (i.e., A-A, SI-SI, and OEE-OEE junctions) but also between cells of different types (i.e., A-SI and SI-OEE junctions). After the secretory phase, the enamel organ differentiates in two stages into a transporting epithelium and becomes involved in the maturation of the enamel. At this time only two cell types can be distinguished, ameloblasts and papillary cells. Gap junctions, now more extensive and complex, exist between ameloblasts, between papillary cells, and between ameloblasts and papillary cells. Garant (1972) suggests that the gap junctions provide a continuous cellular route whereby substances required for enamel synthesis and deposition can be passed from the blood vessels to the ameloblasts. A particularly interesting possibility is that, in the maturation phase, Ca^{2+} is taken up by the papillary cells and is transported via gap junctions to the ameloblasts, which then release it for combination with the enamel matrix. Cell-to-cell transport of Ca^{2+} has also been postulated for bone and is discussed under category 4 (p. 200). Further evidence of a possible physiological role for gap junctions in the enamel organ will be presented later.

Although examples of heterocellular gap junctions are interesting, they are not common. Cells of different types (e.g., epithelial versus connective tissue), even in the early embryo, are commonly separated by basement membranes or more extensive connective tissue elements (Hay, 1968). These structures often prevent contact between the different cell types, and the cases that we have noted may be unusual. However, in this category there is a group of more common examples under the general heading of mosaics. These examples provide especially interesting possibilities for the involvement of gap junctions.

The most established examples in this group are the X-linked mosaics. According to the Lyon hypothesis (Lyon, 1961), at an early developmental stage there is a random inactivation of one or the other X chromosome in somatic cells of mammalian females. The inactivation is permanent, being transmitted to all future daughter cells. The result is that the adult female is a mosaic of cells having either the maternal or the paternal X chromosome activated. Since many of the alleles on the two X chromosomes will differ, that is, the individual will be heterozygous with respect to those genes, X-linked mosaicism clearly provides a mechanism for adjacent cells to differ. The frequency with which adjacent cells will have different X chromosomes in an active state is a function of the "grain" of the mosaicism. The grain in turn depends on many factors, including the number of cells present at the initial time of inactivation and the degree of subsequent mixing of cells due to cell movements. For example (Lyon, 1968), skin pigmentation and coat color generally show "coarse-grained" mosaicism, presumably because single neural crest cells populate large regions of dermis. However, much finer grain can occur in hair follicles (Goldstein et al., 1971) and presumably in other tissues.

The mosaicism in heterozygotes can be detected only if the gene products can be identified and distinguished either directly (e.g., by electrophoretic differences) or indirectly (e.g., by alteration in functional characteristics). It is this second method of detection which leads us to consider some possible influences of low-resistance junctions. From our knowledge about the permeability of junctions, we might expect that functional mosaicism involving coupled cells would be reduced or effectively abolished for X-linked genes whose gene products control concentrations of small molecules. Furthermore, from what we know about the distribution of junctions, we might expect a more definite mosaic pattern in blood cells and most nerve cells, which lack low-resistance junctions, provided that the genes are normally expressed in these cells. (Skeletal muscle cells, which also lack junctions, nevertheless should not generally show the full mosaic phenotype because they are formed by fusion of myoblasts, which might have different active X chromosomes: Mintz and Baker, 1967; see also "Cells without Junctions," p. 202).

Experimental evidence consistent with these ideas comes from two groups of observations. The first group concerns cells from patients having *Lesch-Nyhan syndrome,* an X-linked recessive disease resulting from a faulty enzyme, HGPRT (Seegmiller et al., 1967). This enzyme, which we encountered before in our discussion of "metabolic cooperation" (p. 190), is necessary for the incorporation of hypoxanthine (or guanine) into intracellular polynucleotides. Cells, such as blood cells or fibroblasts, from hemizygotes (males with the defective gene on their X chromosome) incorporate only minimal amounts of tritiated hypoxanthine into their nuclei and cytoplasms. Cells from heterozygous females, however, present an unusual picture. Their epidermal cells *in situ* (Frost et al., 1970), their lymphocytes *in vitro* (Dancis et al., 1968), and their fibroblasts *in vitro* (at high density) (Cox et al., 1970) show uniform incorporation of labeled hypoxanthine, apparently contradicting the Lyon hypothesis.

Part of the answer to this puzzle has come from *in vitro* experiments showing that mutant (HGPRT−) fibroblasts cooperate metabolically with normal (HGPRT+) fibroblasts (Cox et al., 1970). From these and other results, it was reasoned that the uniform incorporation in dense primary cultures of fibroblasts and in the epidermis resulted from transfer of labeled inosine monophosphate (or derivative) from normal to mutant cells. Incorporation into the lymphocytes, however, could not be explained in the same way. For these a developmental selection of the normal over the mutant phenotype was postulated (Dancis et al., 1968). It should be noted that this explanation implies selection after the loss of junctions has occurred. Before that time, metabolic cooperation would have reduced the basis for selection.

The second group of observations concern an X-linked defect in tooth development (see Witkop, 1967). The defect leads to a syndrome termed *hypomaturation amelogenesis imperfecta* and results in softened enamel containing decreased amounts of bound calcium and phosphorus. Hemizygous males have a uniform defect in all teeth, whereas heterozygous females have random vertical bands of deficient and normal enamel. The banded appearance had been interpreted according to the Lyon hypothesis as representing the grain of the mosaicism in amelo-blasts.

In order to obtain further quantitative support for the application of the Lyon hypothesis to this sydrome, Sauk et al. (1972) used electron microprobe analysis to determine the calcium and phosphorus contents of the different bands in heterozygote teeth. The bands with normal appearance had normal levels of calcium and phosphorus. However, it was found unexpectedly that the calcium and phosphorus contents of the abnormal enamel were midway between the normal levels and the levels in the hemizygote. Furthermore, there was a gradual decrease in the ion content of calcium and phosphorus between the normal and abnormal bands, rather than the abrupt change anticipated.

There are a number of possible explanations for the intermediate calcium and phosphorus levels. The one most germane to our discussion, however, is that certain small molecules are transferred between ameloblasts (or other cells in the enamel organ) and alleviate some of the phenotypic expression of the defective gene products. As we have seen, the enamel organ has extensive gap junctions between all cells, and the opportunities for metabolic cooperation are numerous. An intriguing possibility is that the defective gene prevents or reduces the uptake of calcium and/or phosphate by the papillary cells (see, however, Witkop et al., 1973). According to Garant's reasoning, the ameloblasts would have to obtain their calcium and phosphate ultimately from normal papillary cells in an adjacent region. Some cooperation would be possible, but might be insufficient to compensate fully for the defect.

We can carry our discussion in this category one step further by considering the possibility of single allele inactivation in autosomes, paralleling X inactivation. Thus far, only one autosomal locus, the *Ig*-1 locus in mammalian lymphocytes, has been found to demonstrate this effect (see Mintz, 1971, for discussion). Inactivation of one allele at this locus ensures that heterozygous lymphocytes will produce a single type of immunoglobulin. Yet there appear to be no *a priori* arguments against the existence of other examples, and certainly their presence would provide more widespread possibilities for junctions to play roles such as I have postulated for X-linked defects. In fact, one would expect difficulty in finding examples of autosomal inactivation if junctions were effective in obscuring certain mosaic phenotypes in heterozygotes. The mosaic genotype would appear to control a simple recessive (or perhaps semidominant) trait.

Finally, I should mention a working hypothesis advanced by Mintz (1971). She suggests that many, if not most, tissues in multicellular organisms are composed of "phenoclones," groups of cells cloned out on the basis of slight differences in patterns of gene expression. The variations would not be extreme enough to convert, for example, a liver cell into an intestinal epithelial cell, but would nevertheless modify the phenotype of the cells. She looks upon phenoclones as a mechanism for ensuring genetic variability and for providing "fine-focus" control of differentiation. Mintz's model has many important implications for the function of low-resistance junctions. As she discusses briefly, cell interaction could be superimposed on the genetic variations and would give further possibilities of fine control. Low-resistance junctions could provide one important type of interaction.

We can conclude from our discussion of various known and postulated mosaic patterns that low-resistance junctions may aid in reducing expression of the mosaic phenotype in heterozygotes. The main requirements are that the cells have junctions, that the phenotype depends on the concentration of small molecules, and that the mosaic pattern has fine enough grain to give a high frequency of contacts between cells expressing different genes.

3. Different Quantities of Gene Product

Cells having identical genomes and expressing identical genes make identical gene products. The cells will be different, however, if at any given time they have different quantities of gene products. Direct comparisons of the amounts of gene product in individual cells are hard to make, but variations might be expected. I will now discuss some specific examples.

The first example involves the interaction of muscle cells in the vertebrate heart. Most cardiac muscle cells can generate action potentials spontaneously. However, when the cells are put into culture, it can be shown that the frequency of action potentials, and thus of contractions, varies from cell to cell (see DeHaan and Sachs, 1972). The molecular basis for the variation probably is quite complex since frequency depends on many factors, including surface area of the cell, specific membrane resistance and capacitance, and density of active Na^+ and K^+ conductance channels. However, variations in any of these factors ultimately reflect variations in the quantity of some gene product, for example, a membrane protein or lipid. In this case, one role of low-resistance junctions is quite clear: they transmit action potentials from cell to cell. Without junctions in the heart, the cells would beat asynchronously and effective delivery of blood would be impossible.

If some membrane-associated gene products can vary from cell to cell, there is no logical reason to expect others to remain invariant. Yet the consequences of certain variations, without the help of low-resistance junctions, could be nearly as disastrous as the consequence of asynchronous beating. If, for example, the quantity of Na^+-K^+ ATPase per volume of cell varied, each cell would have a different capacity for maintaining internal Na^+ and K^+ concentrations in the presence of the substantial fluxes accompanying the production of action potentials. The presence of junctions would allow the cells with higher enzyme activity to help the cells with lower activity to maintain ionic balance.

Once we accept the possibility of variations in quantities of gene product, we can immediately suggest many examples in which low-resistance junctions might play a vital role in coordination. A particularly dramatic possibility can be developed on the basis of recent evidence implicating cyclic nucleotides in the control of cell growth. It has been suggested that cAMP (3′, 5′-cyclic adenosine monophosphate) inhibits (see Sheppard, 1973) and cGMP (3′, 5′-cyclic guanosine monophosphate) stimulates (Goldberg et al., 1973) cell proliferation. These suggestions have rapidly gained support, though not without controversy. The molecular basis of the proliferative effects is not known. For our present discussion, however, we need only accept the idea that the two agents act antagonistically (Goldberg et al., 1973).

The intracellular concentration of each cyclic nucleotide is controlled independently by two enzymes. One is a specific cyclase, which produces the cyclic nucleotide from the appropriate nucleoside triphosphate, and the other is a specific phosphodiesterase, which catalyzes hydrolysis of the 3′-phosphate bond. The activities of the cyclases, in turn, are regulated by external factors, such as hormones and neurotransmitters, which act via membrane receptors, and perhaps by internal factors, such as divalent cations. Clearly, variations in the quantity of any of these enzymes, or of the specific membrane receptors regulating the cyclases, might result in variations in the concentration of the appropriate cyclic nucleotide.

For the sake of discussion, we might reasonably postulate than any population of dividing cells contains cells with different absolute (and/or relative) amounts of adenylate cyclase and guanylate cyclase. Consider the effects of a growth-promoting hormone that acts via cGMP. The cells with the larger absolute (and/or relative) amounts of guanylate cyclase will be preferentially stimulated. Thus, as the population increases in size, the proportion of cells with greater ability to produce cGMP will increase. If we now add a growth-inhibiting hormone that acts via cAMP (mimicking a normal physiological control mechanism), there will be greater inhibition of cells with greater ability to produce cAMP. However, the inhibited cells are unlikely to be the same cells that were originally stimulated to divide. Thus the inhibition of the population will be incomplete, and the inbalance generated by the nonuniform stimulation will remain. Any additional stimulus to grow will throw the population further out of balance. This change in the population can be obviated, however, if the cells exchange cAMP and cGMP via low-resistance junctions. Although different cells produce different amounts of cAMP and cGMP, the final concentrations will be averaged over the population by loss of the cyclic nucleotides from the "high producers" to the "low producers."

Although these arguments are speculative, they are consistent with certain experimental observations. For example, if mouse embryonic fibroblasts (presumably having low levels of cAMP: Otten et al., 1971) are seeded at low density on a confluent monolayer of 3T3 cells ("normal" fibroblasts with higher cAMP levels), the embryonic fibroblasts stop proliferating (Weiss and Njeuma, 1971). Although there are many possible interpretations of these data, they could indicate inhibition by transfer of cAMP from the 3T3 cells to the embryonic cells.

Contact-dependent stimulation has also been reported (Dewey et al., 1973). When cells synthesizing DNA (S-phase cells) are mixed with cells just about to begin DNA synthesis (late G1 cells), there is an acceleration of the G1 cells into the S phase. This acceleration could be due in part to transfer of cGMP from the S-phase to the G1 cells. Loss of cAMP from the G1 to the S-phase cells would enhance the stimulation of the G1 cells, yet might be unable to inhibit the S-phase cells, which are already committed to DNA synthesis.

In conclusion, the amounts of certain gene products may vary independently from cell to cell. Some of these gene products affect intracellular concentrations of small molecules, leading to possible nonuniformity in growth or other processes. By allowing the small molecules to distribute more evenly, low-resistance junctions can affect the potential heterogeneity and guarantee a more uniform cell population.

4. Different Environmental Conditions

In discussing previous categories, I have neglected the fact that adjacent cells can be exposed to different environmental conditions. Variations in innervation, blood supply, and physical or chemical stress can cause differences even in cells with identical genotypes and phenotypes. In most cases the differences would lead to unequal concentrations of small molecules and thus create situations in which low-resistance junctions might have an important function.

Some interesting examples of variable input occur in mammalian smooth muscle. The contraction of smooth muscle is under nervous (and hormonal) control. There are three general patterns of innervation (M. R. Bennett, 1972). In some types of smooth muscles, each cell receives direct "close-contact" innervation. In other types, there are no close contacts, only small axon bundles which presumably release transmitter into the general extracellular space. In a few types, there is an intermediate pattern consisting of small axon bundles and only a few close contacts. Clearly, the smooth muscles with the intermediate pattern of innervation have cells with different inputs. A more dramatic example, however, is vascular smooth muscle, which itself has no nerves, nerves being present only in the ad-ventitia surrounding the muscle. In this case there is a need to distribute the effect of the nerve input from the muscle cells adjacent to the adventitia to the muscle cells lying more internally, or closer to the lumen. One effect of the input is to produce membrane potential changes, which in turn regulate contraction. Low-resistance junctions could distribute this effect by transmitting the changes in potential to the internal cells. Another effect of the input is to alter cAMP (and perhaps cGMP) concentrations (Andersson, 1973; Goldberg et al., 1973). These alterations are presumably independent of membrane potential, yet are important at least in regulating the metabolic events accompanying contraction. We might propose that the low-resistance junctions also transfer the cyclic nucleotides from the peripheral to the internal cells. Low-resistance junctions may play similar roles in smooth muscles with mixed innervation and even in muscles more densely in-nervated with close contacts, provided that the number of contacts per cell varies.

Many other tissues (e.g., fat and exocrine glands) received autonomic innerva-tion. Although these innervation patterns have not been characterized as completely as those in smooth muscle, it is reasonable to suppose that low-resistance junctions might act to distribute the effects of nerve input in these tissues as well (Sheridan, 1971b). Since the role of potential changes in these tissues is unclear, the ability of junctions to pass molecules other than inorganic ions may be of prime impor-tance. It is interesting that cyclic nucleotides have been implicated in the responses of most of these tissues to autonomic stimulation. Transfer of cyclic nucleotides could therefore be a vital function of the low-resistance junctions in these tissues.

Other forms of external inputs might also vary from cell to cell. For example, some cells may be closer to capillaries and thus may receive a higher concentration of hormone. In such cases low-resistance junctions could act to integrate the re-sponse of the tissue to the hormone. An extreme example of this type appears to exist in bone. Holtrop and Weinger (1972) have reported that osteocytes are tightly adherent to the surrounding bone matrix, especially in the interlacunal cana-liculi. These authors believe that the extracellular space is effectively occluded, leaving no extracellular route for nutrients or other molecules to reach the deeper osteocytes. The osteocytes are, however, connected by extensive "tight junctions" (probably gap junctions in which the "gap" was not resolved). There is thus a possibility for an extensive cell-to-cell system connecting the peripheral and internal regions of the bone. Osteocytes have been proposed as part of a complex calcium regulatory system under the control of parathyroid hormone (PTH). It is thought that PTH stimulates calcium entry from the bone into osteocytes and then into the blood, and part of this effect is apparently mediated by cAMP (Borle, 1972). However, the lack of appreciable extracellular space implies that PTH must exert its direct effect on the peripheral osteoblasts and osteocytes and affect the internal osteocytes indirectly. A strong possibility is that low-resistance junctions, which are proposed to carry nutrients to internal cells (Holtrop and Weinger, 1972), also carry cAMP, thus distributing the effects of PTH. If these suggestions are correct, the calcium resorbed from the bone must be transported from the internal osteo-cytes out to the peripheral cells for release into the circulation. This Ca^{2+} might also move through the junctions.

Further evidence that the osteocytes might act to transport regulatory molecules through the bone comes from a study of bone wounding by Melcher and Accursi (1972). They showed that localized damage of the scapula induced not only formation of new bone on the damaged side, directly opposite the damage site, but also formation of new bone by osteocytes in lacunae between the two sites of periosteal bone formation. The authors suggest that some "osteogenic" stimulus was generated at the site of the damage and was transferred between osteocytes in the bone matrix to the progenitor cells on the opposite side. It is likely that low-resistance junctions play a role in this transfer.

In conclusion, local environmental conditions can vary from cell to cell. The clearest examples are variations in innervation, although other instances are also seen. Low-resistance junctions might play an important role in distributing the effects of the environmental factors over the cell population, providing for a more integrated tissue response.

A Developmental Illustration

I have approached the four categories of cell differences as if each occurred in exclusion of the others. This is not likely, however, and to make this point clear I wish to briefly consider possible junctional involvement in a complex process, embryonic induction. Induction refers to an interaction between two tissues, resulting in a new path of differentiation for one or both of them. Theoretically, junctions could exist within each interacting tissue and between the tissues. In one induction system (notochord and ectoderm in chick embryos), there are definitely low-resistance junctions between notochord cells and between ectodermal cells (Sheridan, 1968). There is no direct evidence, however, for junctions connecting notochord and ectodermal cells, and if they exist, they are few in number.

There are many ways in which low-resistance junctions might be involved in the induction process (for example, see Loewenstein, 1967). I will consider just one for illustration. In this model the inducing tissue produces some extracellular material(s) that reacts with receptors in the membranes of cells in the responding tissue. Part of the response of these cells is a change in the intracellular concentration of small molecules (e.g., cyclic nucleotides or divalent cations). The extracellular material(s) might, according to our previous discussion (category 3, "Different Quantities of Gene Product"), vary from point to point along the interface between the two tissues. This would present different stimulatory input to each cell in the responding tissue (category 4, "Different Environmental Conditions"). Furthermore, the cells in the responding tissue might vary in ability to respond to the extracellular stimulus (category 3). Low-resistance junctions could then play an important role in passing the critical small molecules from cell to cell in the responding tissue, ensuring that each cell was induced with equal efficiency.

Cells without Junctions

If we wish to contend that junctions have a vital function, we must attempt to explain how a few cell types are able to get along in their absence. As I mentioned previously, low-resistance junctions are found only infrequently between nerve cells, never between skeletal muscle cells, and only in unusual circumstances between blood cells. For nerve cells and skeletal muscle cells we can suggest an electrical explanation. The indiscriminate presence of low-resistance junctions between either of these types of cells would not permit the fine degree of control necessary for complex nervous function or for motor coordination. In the case of nerve cells, there is another important consequence of the lack of junctions. In comparison with cells having junctions, individual nerve cells can have more independent regulation of the concentrations of small molecules and thus greater diversity in function, even at the metabolic level. One might even speculate that such independence and diversity are important for "higher-order" functions such as learning and memory. However, such a consequency is not necessary, nor is it likely to occur, in skeletal muscle. Since skeletal muscle cells are generally large cells formed by the fusion of many myoblasts, possible variations in gene product and even in environmental conditions might readily be averaged out in each individual cell, even in the absence of junctions.

Lack of junctions between blood cells is consistent with the necessity for distributing them rapidly throughout the body, ultimately through capillaries roughly the diameter of the cells themselves. However, lack of junctions may serve another purpose, at least in the case of so-called B lymphocytes. These cells have the task of producing specific antibodies. When a foreign antigen enters the body, it stimulates a small percentage of the lymphocytes to divide and then to begin secreting an antibody specific for the antigen. The current idea is that the antigen stimulates only the cells capable of producing the appropriate antibody. Absence of low-resistance junctions would restrict the stimulus to the appropriate cells and prevent indirect stimulation of nonspecific cells by junctional transfer. As in the case of nerve cells, then, the lack of junctions helps to provide for the greater diversity of cellular responses needed for a discriminatory immunological system.

CONCLUDING REMARKS

It has been frequently stated that low-resistance junctions act to coordinate or integrate the activities of populations of cells. In my discussion, I have shown how the opportunity or need for such coordination can arise from cellular heterogeneity at the genotypic, phenotypic, and environmental levels. Such heterogeneity is a common feature of all cell populations, but only metazoan cells have so far been found to have low-resistance junctions.* It is therefore tempting to suggest that the development of junctional communication, with its potential effect of stabilizing cell populations, has been a major factor in the evolution of metazoa from protists. Such a general evolutionary significance, if correct, might far outweigh the importance of any specific function, for example, in electrical transmission or growth control.

---

* I have written this paper with the typical bias of an animal biologist. However, the conclusions regarding the importance of junctional communication may be applicable also to higher plants. Plasmadesmata, which are found in metaphyta, may serve as low-resistance junctions and selective pathways for the transfer of small molecules (Spitzer, 1970).

SUMMARY

We have learned a great deal about the permeability, ultrastructure, and distribution of low-resistance junctions. However, we still do not understand their functions except as electrical synapses between certain excitable cells. Many interesting functions have been suggested, but there have been few attempts to relate these functions in a general way to each other or to the more established role of junctions in electrical transmission. In this paper I discuss some ways in which the various suggested and established functions might fit together. The ideas are based on a consideration of the sources of heterogeneity in cell populations and the predicted ability of cell junctions to compensate for this heterogeneity by passing small ions and metabolites between adjacent cells. More specialized functions, such as regulation of growth and differentiation and transmission of electrical signals, can be considered to be adaptations of the basic homeostatic function.

ACKNOWLEDGMENTS

I wish to thank Dr. R. G. Johnson, Ms. D. Preus, and Ms. M. Hammer for permission to use unpublished figures, Dr. Johnson and Dr. J. Sheppard for comments on the manuscript, Dr. C. Witkop for stimulating discussions, Ms. L. Steere for photographic work, and Ms. G. Busack for preparing the manuscript. Some of the studies reported here were supported in part by U.S. Public Health Service Grant CA11114 and Research Career Award K4-CA-70,388.

REFERENCES

Andersson, R. (1973). Role of cyclic AMP and Ca^{++} in mechanical and metabolic events in isometrically contracting smooth muscle. *Acta Physiol. Scand.* **87**, 84.

Asada, Y., and M. V. L. Bennett (1971). Experimental alteration of coupling resistance at an electrotonic synapse. *J. Cell Biol.* **49**, 159.

Azarnia, R., and W. R. Loewenstein (1971). Intercellular communication and tissue growth. V. A cancer cell strain that fails to make permeable membrane junctions with normal cells. *J. Membrane Biol.* **6**, 368.

Barr, L., W. Berger, and M. Dewey (1965). Propagation of action potentials and the structure of the nexus in cardiac muscle. *J. Gen. Physiol.* **48**, 797.

Bennett, M. R. (1972). *Autonomic Neuromuscular Transmission.* Cambridge University Press, London.

Bennett, M. V. L. (1966). Physiology of electrotonic junctions. *Ann. N.Y. Acad. Sci.* **137**, 509.

Bennett, M. V. L. (1972). A comparison of electrically and chemically mediated transmission. In *Structure and Function of Synapses* (G. D. Pappas and D. P. Purpura, eds.). Raven Press, New York.

Bennett, M. V. L. (1973). Function of electrotonic junctions in embryonic and adult tissues. *Fed. Proc.* (in press).

Bennett, M. V. L., M. E. Spira, and G. D. Pappas (1972). Properties of electrotonic junctions between embryonic cells of *Fundulus. Develop. Biol.* **29**, 419.

Borle, A. B. (1972). Parathyroid hormone and cell calcium. In *Calcium, Parathyroid Hormone and the Calcitonins* (R. V. Talmage and P. L. Munson, eds.). Exerpta Medica, Amsterdam.

Cox, R. P., M. R. Krauss, M. E. Balis, and J. Dancis (1970). Evidence for transfer of enzyme product as the basis of metabolic cooperation between tissue culture fibroblasts of Lesch-Nyhan disease and normal cells. *Proc. Nat. Acad. Sci. U.S.A.* **67**, 1573.

Cox, R. P., M. R. Krauss, M. E. Balis, and J. Dancis (1974). Metabolic cooperation in cell culture: a model for cell to cell communication. In *Cell Communication* (R. P. Cox, ed.), Wiley-Interscience, New York, in press.

Dancis, J., P. H. Berman, V. Jansen, and M. E. Balis (1968). Absence of mosaicism in the lymphocyte in X-linked congenital hyperuricosuria. *Life Sci.* **7**, 587.

DeHaan, R. L., and H. G. Sachs (1972). Cell coupling in developing systems. In *Current Topics in Developmental Biology*, Vol. 7 (A. A. Moscona and A. Monroy, eds.). Academic Press, New York, p. 193.

Dewey, M. D., and L. Barr (1964). A study of the structure and distribution of the nexus. *J. Cell Biol.* **23**, 553.

Dewey, W. C., H. H. Miller, and H. Nagasawa (1973). Interactions between S and G1 cells. *Exp. Cell Res.* **77**, 73.

Friend, D., and N. B. Gilula (1972). Variations in tight and gap junctions in mammalian tissues. *J. Cell Biol.* **53**, 758.

Frost, P., G. D. Weinstein, and W. L. Nyhan (1970). Diagnosis of Lesch-Nyhan syndrome by direct study of skin specimens. *J. Am. Med. Assoc.* **212**, 316.

Furshpan, E. J., and D. D. Potter (1959). Transmission at the giant motor synapses of the crayfish. *J. Physiol.* **143**, 289.

Furshpan, E. J., and D. D. Potter (1968). Low-resistance junctions between cells in embryos and tissue culture. In *Current Topics in Developmental Biology*, Vol. 3 (A. A. Moscona and A. Monroe, eds.). Academic Press, New York, p. 95.

Garant, P. R. (1972). The demonstration of complex gap junctions between the cells of the enamel organ with lanthanum nitrate. *J. Ultrastruct. Res.* **40**, 333.

Gilula, N. B., O. R. Reeves, and A. Steinbach (1972). Metabolic coupling, ionic coupling and cell contacts. *Nature* **235**, 262.

Goldberg, N. D., M. K. Haddox, E. Dunham, E. Lopez, and J. W. Hadden (1973). The Yin-Yang hypothesis of biological control: opposing influences of cyclic GMP and cyclic AMP in the regulation of cell proliferation and other biological processes. In *The Cold Spring Harbor Symposium on the Regulation of Proliferation in Animal Cells* (B. Clarkson and R. Baserga, eds.). Academic Press, New York.

Goldstein, J. C., J. F. Marks, and S. M. Gartler (1971). Expression of two X-linked genes in human hair follicles of double heterozygotes. *Proc. Nat. Acad. Sci. U.S.A.* **7**, 1425.

Goodenough, D. A., and J. P. Revel (1970). A fine structural analysis of intercellular junctions in the mouse liver. *J. Cell Biol.* **45**, 272.

Goodenough, D. A., and W. Stoeckenius (1972). The isolation of mouse hepatocyte gap junctions. *J. Cell Biol.* **54**, 646.

Hand, A. R., and S. Gobel (1972). The structural organization of the separate and gap junctions of *Hydra*. *J. Cell Biol.* **52**, 397.

Hay, E. D. (1968). Organization and fine structure of epithelium and mesenchyme in the developing chick embryo. In *Epithelial-Mesenchymal Interactions* (R. Fleishmajer and R. Billingham, eds.). Williams and Wilkins Co., Baltimore.

Holley, R. W. (1972). A unifying hypothesis concerning the nature of malignant growth. *PNAS,* **69**, 2480.

Holtrop, M. E., and J. M. Weinger (1972). Ultrastructural evidence for a transport system in bone. In *Calcium, Parathyroid Hormone and the Calcitonins* (R. V. Talmage and P. L. Munson, eds.). Exerpta Medica, Amsterdam.

Hülser, D. F., and J. H. Peters (1972). Contact cooperation in stimulated lymphocytes. II. Electrophysiological investigations on intercellular communication. *Exp. Cell Res.* **74**, 319.

Johnson, R. G., and J. D. Sheridan (1971). Junctions between cancer cells in culture: Ultrastructure and permeability. *Science* **174**, 717.

Johnson, R., W. Herman, and D. Preus (1973). Homocellular and heterocellular gap junctions in *Limulus. J. Ultrastruct. Res.* **43**, 298.

Kaneko, A. (1971). Electrical connexions between horizontal cells in the dogfish retina. *J. Physiol.* **213**, 95.

Kanno, Y., and W. R. Loewenstein (1966). Cell-to-cell passage of large molecules. *Nature* **212**, 629.

Loewenstein, W. R. (1966). Permeability of membrane junctions. *Ann. N.Y. Acad. Sci.* **137**, 441.

Loewenstein, W. R. (1967). On the genesis of cellular communication. *Develop. Biol.* **15**, 503.

Loewenstein, W. R. (1968). Communication through cell junctions: Implications in growth control and differentiation. *Develop. Biol., Suppl.* **2**, 151.

Loewenstein, W. R. (1973). Membrane junctions in growth and differentiation. *Fed. Proc.* **32**, 60.

Loewenstein, W. R., S. J. Socolar, S. Higashino, Y. Kanno, and N. Davidson (1965). Intercellular communication: renal, urinary bladder, sensory and salivary glands. *Science* **149**, 295.

Loewenstein, W. R., M. Nakas, and S. J. Socolar (1967). Junctional membrane uncoupling—permeability transformations at a cell membrane junction. *J. Gen. Physiol.* **50**, 1865.

Lyon, M. F. (1961). Gene action in the X-chromosome of the mouse (*Mus musculus L.*). *Nature* **190**, 372.

Lyon, M. F. (1968). Chromosomal and subchromosomal inactivation. *Ann. Rev. Genet.* **2**, 31.

Martinez-Palomo, A. (1971). Intercellular junctions in normal and malignant cells. In *Pathobiology Annual*, p. 271.

McNutt, S., and R. Weinstein (1973). Membrane ultrastructure at mammalian intercellular junctions. *Prog. Biophys. Mol. Biol.* **26**, 45.

Melcher, A. H., and G. E. Accursi (1972). Transmission of an "osteogenic message" through intact bone after wounding. *Anat. Rec.* **173**, 265.

Michalke, W., and W. R. Loewenstein (1971). Communication between cells of different type. *Nature* **232**, 121.

Mintz, B. (1971). Genetic mosaicism *in vivo*: Development and disease in allophenic mice. *Fed. Proc.* **30**, 935.

Mintz, B., and W. Baker (1967). Normal mammalian muscle differentiation and gene control of isocitrate dehydrogenase synthesis. *Proc. Nat. Acad. Sci. U.S.A.* **58**, 592.

Oliveira-Castro, G. M., and W. R. Loewenstein (1971). Junctional membrane permeability: Effects of divalent cations. *J. Membrane Biol.* **5**, 51.

Otten, J., G. S. Johnson, and I. Pastar (1971). Cyclic AMP levels in fibroblasts: Relationship to growth rate and contact inhibition of growth. *Biochem. Biophys. Res. Commun.* **44**, 1192.

Pardee, A. B. (1964). Cell division and a hypothesis of cancer. *Nat. Cancer Inst. Monogr.* **14**, 7.

Payton, B. W., M. V. L. Bennett, and G. D. Pappas (1969). Permeability and structure of junctional membranes at an electrotonic synapse. *Science* **166**, 141.

Revel, J. P., and M. J. Karnovsky (1967). Hexagonal array of subunits in intercellular junctions of the mouse heart and liver. *J. Cell. Biol.* **33**, C7.

Revel, J. P., A. G. Yee, and A. J. Hudspeth (1971). Gap junctions between electrotonically coupled cells in tissue and in brown fat. *Proc. Nat. Acad. Sci. U.S.A.* **68**, 2924.

Rose, B. (1971). Intercellular communication and some structural aspects of membrane junctions in a simple cell system. *J. Membrane Biol.* **5**, 1.

Satir, P., and N. B. Gilula (1973). The fine structure of membranes and intercellular communication in insects. *Ann. Rev. Entomol.* **18**, 143.

Sauk, J. J., H. W. Lyon, and C. J. Witkop (1972). Electron optic microanalysis of two gene products in enamel of females heterozygous for X-linked hypomaturation amelogenesis imperfecta. *Am. J. Human Genet.* **24**, 267.

Seegmiller, J. E., F. M. Rosenbloom, and W. N. Kelley (1967). Enzyme defect associated with a sex-linked human neurological disorder and excessive purine synthesis. *Science* **155**, 1682.

Sheppard, J. R. (1974). Cyclic AMP and cell division. In *Molecular Pathology* (R. Good and S. Day, eds.). Charles C Thomas, Springfield, Ill.

Sheridan, J. D. (1968). Electrophysiological evidence for low-resistance intercellular junctions in the early chick embryo. *J. Cell Biol.* **37**, 650.

Sheridan, J. D. (1971a). Dye movement and low resistance junctions between reaggregated embryonic cells. *Develop. Biol.* **26**, 627.

Sheridan, J. D. (1971b). Electrical coupling between fat cells in newt fat body and mouse brown fat. *J. Cell Biol.* **50**, 795.

Sheridan, J. D. (1973). Functional evaluation of low-resistance junctions: Influence of cell shape and size. *Am. Zool.* **13**, 1119.

Sheridan, J., and R. Johnson (1974). Cancer cell junctions. In *Molecular Pathology* (R. Good and S. Day, eds.). Charles C Thomas, Springfield, Ill.

Slack, C., and J. P. Palmer (1969). The permeability of intercellular junctions in the early embryo of *Xenopus laevis*, studied with a fluorescent tracer. *Exp. Cell Res.*, **55**, 416.

Spira, A. W. (1971). The nexus in the intercalated disc of the canine heart: Quantitative data for an estimation of its resistance. *J. Ultrastruct. Res.* **34**, 409.

Spitzer, N. C. (1970). Low resistance connections between cells in the developing anther of the lily. *J. Cell Biol.* **45**, 565.

Subak-Sharpe, H., R. Bürk, and J. Pitts (1966). Metabolic cooperation by cell-to-cell transfer between genetically different mammalian cells in tissue culture. *Heredity* **21**, 342.

Temin, H. M. (1972). The RNA tumor viruses—background and foreground. *Proc. Nat. Acad. Sci. U.S.A.* **69**, 1016.

Todaro, G. J., and R. J. Huebner (1972). The viral oncogene hypothesis: New evidence. *Proc. Nat. Acad. Sci. U.S.A.* **69**, 1009.

Tupper, J. T., and J. W. Saunders, Jr. (1972). Intercellular permeability in the early *Asterias* embryo. *Develop. Biol.* **27**, 546.

Weiss, R. A., and D. L. Njeuma (1971). Growth control between dissimilar cells in culture. In *Growth Control in Cell Cultures* (G. E. W. Wolstenholme and J. Knight, eds.). Churchill Livingstone, London.

Witkop, C. J. (1967). Partial expression of sex-linked recessive amelogenesis imperfecta in females compatible with the Lyon hypothesis. *Oral Surg.* **23**, 174.

Witkop, C. J., W. Kuhlmann, and J. J. Sauk (1973). Autosomal recessive pigmented hypomaturation amelogenesis imperfecta: Report of a kindred. *Oral Surg.* **36**, 367.

Woodbury, J. W., and W. E. Crill (1970). The potential in the gap between two abutting cardiac muscle cells: A closed solution. *Biophys. J.* **10**, 1076.

CHAPTER ELEVEN

The Cholinergic Receptor Protein: Functional Properties and Its Role in the Regulation of Developing Synapses

JEAN-PIERRE CHANGEUX

Since the discovery that the cholinergic (nicotinic) receptor protein can be selectively labeled by snake venom α-toxins (Lee, 1970; Changeux, Kasai, and Lee, 1970; Miledi et al., 1971) and solubilized by mild detergents without loss of its ability to bind cholinergic agonists and antagonists (Changeux, Kasai, and Lee, 1970; Miledi et al., 1971; Changeux, Kasai, Huchet, and Meunier, 1970; Changeux, Meunier, and Huchet, 1971), an increasing number of studies have been undertaken (Meunier et al., 1971; Meunier et al., 1972; Raftery et al., 1971; Eldefrawi and Eldefrawi, 1972; Fulpius et al., 1972; Karlsson et al., 1972; Olsen et al., 1972; Hall, 1972) on this important *regulatory* protein. (Changeux, 1966; Changeux et al., 1967; Karlin, 1967; Changeux et al., 1969). In this report, we shall summarize the results obtained *in our laboratory* with the receptor protein from the electric organs of *Electrophorus electricus* and *Torpedo marmorata* and present and discuss some recent findings on the critical role played by the receptor protein in the development of a neuromuscular synapse.

ISOLATION AND PURIFICATION

The major requirements for the isolation and purification of a receptor protein are (*a*) a tissue particularly rich in receptor, (*b*) a set of specific ligands, and (*c*) the possibility of correlating physiological response and *in vitro* properties of the isolated molecule. The electric organ of *Electrophorus* satisfies these three requirements.

The electric organ of an average-size eel (1 meter long) weighs up to 1 kg and contains about 10^{10} identical synaptic contacts. These synapses are cholinergic and sensitive to typical nicotinic agents. The most effective agonists are acetylcholine, carbamylcholine, and decamethonium. Antagonists like *d*-tubocurarine, flaxedil, and hexamethonium block the effect of these agonists in a "competitive" manner; local anesthetics like tetracaine and procaine, in a "noncompetitive" way. In addition, snake venom α-toxin like that from *Bungarus multicinctus* (74 amino acids, 5 disulfide bridges, MW = 8000) (Lee, 1970) or *Naja nigricollis* (61 amino acids, 4 disulfide bridges, MW = 6800) (Karlsson et al., 1966; Lee, 1970), behave in this system (Changeux, Kasai, and Lee, 1970; Changeux et al., 1971) as powerful and slowly reversible curarizing agents. The α-toxin from *N. nigricollis* was tritiated by the method of Fromageot, that is, iodination followed by catalytic dehalogenation in the presence of tritium gas (Menez et al., 1971). The tritiated α-toxin, with a specific radioactivity of $10-14$ Ci mmole^{-1}, presents the same structure as the native α-toxin (except that one hydrogen atom is replaced by a tritium atom) and possesses all its pharmacological properties (Meunier et al., 1972). It has been used to label and assay the receptor protein.

The physiological response to cholinergic agents can be recorded by electrophysiological techniques applied to single electroplax dissected from the electric organ (Schoffeniels and Nachmansohn, 1967; Higman et al., 1963). Decrease in membrane potential as a function of concentration of bath-applied agonists gives dose-response curves and "apparent" dissociation constants. A more reliable method consists in measuring steady-state slope conductances by voltage clamp (Changeux and Lester, manuscript in preparation). With carbamylcholine the apparent dissociation constant is $2.5 \pm 0.5 \times 10^{-5}$ M by the first method and around 3×10^{-4} M by the second one, depending on the value of the clamping potential (Higman et al., 1963; Changeux and Lester). With decamethonium the values obtained by both techniques fall between 10^{-6} and 10^{-5} M.

As shown by Kasai and Changeux (1971), the pharmacological and ionic properties of the response are preserved, *in vitro*, in excitable membrane fragments or microsacs purified from crude homogenates of electric tissue. The passive permeability of these vesicles to Na$^+$, K$^+$, and Ca^{2+} increases in the presence of agonist. The same vesicles bind ^{3}H-decamethonium and ^{3}H-α-toxin (Kasai and Changeux, 1971; Meunier et al., 1972). "Reduction" of the system from the cellular to the subcellular level, therefore, does not result in a dramatic alteration of the physiological function.

Mild detergents like sodium deoxycholate or cholate, Triton X-100, and Emulphogen completely solubilize excitable microsacs without loss of the ability to bind cholinergic ligands and snake α-toxins (Changeux, Kasai, and Lee, 1970; Miledi et al., 1971). With the solubilized material, as well as with the membrane fragments, the amounts of bound ^{3}H-decamethonium displaced by unlabeled α-toxin and, conversely, of bound ^{3}H-α-toxin displaced by an excess of unlabeled decamethonium were considered to be specifically associated with the cholinergic receptor protein (Changeux, Kasai, and Lee, 1970; Changeux, Kasai, Huchet, and Meunier, 1970; and Changeux, Meunier, and Huchet, 1971).

Several assays for the solubilized toxin-binding material have been used, for example, ammonium sulfate precipitation (Meunier, Olsen, Menez, Fromageot et al., 1972; Franklin and Potter, 1972) and adsorption on DEAE cellulose paper (Fulpius, Klett et al., 1971; Schmidt and Raftery, 1973). The most reliable and fast method was, in our hands, precipitation in the absence of detergents, followed by Millipore filtration (Olsen et al., 1972).

Massive Triton X-100 extraction of a crude membrane preparation from *Electrophorus* electric organ gives a convenient starting material for purification (Olsen et al., 1972). Successful purification of the toxin-binding material was achieved by affinity chromatography on a column of Sepharose derivative with cholinergic arms resembling flaxedil, followed by chromatography on DEAE cellulose and sucrose gradient centrifugation (Olsen et al., 1972; Meunier, Olsen, Sealock, and Changeux, 1974).

The material purified in a satisfactory yield (30% from crude extract) gives one major band by sucrose gradient centrifugation and gel electrophoresis in the presence of Na cholate and Emulphogen. In both cases a reasonable superimposition of protein and toxin binding was achieved. The specific activity corresponds to 1 mole of α-toxin bound for 150,000–180,000 molecular weight of protein. The purified preparation contained less than one catalytic site of acetylcholinesterase (assuming 65,000 g per site) per 100 ^{3}H-α-toxin binding sites.

The amino acid composition does not deviate from that of standard globular protein. The "polarity" (Capaldi and Vanderkooi, 1972) of the protein is close to 0.47. In addition, the purified protein is precipitated by concanavalin A and thus carries carbohydrate residues.

Disk gel electrophoresis in SDS reveals two bands having apparent molecular weights of 43,000 and 48,000–52,000 daltons. It is not yet known whether these two bands represent two distinct subunits or result from a proteolytic attack. These values are in the range of those found by Karlin and his associates, using a covalent affinity label (Reiter et al., 1972).

Electron microscopy on the purified fraction (Cartaud et al., 1973) shows, after negative staining by uranyl acetate, a homogeneous population of particles with an average diameter of 80–90 Å, a characteristic electron-dense center, and a subunit pattern of five to six subunits with a diameter of 30–40 Å.

Hydrodynamic experiments and gel electrophoresis in the presence of SDS after treatment by cross-linking reagents suggest that the molecular weight of the protein would be in the range of 250,000. In any case the detergents extract a molecular form of the receptor protein which presents an oligomeric structure.

THE RECEPTOR PROTEIN AS AN INTEGRAL MEMBRANE PROTEIN

Distilled water and high (0.8 *M*) and low (0.02 *M*) ionic strength buffers do not extract the receptor protein from membrane fragments of *Electrophorus* electric organ. Only after treatment by detergents like Triton X-100, Emulphogen and Na cholate is the receptor found in soluble form. It is quite clear that the receptor protein is strongly membrane bound. By contrast, acetylcholinesterase, whose binding properties present some analogies with those of the receptor, is released from the membrane during a wash with 0.8 M NaCl. According to Singer and Nicolson (1972), acetylcholinesterase would be a "peripheral" protein, and the receptor protein an "integral" one.

In the presence of detergent the receptor protein shows unusual hydrodynamic properties (Meunier et al., 1972). On Sepharose 6B columns equilibrated with buffers containing Triton X-100, the receptor has an apparent Stokes radius of 73 Å (slightly larger than that of β-galactosidase, MW 550,000). On the other hand—and this was confirmed by Reich's (Fulpius et al., 1972) and Raftery's (Raftery et al., 1971) groups—upon sedimentation in sucrose gradients in the presence of Triton X-100, the receptor has an apparent sedimentation constant of 9.5 S, considerably less than the 16 S found for β-galactosidase. Sedimentation in D_2O instead of H_2O (Meunier, Olsen and Changeux, 1971) shows that the receptor has an unusually low density or high $\bar{v}$ ($\bar{v}=0.78$), while $\bar{v}$ calculated from the amino acid composition (0.73–0.74) is typical of proteins in general (Meunier et al., 1974).

This low density probably arises from the binding of considerable quantities of Triton X-100 ($\bar{v}=0.99$) and accounts to large degree for the low sedimentation constant observed. A possible contribution from asymmetry of the receptor protein cannot be eliminated, however, on the basis of these data (Meunier et al., 1974).

The state of aggregation of the receptor particle varies with the presence or absence of detergent. Removal of the detergent, in general, causes aggregation of the receptor protein. Passing a solution of receptor in ^{14}C-cholate through a G-75 Sephadex column removes the bulk of unbound cholate but leaves enough detergent tightly associated with the protein to keep it in solution (Meunier et al., 1974).

One simple interpretation of the solubility properties and of the high affinity for detergents is that after extraction from the membrane the receptor protein possesses hydrophobic areas exposed to the solvent. In its membrane-bound state these areas would anchor the receptor protein to the apolar phase of the membrane. An interesting problem is then raised by the fact that the amino acid composition of the receptor protein, like that of several other integral proteins, does not differ markedly from that of water-soluble proteins. As already discussed, one explanation would be that regions of the protein are folded in such a way that hydrophilic groups face the inside of the protein, and hydrophobic ones the outside (Changeux, 1972). In these regions the tertiary structure would be "inverted" as compared to that of water-soluble proteins. An alternative is that, as in cytochrome b_5 (Spatz and Strittmatter, 1971) or erythrocyte major glycoprotein (Marchesi et al., 1972), the protein possesses "segments" or "tails" particularly rich in hydrophobic amino acids, which would preferentially interact with the aliphatic chains of membrane lipids.

In any case, since the protein acts as a pharmacological receptor, at least the region carrying the receptor site is not occluded within the membrane phase and exposed to the physiological medium. Indeed, this is confirmed by freeze-etching pictures of receptor-rich membrane fragments from *Torpedo* (Cartaud et al., 1973; Nickel and Potter, 1973), which reveal on the membrane surface structures resembling the purified *Electrophorus* receptor.

In an attractive but still highly hypothetical structure of the receptor molecule, some subunits or "globules" in direct contact with the physiological medium carry the receptor site and possess the usual globular structure, while others, with an

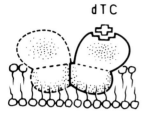

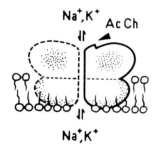

Figure 1. A model for the control of ionic translocation by the cholinergic receptor protein.

"inverted" structure, are more directly involved in ionic translocation and are deeply buried in the hydrocarbon phase of the membrane (Fig. 1).

THE CHOLINERGIC RECEPTOR PROTEIN AS A REGULATORY PROTEIN

As already extensively discussed (Podleski and Changeux, 1970), the cholinergic receptor protein can be viewed as a regulatory protein which controls the selective translocation of small cations through the membrane. To account for the coupling between the receptor site for agonists and the site for ion translocation, the receptor-ionophore complex was postulated to exist under at least two conformational states: the "resting" state, with a low permeability for cations but a high affinity for antagonists, and the "active" state, with a high permeability and a high affinity for agonists. Although these ideas are still hypothetical, several experimental results make them plausible.

The yield of the purification procedure is high enough to study the binding of cholinergic agonists and antagonists to the receptor protein by equilibrium dialysis. In agreement with our initial results on crude extracts, decamethonium binds to the purified protein and is completely displaced by *N. nigricollis* α-toxin and by the cholinergic antagonist flaxedil. Interestingly, the affinity of the receptor protein for the agonists tested, including acetylcholine, is 1–2 orders of magnitude *larger* than the *apparent* affinity of the same agonists with either the isolated electroplax or the excitable microsacs (Meunier and Changeux, 1973). On the other hand, no significant difference is found with the antagonists (Meunier and Changeux, 1973). Among the several interpretations which can be proposed for this phenomenon, one is that solubilization by detergents and purification release a membrane "constraint" created by either membrane lipids or proteins or by both and stabilize the molecule in an "active" conformation exhibiting high affinity for agonists. The limited changes of affinities for the antagonists would be caused by their *nonexclusive* binding to both the "active" and the "resting" conformations.

A sigmoid shape of the dose-response curve of the electroplax to cholinergic agonists was noticed in the early work of Higman et al. (1963) and subsequently confirmed for a variety of agonists with the same system (Hall, 1972; Changeux and Podleski, 1968) or excitable microsacs (Kasai and Changeux, 1971). Isolation and purification of membrane fragments from *Torpedo* electric tissue particularly rich in receptor protein (Cohen et al., 1972) favored accurate analysis of the *binding* curve of acetylcholine to the cholinergic receptor site (Weber and Changeux, 1974). Clear-cut deviation from the Langmuir isotherm has been observed: the Hill coefficient of the binding curve ranges from 1.3 to 1.5. However, the binding curve of the same ligand to the protein purified from *Electrophorus* follows a hyperbola (Meunier and Changeux, 1973). *Cooperative binding* of acetylcholine, therefore, occurs in the membrane-bound state of the receptor protein (Weber and Changeux, 1974). It is not known yet whether these cooperative effects are associated with the oligomeric structure of the receptor protein (interaction between subunits) or are relevant to its organization into a lattice structure (interaction between oligomers). It should be mentioned, however, that a lattice organization of the cholinergic receptor protein, with a center-to-center distance of approximately 90 Å, has been demonstrated by freeze-etching (Cartaud et al., 1973) and X-ray diffraction (Dupont, Cohen and Changeux, 1974) with the same membrane fragments.

Finally, recent studies with the dansylated cholinergic ligand introduced by Weber et al. (1971) provide some evidence for a structural transition of the receptor protein upon binding of agonists (Cohen and Changeux, 1973):

$$CH_3 \qquad CH_3$$
$$N$$

$$\overset{+}{SO_2 - NH - CH_2 - CH_2 - N(CH_3)_3}$$

DNS-chol

When added to a suspension of *Torpedo* membrane fragments rich in receptor protein, the excitation and emission spectra of DNS-chol change notably. In the excitation spectrum observed at 550 nm, the 340-nm band characteristic of DNS-chol in aqueous solution persists but a new band at 280 nm appears. The emission spectrum changes as well: at 340-nm excitation wavelength the maximum emission shifts from 560 to 550 nm; at 287 nm the shift reaches 535 nm. DNS-chol interacts with the membrane fragments, and probably an energy transfer takes place between membrane proteins and DNS-chol. Cholinergic agonists and antagonists, as well as α-toxin from *N. nigricollis*, reduce by approximately 80% the light emitted at 550 nm by DNS-chol in the presence of membrane fragments upon illumination at 287 nm. This observation and others indicate that DNS-chol binds to at least two classes of membrane sites, one of them, the α-toxin-sensitive type, possessing the specificity of the cholinergic receptor site. In the course of these experiments it was shown that some cholinergic ligands not only provoke a decrease of fluorescence intensity, associated with the displacement of DNS-chol from the cholinergic receptor site, but also cause a shift of the wavelength of maximal emission from 537 to 522 nm. This blue shift takes place when the cholinergic agent used acts as an agonist on *Electrophorus* electroplax; it is absent in the presence of antagonists or α-toxin. The effect is interpreted as indicating a change in the spectral properties of DNS-chol bound to a site distinct from the cholinergic receptor site. There, DNS-chol would be sensitive to the agonist or antagonist character of the compound bound to the receptor site and would therefore monitor a structural transition associated with the physiological effect of cholinergic ligands (Cohen and Changeux, 1973).

These still fragmentary results are consistent with the hypothesis that the receptor protein is a regulatory protein, although final demonstration of this requires more extensive documentation on the structural properties of the protein.

IMMUNOLOGICAL STUDIES WITH THE RECEPTOR PROTEIN

Among the several hypotheses which should be mentioned concerning the changes of affinity observed after purification, one is that the purified protein is *not* the macromolecule which accounts for the electrogenic action of acetylcholine! Although the arguments based on the binding specificity of the isolated macromolecule for both cholinergic ligands and snake α-toxins are rather convincing, immunological studies with the purified protein have brought additional support that this protein carries the physiological receptor site for acetylcholine (Sugiyama et al., 1973).

Antibodies directed against the purified protein were raised in rabbits by injecting 0.5 mg of protein in Freund's adjuvant per rabbit. After 3 weeks a booster injection of the same amount of purified protein was given. Four days later the rabbits developed a flaccid paralysis and died rapidly. A similar effect was observed as well by Patrick and Lindström (1973) with a preparation of protein prepared with a different affinity column and interpreted by them as an autoimmune response of the rabbit to the receptor protein from *Electrophorus*. The antibody precipitates the toxin-binding protein from both a crude extract and a purified preparation and cross-reacts, but with a different ratio of antigen to serum at equivalence, with the receptor protein from *Torpedo* and chick embryo. The serum gives a single precipitation band by Ouchterlony's double-diffusion technique against the purified protein. Interestingly, a tenfold dilution of the serum applied to the electroplax for 20 minutes blocks in an irreversible manner the response to bath-applied carbamylcholine. Therefore, the protein which, *in vitro*, binds cholinergic agonists with a very high affinity is that which mediates the electrical response of the electroplax to the same agonists, but with an "apparent" affinity 10–100 times smaller.

RECONSTITUTION OF A CHEMICALLY EXCITABLE MEMBRANE

Isolation and purification of the protein which carries the cholinergic receptor site does not bring per se any information on the structural element involved in the selective translocation of ions, which is under its control: the cholinergic ionophore. Classical electrophysiological experiments and flux measurements with isolated microsacs (Kasai and Changeux, 1971) have shown that the cholinergic ionophore is selective for small metallic cations, Na^+, K^+, or Ca^{2+}. The exact magnitude of the ionic flux through a single active ionophore is still unknown; on the other hand, the average conductance under the control of a single α-toxin binding site ranges between 10^{-13} and $10^{-12}\Omega^{-1}$ (Changeux and Lester, in preparation).

Our approach to identify this element is as follows (Changeux, Huchet, and Cartaud, 1972; Hazelbauer and Changeux, in press). Microsacs of excitable membrane could be dissociated into their elementary components and subsequently reconstituted in a form which exhibits both the characteristic selective permeability to cations and the sensitivity to acetylcholine of native microsacs. Then, fractionation before reconstitution might lead to identification of the components responsible for the selective permeability change caused by the cholinergic agents.

The first step has been reached. Solubilization of excitable microsacs by Na cholate or deoxycholate under adequate conditions occurs without loss of the specific binding capacity of the cholinergic receptor for α-toxin and of the catalytic activity of acetylcholinesterase. Extensive dialysis, removing detergents and introducing divalent cations, yields membrane fragments which bind ^3H-α-toxin with a high specific activity and contain large amounts of acetylcholinesterase (Changeux, Huchet, and Cartaud, 1972). Recently, conditions have been defined in which the reconstituted microsacs have an apparent volume for $^{22}Na^+$ and are sensitive to carbamylcholine (Hazelbauer and Changeux, in press). Reconstitution of an excitable membrane can thus be achieved. Experiments are in progress to identify the ionophore in the solubilized extracts and, in particular, to determine its relation to the receptor protein.

Several plausible mechanisms have already been presented for the relationship between the conformational transition of the receptor protein and the functioning of the ionophore (Changeux, Podleski et al., 1970). In the one presented in Fig. 1, the conformational transition would be associated with a movement of the receptor protomer within the thickness of the membrane. For instance, in the permeable state, the receptor-ionophore complex would span the membrane; at rest, it would restrict its position to one face of the membrane, leaving one layer of lipids to seal the pore on the other face.

DISTRIBUTION OF THE CHOLINERGIC RECEPTOR SITES IN ELECTROPHORUS ELECTROPLAX AND CONSEQUENCES OF DENERVATION ON THIS DISTRIBUTION

Since the snake α-toxins constitute highly specific and slowly reversible markers of the receptor site, they can conveniently be used to localize the receptor at the cellular and subcellular level. In a first series of experiments we used the unlabeled α-toxin from *N. nigricollis,* and the bound toxin was revealed by a combination of rabbit antiserum directed against the toxin and fluorescent α-globulins directed against the rabbit antibodies. Only the innervated membrane became fluorescent, suggesting an exclusive localization of the receptor protein on this face of the cell (Bourgeois et al., 1971).

This finding was confirmed by autoradiography, using the tritiated α-toxin. After exposure of a slice of electroplax to the ^3H-α-toxin, more than 99% of the grains were present on the innervated face. In addition, as expected for a specific binding to the receptor site, no radioactivity was bound after exposure to an irreversible antagonist or after contact with an excess of unlabeled toxin (Bourgeois et al., 1972).

Fine localization of the α-toxin binding sites was achieved by high-resolution autoradiography with the electron microscope (Bourgeois et al., 1972). After exposure of the whole electroplax to ^{3}H-α-toxin, grains appear much more numerous under than between the synapses. The absolute density of receptor sites per unit of surface area was calculated on the basis of several assumptions about the actual area of the surface of the cytoplasmic membrane (the subsynaptic membrane constitutes between 1 and 2% of the total surface of the cell), the yield of dpm to grains, the thickness of the fine section, and so forth. Between 30,000 and 60,000 toxin binding sites per square micron are found under the synapses and approximately 100 times less between the synapses, the density on the innervated side being a further 10 times smaller. A similar difference in density between subsynaptic and extrasynaptic areas has also been reported for muscle (Miledi and Potter, 1971; Barnard et al., 1971). If all our assumptions are correct, the subsynaptic membrane would be occupied almost exclusively by the receptor protein. Little space would be left for acetylcholinesterase unless this enzyme is integrated to the membrane in a manner different from that for the receptor protein. The loose attachment of the esterase to the cytoplasmic membrane gives some support to this view.

In order to characterize the factors involved in the regulation of receptor distribution, *Electrophorus* electric organs were denervated by destruction of the caudal part of the spinal cord. Degeneration of the nerve terminals was monitored by following the "neurally evoked" action potential and the ultrastructure of the synaptic boutons. After 2 days, the neurally evoked action potential disappears, as well as most nerve endings. After 8 days, the number of synaptic contacts becomes reduced to less than 5% of the number in the intact cell. After a fortnight, none of them persists. However, both the total number and the distribution of cholinergic receptor sites remain constant up to 142 days. In particular patches with a high density of receptor sites corresponding to former subsynaptic membranes persist, freely exposed, for several weeks, without any sign of lateral diffusion of the receptor molecule (Bourgeois et al., 1973).

There is, thus, no evidence, at least at the level of the subsynaptic membrane, for a "fluidity" as high as that expected from experiments developed with spin-labeled lipids and several artificial and biological membranes (McConnel et al., 1972). On the other hand, if one still wishes to consider "fluidity" as a *general* property of biological membranes [this, in fact, has never been our opinion as far as membrane proteins are concerned (Wahl et al., 1971)], "special" mechanisms might be invoked to prevent lateral diffusion of the cholinergic receptor protein: a tendency to form lattice structures, the presence of "cleft substances," for instance, polysaccharides, which would cross-link neighboring receptor molecules, and so forth.

The consequences of denervation in *Electrophorus* electroplax differ from those observed with striated muscle: no supersensitivity or increase of receptor number in extrasynaptic areas after denervation occurs. The difference in zoological origin, the absence of contractile apparatus, and the eventual "induced" depolarization of the denervated cells by the electric field developed in the nondenervated region of the organ are some of the factors that might explain the particular behavior of the electroplax.

In any case, the differential and stable distribution of the receptors between and under the synapses remains to be explained. Several hypotheses such as the following might be proposed for the local enrichment in receptor protein under the nerve terminals:

1. The synthesis and liberation by nerve terminals of additional receptor molecules, which become integrated in the synaptic membrane.

2. The already mentioned tendency of the receptor protein to aggregate and produce a lattice organization or the cross-linking of neighboring receptor molecules under the synapse which prevents their lateral diffusion.

3. The differential turnover of receptor proteins in subsynaptic and extrasynaptic areas. In extrasynaptic areas the turnover might be enhanced by the ionic currents or the potential inversion associated with spike propagation (Stent, 1973; Changeux et al., 1973).

Mechanism 3 might be accounted for by either a different structure or a different environment of the extrasynaptic and subsynaptic receptor molecules. No significant differences were revealed by immunoprecipitation between the receptor proteins from soluble extracts of extrasynaptic and subsynaptic areas of normal and denervated rat diaphragms (Sugiyama and Changeux, 1973). But covalent modifications such as phosphorylation, adenylation, glycosylation, or mild proteolytic attack might remain undetected by the immunological test. In this case the enzymes involved might be liberated by the presynaptic nerve terminal simultaneously with acetylcholine, or acetylcholine itself might activate some postsynaptic enzyme. In addition, it is known that cholinergic vesicles present in the nerve terminals of *Torpedo* electric tissue contain, like the chromaffin granules, high quantities of ATP (Whittaker, 1973). As a consequence of stimulation, the ATP content of these vesicles decreases (Whittaker, 1973) and ATP is released into the synaptic cleft (Hubbard et al., 1972). The liberated ATP might serve as an energy source for covalent reactions occurring in the cleft, particularly those which modify the stability of the receptor protein.

ROLE OF THE RECEPTOR PROTEIN IN THE DIFFERENTIATION OF THE MYONEURAL JUNCTION

Establishment of a synapse involves, in general, two major steps: (1) recognition of the target cell, and (2) stabilization or "consolidation" of the adequately functional contact and its further maturation into a typical synapse, exemplified by the development of the subneural apparatus in the myoneural junction.

In order to investigate whether the cholinergic receptor protein is involved in any one of these steps in the myoneural junction, chronic injections of N. *nigricollis* α-toxin were given to developing chick embryos (Giacobini et al., 1973), which survive high doses (hundreds of micrograms) of α-toxin. However, embryos injected after 3, 8, and 12 days of incubation and observed on the 16th day of embryonic development show marked atrophy of the skeletal muscles. The atrophic muscle presents signs of delayed differentiation (abundance of myotubes) or even regression (increased connective tissue, phagocytic histiocytes, degeneration of muscle fibers), as in surgically denervated muscle. Nor are the effects of the α-toxin restricted to muscle: in 16-day embryos treated with the α-toxin, typical motor end plates revealed by the Koelle reaction are almost completely absent, and the sciatic nerve, the spinal roots, and sensory ganglia are markedly reduced in size; the specific and total activity of choline acetyltransferase, a typical presynaptic enzyme, decreased by more than 50%. The α-toxin interferes with both the differentiation of the muscle and that of its innervation.

The α-toxin, at the doses employed, is neither lethal nor teratogenic; its effects seem limited to systems involving typical nicotinic receptors. In addition, most present evidence indicates that the α-toxin affects neither central nervous activity nor peripheral propagation of impulses. Probably the observed effects arise from the blockade by the α-toxin of the cholinergic receptor site present on the embryonic muscle. In agreement with several recent observations (Lømo and Rosenthal, 1971), all the postsynaptic effects could be caused simply by the lack of receptor activation by acetylcholine and therefore of end-plate potential genesis, spike propagation, and muscle contraction.

The existence of presynaptic effects raises two questions. First, how does an essentially postsynaptic block influence the motor nerve terminal? One possibility is that the cholinergic receptor participates to step 1: it contributes by some characteristic structural feature to the recognition between nerve and muscle. The α-toxin might interfere with this recognition step. In embryos injected at the 4th day of incubation, primitive junctions do not differ significantly from those in the noninjected controls until the 12th day of incubation. The α-toxin does not seem to interfere with the establishment of the early myoneural contacts. It remains possible, however, that some elements of structure involved in step 1 are present on the receptor protein but are not blocked by the α-toxin. In this respect one might mention that the receptor protein possesses a glycosyl moiety which does not seem to be directly involved in the regulation of ionic transport by the cholinergic receptor site.

There is thus strong evidence that the α-toxin interferes with step 2: the stabilization and maturation of a functional endplate. This would mean, first, that this second step depends strongly on the state of activity of the synapse. Functioning stabilizes the structure of the synapse during its development. Spontaneous neural activity and mobility of the embryos would be required for the maturation of the myoneural junction.

Then one asks a second question: How is the nerve terminal informed of the state of activity of the cholinergic receptors located on the postsynaptic membrane? One channel could be the sensory pathway. The lack of a typical sensory innervation in oculomotor muscles, which nevertheless do not show end plates in α-toxin-treated embryos, tends to rule out the contribution of this innervation, at least in oculomotor muscles. One is therefore left with the *retrograde* transfer of a signal directly from the postsynaptic to the presynaptic cell in a direction opposite to that of the transmission of impulse. We still ignore the question of whether the retrograde signal derives simply from changes of ionic concentration in the synaptic cleft consecutive to the activation of the cholinergic receptors, or its genesis requires the propagation of action potentials in the muscle and its contraction. In any case, whatever its exact nature, this retrograde signaling would constitute a critical regulatory element in the functional stabilization of a developing synapse.

The implications of this conclusion have been extended to the problem of the determinism of the end-plate *position* on the muscle fiber. In most but not all categories of striated muscle fibers (fast muscle fibers in higher vertebrates) the adult end plate is located in the middle of the fiber. Since significant fluctuations occur during embryonic nerve growth, however, the precise position of the end plate is not expected to be genetically programmed. At the early stage of development, growing nerve terminals, referred to as "exploratory fibers," appear to establish multiple rudimentary but functional synaptic contacts. Maturation of the end plate appears to result from the *selection* of only one of these early contacts.

A theoretical analysis of this question has recently been made as an application of a more general theory on the epigenesis of neuronal networks (Changeux, Courrege, and Danchin, 1973). According to the theory, the selection of the central synapse would require the actual functioning both of the early synaptic contacts and of the muscle fibers. Transfer of signals through four different channels appears necessary for the system to evolve from a multisynaptic to a monosynaptic state: (*a*) the transmission of the nerve impulse from the nerve terminal to the muscle; (*b*) the propagation of a centrifugal signal, for instance, the action potential on the muscle fiber; (*c*) the propagation of a centripetal signal from both ends of the muscle fiber, for instance, contraction or a related biochemical process; and (*d*) the transfer, in a *retrograde* manner, of a stabilizing signal from the muscle to the nerve terminal at the position where the confluence of two centripetal signals reaches a certain threshold (Fig. 2).

This theory permits several interesting predictions. One is that the topology of the stabilized synapses depends on the functioning of the developing myoneural system during a critical period. An "optimal" activity is required for the stabilization of the central synapse. Too much activity would stabilize all the early synaptic contacts; not enough would lead to the degeneration of all the contacts. The

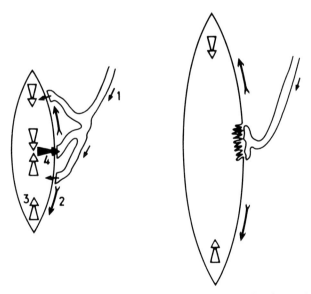

Figure 2. Diagrammatic representation of the evolution, during development, of the innervation of a striated muscle fiber. The four classes of arrows indicate the four classes of signals postulated by the theory.

spontaneous electrical activity of the spinal cord in developing embryos might provide the optimal dose of stabilizing impulses. Blocking of this activity or its enhancement by chronic stimulation should therefore lead to a critical test of the theory.

ACKNOWLEDGMENTS

This work was supported by funds from the Centre National de la Recherche Scientifique, the Délégation Générale à la Recherche Scientifique et Technique, the Collège de France, the Commissariat à l'Energie Atomique, and the National Institutes of Health.

REFERENCES

Barnard, E., J. Wieckowski, and T. H. Chiu. *Nature* **234**, 207 (1971).

Bourgeois, J. P., J. L. Popot, A. Ryter, and J. P. Changeux. *Brain Res.* **62**, 557 (1973).

Bourgeois, J. P., A. Ryter, A. Menez, P. Fromageot, P. Boquet, and J. P. Changeux, *FEBS Lett.* **25**, 127 (1972).

Bourgeois, J. P., S. Tsuji, P. Boquet, J. Pillot, A. Ryter, and J. P. Changeux. *FEBS Lett.* **16**, 92 (1971).

Capaldi, R. A., and Vanderkooi. *Proc. Nat. Acad. Sci. U.S.A.* **69**, 930 (1972).

Cartaud, J., L. Benedetti, J. B. Cohen, J. C. Meunier, and J. P. Changeux. *FEBS Lett.* **33**, 109 (1973).

Changeux, J. P. *Mol. Pharmacol.* **2**, 369 (1966).

Changeux, J. P. *CIBA Foundation Symposium*, Vol. 7, 1972, p. 289.

Changeux, J. P., P. Courrège, and A. Danchin. *Proc. Nat. Acad. Sci. U.S.A.* **70**, 2974 (1973).

Changeux, J. P., M. Huchet and J. Cartaud. *C. R. Acad. Sci. Paris* **274D**, 122 (1972).

Changeux, J. P., M. Kasai, and C. Y. Lee. *Proc. Nat. Acad. Sci. U.S.A.* **67**, 1241 (1970).

Changeux, J. P., M. Kasai, M. Huchet, and J. C. Meunier. *C.R. Acad. Sci. Paris* **270D**, 2864 (1970).

Changeux, J. P., and H. Lester.—Manuscript in preparation.

Changeux, J. P., J. C. Meunier, and M. Huchet. *Mol. Pharmacol.* **7**, 538 (1971).

Changeux, J. P., J. C. Meunier, R. W. Olsen, M. Weber, J. P. Bourgeois, J. L. Popat, J. B. Cohen, G. L. Hazelbauer, and H. A. Lester, *Drug receptors* (H. P. Rang, ed.). Macmillan 1973, p. 273.

Changeux, J. P., and T. R. Podleski. *Proc. Nat. Acad. Sci. U.S.A.* **59**, 944 (1968).

Changeux, J. P., T. R. Podleski, M. Kasai, and R. Blumenthal. *Proceedings of 5th International Meeting of Neurobiologists* (P. Andersen and J. Jansen, eds.). Universitets Forlaget, 1970, p. 123.

Changeux, J. P., T. Podleski, and J. C. Meunier. *J. Gen. Physiol.* **54**, 225 (1969).

Changeux, J. P., J. Thiery, Y. Tung, and C. Kittel. *Proc. Nat. Acad. Sci. U.S.A.* **57**, 335 (1967).

Cohen, J., and J. P. Changeux. *C.R. Acad. Sci. Paris*, **277D**, 603–606 (1973).

Cohen, J., and J. P. Changeux. *Biochemistry* **12**, 4855 (1973).

Cohen, J. B., M. Weber, M. Huchet, and J. P. Changeux. *FEBS Lett.* **26**, 43 (1972).

Dupont, Y., J. B. Cohen, and J. P. Changeux. *FEBS Lett.* **40**, 130 (1974).

Eldefrawi, M. and A. Eldefrawi. *Proc. Nat. Acad. Sci. U.S.A.* **69**, 1776 (1972).

Franklin, G. I., and L. T. Potter. *FEBS Lett.* **28**, 101 (1972).

Fulpius, B., S. Cha, and E. Reich. *FEBS Lett.* **24**, 323 (1972).

Fulpius, B., R. Klett, D. Cooper, and E. Reich. *5th International Congress of Pharmacology*, San Francisco, July 1971, p. 107.

Giacobini, G., G. Filogamo, M. Weber, P. Boquet, and J. P. Changeux. *Proc. Nat. Acad. Sci. U.S.A.* **70**, 1708 (1973)

Hall, Z. W. *Ann. Rev. Biochem.* **41**, 925 (1972).

Hazelbauer, J., and J. P. Changeux. *Proc. Nat. Acad. Sci. U.S.A.* (in press).

Higman, H., T. R. Podleski, and E. Bartels. *Biochim. Biophys. Acta* **75**, 187 (1963).

Hubbard, J. L., J. Musick, and E. Silinsky. 1972, *INSERM Colloque: La transmission cholinergique de l'excitation*, pp. 187–195.

Karlin, A. *J. Theor. Biol.* **16**, 306 (1967).

Karlsson, E., D. Eaker and J. Porath. *Biochim. Biophys. Acta* **127**, 505 (1966).

Karlsson, E., E. Heilbronn, and L. Widlund, *FEBS Lett.* **28**, 107 (1972)

Kasai, M., and J. P. Changeux. *J. Membrane Biol.* **6**, 1 (1971).

Lee, C. Y. *Clin. Toxicol.* **3**, 457 (1970).

Lφmo, T., and J. Rosenthal. *J. Physiol.* **216**, 52 (1971).

Marchesi, V. T., T. W. Tillack, R. L. Jackson, J. P. Segrest, and R. E. Scott. *Proc. Nat. Acad. Sci. U.S.A.* **69**, 1445 (1972).

McConnel, H., P. Devaux, and C. Scandella. *Membrane Research.* Academic Press, New York, 1972, p. 27.

Menez, A., J. L. Morgat, P. Fromageot, A. M. Ronseray, P. Boquet, and J. P. Changeux. *FEBS Lett.* **17**, 333 (1971).

Meunier, J. C., and J. P. Changeux. *FEBS Lett.* **32**, 143 (1973).

Meunier, J. C., R. N. Olsen, and J. P. Changeux. *FEBS Lett.* **24**, 63 (1971).

Meunier, J. C., R. W. Olsen, A. Menez, P. Fromageot, P. Boquet, and J. P. Changeux. *Biochemistry* **11**, 1260 (1972).

Meunier, J. C., R. W. Olsen, A. Menez, J. L. Morgat, P. Fromageot, A. M. Ronseray, P. Boquet, and J. P. Changeux. *C.R. Acad. Sci. Paris* **273D**, 595 (1971).

Meunier, J. C., R. Sealock, R. Olsen, and J. P. Changeux. *Eur. J. Biochem.* (1974) in press.

Miledi, R., P. Molinoff, and L. T. Potter. *Nature* **229**, 554 (1971).

Miledi, R., and L. Potter. *Nature* **233**, 599 (1971).

Nickel, E., and L. T. Potter. *Brain Res.* **57**, 508 (1973).

Olsen, R. W., J. C. Meunier, and J. P. Changeux, *FEBS Lett.* **28**, 96 (1972).

Patrick, J., and J. Lindström. *Science* **180**, 871 (1973).

Podleski, T. R., and J. P. Changeux. In Fundamental concepts of drug-receptor interactions, *Proceedings of 3rd Annual Buffalo-Milan Symposium on Molecular Pharmacology, 1968* (Danielli, Moran, and Triggle, eds.). Academic Press, New York, 1970, p. 93.

Raftery, M. A., J. Schmidt, D. G. Clark, and R. G. Wolcott. *Biochem. Biophys. Res. Commun.* **45**, 1622 (1971).

Reiter, M. J., D. M. Cowburn, J. M. Prives, and A. Karlin. *Proc. Nat. Acad. Sci. U.S.A.* **69**, 1168 (1972).

Schoeffeniels, E., and D. Nachmansohn. *Biochim. Biophys. Acta* **26**, 1 (1967).

Schmidt, J., and M. A. Raftery. *Biochemistry* **12**, 852 (1973).

Singer, S. J. and G. L. Nicolson. *Science* **175**, 720 (1972).

Spatz, L., and P. Strittmatter. *Proc. Nat. Acad. Sci. U.S.A.* **68**, 1042 (1971).

Stent, G. *Proc. Nat. Acad. Sci. U.S.A.* **70**, 997 (1973).

Sugiyama, H., P. Benda, J. C. Meunier, and J. P. Changeux. *FEBS Lett.* **35**, 124 (1973).

Sugiyama, H., and J. P. Changeux. Unpublished results, 1973.

Wahl, P., M. Kasai, and J. P. Changeux. *Eur. J. Biochem.* **18**, 332 (1971).

Weber, G., D. Borris, E. De Robertis, F. Barrantes, J. La Torre, and M. De Carlin. *Mol. Pharmacol.* **7**, 530 (1971).

Weber, M., and J. P. Changeux. *Mol. Pharmacol.* **10**, 1 (1974).

Whittaker, V. P. *Naturwissenschaften* **60**, 280 (1973).

CHAPTER TWELVE

Contact Interaction Among Developing Mammalian Brain Cells

RICHARD L. SIDMAN

A mosaic organization of nerve cell surfaces in adult vertebrates has not been demonstrated by direct chemical evidence, but is inferred from other data. Electron microscopy provides the most direct evidence (Peters et al., 1970). Obvious surface specializations are seen at synapses, where the surface membrane of the post-synaptic cell or of both pre- and postsynaptic cells commonly appears thickened at the cytoplasmic side of the membrane, and an amorphous material often is seen in the synaptic cleft. Freeze-fracture electron microscopy reveals additional surface specializations within the area of a given synapse (Akert et al., 1972; Sandri et al., 1972; Akert, 1973) and at paranodal contact points between glial cell and axon (Livingston et al., 1973). The latter resemble the rarely encountered interaxonal septate junctions (Sotelo and Llinas, 1972). At the initial segment of the axon, the surface membrane has an undercoating of amorphous material not seen along the main axonal shaft, and this undercoating has been suggested as indicative of local specialization of the cell surface (Palay et al., 1968; Peters et al., 1968). A similar-appearing material lines the inside of the axonal membrane focally at central nodes of Ranvier and sometimes for a short distance along axonal branches arising at such nodes (Waxman, 1974). The electrical properties are probably different at axonal surface sites with and without undercoats (Bennett, 1970; Waxman et al., 1972). Further surface differences at the molecular level are implied by physiological evidence that ionophores for Na^+ ions are differently distributed from ionophores for Ca^{++} ions (Katz and Miledi, 1965, 1967; Rahaminoff, 1974). Additional areas of specialized surface membrane are inferred to exist on the neuron soma on the basis of a focal distribution of subsurface cisterns of endoplasmic reticulum (Rosenbluth, 1962). Pharmacological and histochemical evidence strongly suggests that receptors for monoamines and other agents are distributed differentially (Fuxe et al., 1970; Iversen, 1971). On the negative side, it must be pointed out that not all receptors have a surface distribution, nor is there evidence at present for differential distribution of matrix materials on or between nervous system cells.

The ontogeny of specialized surface properties on neural cells is even more obscure, and some of the pertinent questions are just beginning to come into focus. For example, do the signaling molecules used at mature synapses play an ontogenetic role? The early presence of monoamines in selected central neurons (Olson and Seiger, 1972b; Tennyson et al., 1972), well in advance of synaptic function, suggests that they might do so. Or are there signaling agents and receptor mechanisms unique to particular developmental stages? Thus far the developmental approach has been restricted mainly to the analysis of the behavior and interactions of whole cells, both under normal conditions and under the influence of mutations and various exogenous perturbing influences. There has been considerable progress at the cellular level in the last few years, and these studies form the basis for the main ideas expressed in this review. Also, new methods are becoming available for a more direct approach to the characterization of surface properties of developing cells of the nervous system, and some of these will be indicated at the end of the chapter.

CELL SURFACE FEATURES DURING NORMAL DEVELOPMENT

Increases in Surface Area and Cell Volume

Of all cell types, nerve cells show the largest increment in surface area and volume during development. Examination of this extreme case may illuminate the general issue of surface-mediated ontogenetic reactions, and justifies the inclusion of this admittedly difficult and poorly studied topic in a symposium on cell surfaces in development.

An example of changes in surface area and volume during neuronal differentiation is given in Table 1 and Fig. 1. The cerebellar Purkinje cell is flattened almost to a two-dimensional shape that simplifies quantitative measurement of changes in surface area and volume during development. Like other neurons generated in the ventricular zone of the early embryonic central nervous system (Hinds and Ruffett, 1971), the Purkinje cell is an essentially round cell immediately after the final mitosis of its precursor. The subsequent growth of an elongate radial process and the outward transposition of the nuclear region have been documented for a few cell types but have not been shown specifically in the case of the Purkinje cell. In the rhesus monkey fetus, at around 70 days of gestation, cells that have taken the form of simple-appearing bipolar elements participating in the formation of the incipient cerebellar cortex are assumed to be Purkinje cells (Fig. 1). Their further dramatic remodeling (Fig. 1) involves an increase of more than eightfold in the area and seventeenfold in volume (Table 1). Between the 95th day of gestation and maturity, the surface of the soma becomes smooth and increases by an

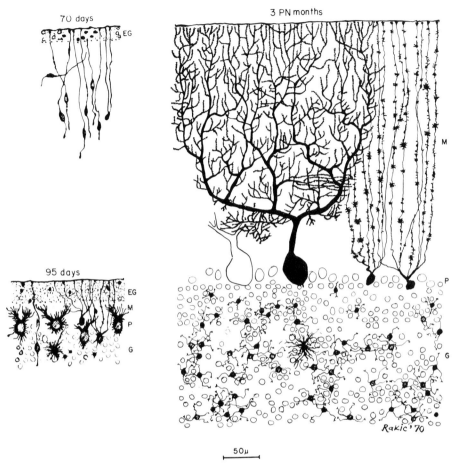

Figure 1. Purkinje cell development in the fetal monkey cerebellum. At 70 days of gestation, Purkinje cells have entered the cerebellar cortex but still maintain a bipolar shape. At 95 days the soma has enlarged, somatic spines project in all directions, and an apical dendritic tree is developing. At 3 postnatal months the cells show the adult configuration, with smooth-surfaced soma and an elaborate apical dendritic tree oriented in the plane transverse to the cerebellar folium. All figures are at the same magnification, indicated by the 50-μm bar at the bottom. (For further details, see Rakic, 1971.)

increment compatible with simple incorporation of the somatic spines into that smooth surface. In the same time period, the dendritic area increases more than fortyfold. The axonal area rises only three- to fourfold but accounts for most of the surface area of the adult cell, even without including collateral branches or terminal sprouts in the calculations.

Such measurements, crude as they are, have rarely been obtained for the development of other classes of neurons, most of which are even more difficult to trace because they have processes extending irregularly in all directions. During development it may be difficult even to visualize the full size of such cells by light microscopy, for only the Golgi procedure demonstrates the ramification of cellular processes. Apparently the impregnations by this staining method are rarely complete or the specimens are not examined at sufficiently high magnifications, if one takes

TABLE 1. QUANTITATIVE PARAMETERS OF SELECTED DEVELOPING AND MATURE NEURONS IN THE RHESUS MONKEY

Neuron	Surface Area (μm^2)	Volume (μM^3)	Approximate Surface/Volume
I. Cerebellar Purkinje neuron			
A. Postmitotic ventricular cell	**250**	**200**	**1.25**
B. Bipolar migrating cell, at 70 days of gestation	**32,000**	**16,000**	**2.0**
C. Partially modeled cell, at 95 days of gestation	**70,000**	**35,000**	**2.0**
1. Soma	800	2,000	0.4
2. Somatic spines	3,700	1,500	2.5
3. Developing dendrites	1,200	800	1.5
4. Axon	63,000	31,000	2.0
D. Mature cell	**279,000**	**285,000**	**1.0**
1. Soma	4,000	16,000	0.25
2. Primary and secondary dendritic branches	20,800	15,500	1.3
3. Tertiary and smaller dendritic branches	26,000	13,000	2.0
4. Dendritic spines	16,700	2,000	8.0
5. Axon	211,000	238,000	0.9
II. Cerebral cortical large pyramidal neuron			
A. Bipolar migrating cell, at 58 days of gestation	**267**	**201**	**1.3**
B. Mature cell	**270,000**	**257,000**	**1.0**
1. Soma	700	3,000	0.2
2. Dendrites	19,000	3,600	5.0
3. Axons	250,000	250,000	1.0

It must be emphasized that the above numbers are previously unpublished crude estimates calculated from drawings of Golgi images (unless otherwise stated) and based on samples of only a few cells. The numbers and related assumptions are as follows: IA: From Zwann and Hendrix (1973); ventricular cells in the rhesus monkey appear similar in size to the chick lens epithelial cells that they measured. IB: It is assumed that externally directed and internally directed processes reach the respective cerebellar surfaces. The cell may actually be smaller. IC: The axon length was estimated to be about 1 cm. The exact value depends on the position of the particular Purkinje cell soma in the cerebellar cortex and the site of termination of its axon. ID: The total length of tertiary dendrites was measured from Golgi images, and the dendritic spine density and average surface area were taken from Fox and Barnard (1957). Axon diameter was estimated to average 4.5 μm, and length to average 1.5 cm; no collateral branches were included in the calculation. IIA: This is cell C from Rakic et al. (1974). IIB: Data on soma volume are for human pyramidal cells (Table 209, p. 391, in Blinkov and Glezer, 1968). The surface area was calculated as that of a right angle cone of the same volume. The axon diameter was estimated at 4.0 μm and length at 2.0 cm.

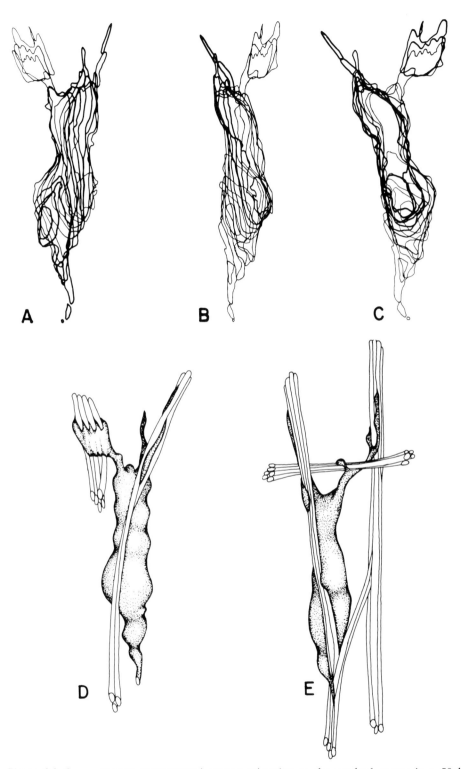

Figure 2A–C, a computer-reconstructed neuron migrating to the cerebral cortex in a 58-day monkey fetus. The computer programmed rotation angles are 0° (Fig. 2A), 135° (2B), and 180° (2C). Figure 2D is a rendition of the same cell, with the addition of segments of the contiguous radial guides. Figure 2E shows another migrating neuron from the same specimen. (From Rakic et al., 1974.)

as a standard the excellent illustrations of Morest (1968, 1969), which show extra-ordinarily extensive and delicate fingerlike filopodial extensions from the somas and growing dendrites, and sheetlike foliopodial extensions from growing axons of immature neurons. Reliable reconstruction of a given cell can be made by tracing its profile in serial electron micrographs (e.g., Hinds and Hinds, 1972; Skoff and Hamburger, 1974) and especially by feeding the data to an appropriate computer for visual display or quantitative analysis (e.g., Rakic et al., 1974). Two such cells migrating to the cerebral isocortex in a monkey fetus are illustrated in Fig. 2 and the quantitative parameters of one of them as determined by computer analysis are summarized in Table 1. Table 1 also includes estimated measurements of such a neuron at adult stages. Though the data are sketchy, it is clear that the develop-mental changes are as dramatic as in the case of the cerebellar Purkinje neuron.

Inferred Interactions between Contiguous Cells

The major generalization to emerge from the "H³-thymidine autoradiographic era" of vertebrate neuroembryology ushered in by the independent publications of Sauer and Walker (1959) and Sidman et al. (1959) was that cells generally arise in different sites from where they will reside in the mature central nervous system. The movement of a cell from one position to another typically involves active mi-gration (Sidman and Rakic, 1973). With the more recent advent of electron micro-scopic study of well-fixed immature specimens, another generalization has become clear: young neurons migrating to cortices are closely apposed to radially oriented cell processes that appear to establish the vector of migration by serving as contact guides (Rakic, 1971, 1972a). These events merit closer examination because of the implication that, in the nervous system, differentiating cells which are in close contact may interact with critical consequences for further development, just as contiguous cells are thought to influence one another in other embryonic rudiments. Let us examine some recently described examples of this in the mammalian central nervous system.

Migrating Young Neurons and Radial Fiber Guides in the Cerebral Cortex

At early stages of development the central nervous system is a tubular structure with the walls composed of a pseudostratified columnar epithelium in which cells round up and divide asynchronously at the inner surface (Boulder Committee, 1970). Any particular postmitotic cell about to elongate is in direct contact with several columnar cells which constrain its growth to the radial direction. At later stages some of these undifferentiated columnar epithelial cells are replaced by im-mature glial cells or radially directed processes of other cells that also span almost the full thickness of the walls of the central nervous system. These in turn may constrain or guide migrating neurons. The first example of such guidance to be recognized was a particularly vivid one—the guidance of future cortical neurons in the cerebrum of the fetal rhesus monkey across a distance of several thousand micrometers to the cortical plate (Rakic, 1972a). Careful three-dimensional re-construction of such bipolar-shaped cells shows that their leading processes, and usually their trailing ones as well, extend only a few hundred micrometers from the nuclear part of the cell (Rakic et al., 1974), so that these cells are actually migrating and are not merely translocating their nuclear regions within a relatively fixed cylinder of cytoplasm, as had been suggested by Berry and Rodgers (1965) and Morest (1970).

The contact relationships between migrating neuron and radial process suggest some kind of cell surface-mediated action or interaction. The leading processes of neurons migrating to the cerebral cortex display complex and very variable forms, with several terminal branches each of which may contact a radial fiber (Fig. 2). It was concluded (Rakic et al., 1974) that these branches are being actively extended and withdrawn as the young neuron migrates, like the branches at the tip of an elongating axon observed directly at multiple time points in tissue culture (e.g., Bunge, 1973). Otherwise a cell such as the one reconstructed in Fig. 2E could not advance beyond the potentially obstructing fiber that lies in its path and occupies the crotch between its two main leading processes. If, then, the leading processes of migrating neurons are "feeling their way," what specifies that they will grow along radial guides and not along processes of any other class, such as axons, dendrites, or blood vessels that cross their path? No answer is available, but whether the signals are fixed on or near cell surfaces, or diffuse as a gradient in the narrow extracellular space, it is difficult to escape the conclusion that the specificity is mediated at least in part by the surfaces of the migrating cells, the guiding cells, or both.

Acquisition of Cortical Addresses

The autoradiographic method shows that neurons take positions in the mammalian cerebral cortex that are statistically predictable and, in general, "inside out" with reference to their time of origin (reviewed in Sidman, 1970, 1974). The pattern is particularly clear in the rhesus monkey visual cortex, where the fetal time span of cell genesis is prolonged relative to that in most previously studied animals (Rakic, 1974). In all species studied it appears that the earliest-generated cells take up positions in the cortical plate just internal to the cell-sparse marginal zone. Most of the cells in each subsequent wave bypass the earlier-arrived cell bodies and likewise take positions immediately internal to the marginal zone, so that the earlier cells come to lie in progressively deeper cortical strata.

What cues provide the addressing information for cortical neurons? No answer is available, but a few possibilities can be considered. One is that the information is already built into the young neuron at the time that it is generated near the ventricular surface or soon thereafter. Another is that radial fibers influence not only the path of migration but also the final stopping point in the cortical plate. While these possibilities cannot be ruled out, neither would seem to have the requisite precision for a mechanism that operates in essentially the same way in species with widely varying time spans of development and migration path lengths. It seems intuitively more probable that cell surface-mediated reactions at or near the cortical plate itself would control the final positions of neuron somas in the cortex. The critical interaction might occur between newly arriving neurons and their predecessors or between migrating neurons and incoming axons extended from cell bodies that reside elsewhere in the nervous system.

The latter possibility becomes particularly intriguing with the realization that afferents are already present, either along the migration pathway or in the incipient cortex itself, virtually as early as a cortical plate can be recognized (Poliakov, 1961; Morest, 1970; Marin-Padilla, 1971; Hinds, 1972). In cat, dog and man, incoming fibers have made specialized contacts which display at least some of the morphological features of synapses. The contacts are found on the processes of cortical neurons immediately external and internal to the cortical plate by the time that the first few rows of neuron somas have constituted a recognizable plate (Cragg, 1972; Molliver and Van der Loos, 1972; Molliver et al., 1973). It remains to be explored whether these remarkably early contacts influence either the addresses or other properties of differentiating neurons.

Cerebellar Granule Cell: Surface Mosaicism at the Translocation Stage

The granule cell neuron of the mouse cerebellar cortex is first recognized after final mitosis as a relatively round cell, with protrusions of somatic filopodia; it lies in the deep strata of the external granular layer. Within hours of its origin, the cell takes on a bipolar shape, its processes extending horizontally (parallel to the external surface of the cerebellum) in the transverse plane (longitudinally within each cerebellar folium). The cell then develops a radially oriented, smoothly tapered cytoplasmic process that pierces into the developing molecular layer. Remarkably, the nuclear portion of the cell then becomes transposed inward inside this elongating cytoplasmic cylinder (Ramon y Cajal, 1960; Mugnaini and Forstrønen, 1967; Rakic, 1971). Within 48–72 hours of the time when the cell is generated, its somatic portion clears the full width of the molecular layer and enters the developing granular layer deep to the row of Purkinje cell somas. The granule cell soma then extends another transient set of filopodia and, finally, a definitive set of four or five short dendrites. The original bipolar horizontal processes and the vertical process that are trailed out behind the transposing soma meanwhile acquire the cytoplasmic features of an axon.

The striking differences in the behavior of the growing horizontal and vertical processes of the young granule cell imply that there may be an underlying chemical difference between them. The horizontal, bipolar processes begin to grow first, and from their onset are oriented in the longitudinal axis of the folium, at right angles to the Bergmann glial fibers and the elongating Purkinje cell dendrites. Virtually nothing is known about the factors determining this consistent orientation of most later-forming components of the cerebellar cortex. Whatever its basis may be, the growing tip of the vertical process must be different, for it extends preferentially along a Bergmann fiber. The difference could be determined even before the rounded granule cell emits any of its processes, when one region of its surface may already be specialized through contact with a Bergmann fiber, or it may become established at any later time. Another unsolved question is whether, by the later stage when the horizontal and vertical processes have become segments of a simple differentiated axon, the imputed differences still persist.

Interneurons of the Cerebellar Molecular Layer

A particularly illuminating example of developmental control through interaction between contiguous cells is represented by the basket and stellate cell interneurons of the molecular layer (Rakic, 1972b, 1973). These cells are first detected when they lie as rounded postmitotic cells at the interface between the external granular layer and the molecular layer of the immature cerebellar cortex. They, like granule cell neurons, are thought to arise from precursors in the external granular layer, but from a different clone or clones. The best evidence comes from a kinetic analysis showing that granule cells in the fetal monkey cerebellum are generated approximately at a logarithmically increasing rate, while interneurons of the molecular layer are produced over a somewhat shorter total time span at a fairly steady rate of about 300,000 cells per day (Rakic, 1973). Additional suggestive evidence comes from the observations that granule cell neurons die postnatally, while the interneurons survive and differentiate in the mutant mice weaver (Rakic and Sidman, 1973a–c) and staggerer (Sidman, 1972; Landis and Sidman, unpublished observations), as described in a later section.

The point of particular interest in the context of the present discussion is that granule cells and interneurons appear to respond differently to the same local environment. At the very time when granule cells are migrating in relation to Bergmann glial fibers and are forming their unusual T-shaped axons in the molecular layer, the interneurons are generating both axonal and dendritic arbors in the molecular layer. The interneuron dendrites appear to acquire their three-dimensional orientation and total mass on the basis of cues from the matrix of horizontal segments of granule cell axons (parallel fibers) in which they are embedded immediately after they take positions at the interface between external granular and molecular layers (see Fig. 15 in Rakic, 1973). The orientation of dendrites, down to small tertiary twigs, is predominantly in the sagittal plane, as though their growth is obeying the rule to maximize the number of parallel fibers contacted per unit length of interneuron dendrite (Rakic, 1973). The additional orientation of these dendrites in the radial direction could be determined by the degree of "synaptic saturation" of the parallel fibers. Thus, dendrites of basket cells, the earliest interneurons to be generated, grow outward through the full width of the molecular layer as it becomes progressively thicker through accretion of new layers of parallel fibers; the volume occupied by the dendritic tree of each interneuron has the shape of a hemiellipsoid, flattened in the sagittal plane, with its dome directed externally. Stellate cell interneurons, generated later, send dendritic branches inward among the "partially saturated," earlier-generated parallel fibers and also outward as new "naive" parallel fibers are generated subsequently; these interneurons are smaller, and their dendritic branches project both inward and, even more extensively, outward. The last-generated (and most superficially positioned) interneurons are the smallest, and their dendrites pass inward along the underlying, partially saturated parallel fibers to occupy a volume shaped as a flattened but inverted hemiellipsoid. The progression in the times of interneuron genesis is matched by a gradient of near mathematical precision in cell size and dendritic mass (Rakic, 1972b).

Another parameter of interneuron development is the latency between the time of final cell division, as determined by H^3-thymidine autoradiography, and the onset of dendrite formation, as determined from Golgi preparations and from the changing position of the cell soma relative to the layer of Purkinje somas (Rakic, 1973). The earliest interneurons begin to form dendrites soon after cell genesis, and thereby become fixed permanently in position in the matrix of parallel fibers. Interneurons that are generated progressively later show progressively longer latencies before they begin to form dendrites. During the latency period the interneuron changes its position by "floating" on the thickening bed of parallel fibers, and becomes permanently set only when dendrite formation is initiated. In the monkey fetus the last interneurons show a latency of more than 2 months between the time of cell genesis and the onset of overt differentiation. The stimulus for dendrite formation appears to derive from the local milieu, either from parallel fibers (whose information becomes progressively more out of phase with interneuron formation because of the differences in the kinetics of cell genesis between granule cells and interneurons of the molecular layer) or from other axonal inputs that must pierce through the thickening molecular layer from below.

Transynaptic Interactions during Development

From the above examples we have seen that the radial orientation of future cortical neurons is maintained during the phases of cell proliferation and migration by guidance systems based on cell contact relationships. It also appears probable that the often extraordinary increases in cell surface areas, the final positions of cells, and the orientations of their processes may be influenced by additional short-range surface-mediated interactions. We shall now consider developmental controls exerted later in development via intercellular contacts at highly restricted surface sites, the sites of synapse formation.

While the exact time of the initial contact between an afferent axon and its target neuron is difficult to establish (see "Acquisition of Cortical Addresses," p. 227), all the available evidence indicates that synapses become recognizable morphologically only after the target neurons are in approximately their final positions. There is a further distinction to be drawn between the time when the first synapses form in a given region and the time when synaptogenesis nears completion. In quantitative terms synaptogenesis proceeds curiously late, even in some systems of neurons that are generated quite early in development. Postganglionic sympathetic neurons, for example, are usually considered to be relatively early-generated derivatives of the neural crest (Yntema and Hammond, 1947), and yet the number of synapses in the superior cervical sympathetic ganglion in rodents is very small at birth and rises several hundredfold in the next 2 weeks (Black et al., 1971). In mice, and presumably also in rats, neurons in virtually all parts of the visual system (with the exception of most photoreceptor cells, many bipolar cells, and some ganglion cells in the retina) are generated prenatally (Sidman, 1961; Angevine and Sidman, 1961; DeLong and Sidman, 1962; Angevine, 1970), but almost all synapses, from the level of the retinal external plexiform layer to the visual cortex, form postnatally (Olney, 1968; Lund and Lund, 1971; Cragg, 1967, 1969, 1972). There need be no simple relationship between the times of cell genesis and of synapse formation. In the mouse retina, for example, synapses form later between the early-generated amacrine and ganglion cells than between the late-arising photoreceptor and bipolar cells (Olney, 1968). In the human fetal retina, synapse formation begins relatively early (Spira and Hollenberg, 1973).

Descriptions of the morphology of developing synapses (Bloom, 1972), like the available data on timing, give little insight into causal relationships. One would like to know whether the cell surface destined to be postsynaptic is already focally specialized before the arrival of its inputs, whether it is already making specialized products but comes to distribute them differentially at the cell surface in response to inputs, or whether differentiation, in the sense of initiation of synthesis of special products, is provoked by the inputs. Apart from a rapidly expanding body of data on the distribution of acetylcholinesterase and acetylcholine receptor on the surface of the skeletal muscle cell in relation to innervation and denervation (e.g., Miledi and Potter, 1971; Fischbach and Cohen, 1973; Hartzell and Fambrough, 1973; Shimada and Fischman, 1973), this issue has not been accessible for direct study. The fact that Purkinje neurons of the cerebellum are responsive to iontophoretically applied neurotransmitters even before noradrenergic synapses have become recognizable suggests that receptor must be present in advance of synapse formation (Woodward et al., 1971). Also there is a fast-growing body of evidence that functional innervation regulates ontogeny at sympathetic ganglion synapses and may have a selective effect on the neurohumor-synthesizing enzymes (Thoenen, 1972, 1974; Hendry, 1973).

Whatever spatial and temporal factors bring potential pre-and postsynaptic surfaces into apposition, there must be ensuing series of interactions beween them. It may be useful to think of the series of steps involved in maturation of the synapse and the sequences of functional synaptic activities as a continuum, a simple set of interdigitated events. With tongue in cheek, we may say that, from the conventional physiologist's point of view, development constitutes a mere prelude to the important issue, which is the functioning of the mature organ, while from the vantage point of some developmental biologists, the machinery of the adult organ is an inevitable but less interesting product of the eventful formative phases. For the nervous system, and perhaps for some other organs also, it might be profitable to consider the hypothesis that development continues without time limit, both producing function and being dependent on it. With reference to synapses, this hypothesis implies a dynamic equilibrium rather than a rigid relationship between presynaptic and postsynaptic elements. It also leads to the view that Ramon y Cajal's powerful tenet concerning functional polarization of the neuron, which is generally accepted as part of the neuron doctrine (reviewed in Ramon y Cajal, 1937, and Peters et al., 1970), should not blind us to the likelihood of a two-way rather than a one-way stream of information transfer across the synaptic cleft.

A given postsynaptic site appears to be incompletely specified, that is, it may rearrange the precise terminal distribution of a given input, and may accept a new and different input if it loses the original one. Dramatic rearrangements have been documented in the case of skeletal muscle, which is probably contacted transiently by axonal branches at a number of surface sites before end plates begin to mature in a more restricted distribution (Redfern, 1970). As another example, the cerebellar Purkinje cell receives climbing fiber axonal contacts initially on the cell soma and the tips of transient somatic spines, but then the inputs become redistributed when collateral axonal branches come into contact with the growing dendritic shafts of the Purkinje cell; the somatic spines are withdrawn or incorporated into the enlarging Purkinje cell surface, and newly arriving basket cell axons take over as inputs to the Purkinje cell soma (Kornguth and Scott, 1972). The interesting further suggestion has been made that axodendritic synapses in some regions may form initially on filopodia of growing dendrites and then be repositioned as the surface membrane becomes incorporated into more proximal parts of the dendrite during further growth (Skoff and Hamburger, 1974). Finally, when certain axons are removed in the adult nervous system, residual intact axons may sprout pre-terminal collateral branches (Raisman, 1969; Katzman et al., 1971) that appear to take over the vacated synaptic sites (Edds, 1953; Lund and Lund, 1971; Raisman and Field, 1973; Lynch et al., 1972; Bernstein and Bernstein, 1973). The mechanisms behind these phenomena are not likely to be understood until the distribution and turnover of specific cell surface components can be plotted. However, it is already clear that synaptic organization is more modifiable than was at one time recognized.

These observations can be encompassed by an *active competition hypothesis of synaptogenesis*: there is a hierarchy of effectiveness in the recognition of apposed elements at a potential synapse, with properties such that (1) formation of a synapse is more probable than no synapse, provided that the apposed elements reach a recognition threshold, (2) elements with greater recognition (or "fit") for one another will attain synapsis at the expense of elements with less effective recognition, and (3) the matching may change wih time, as new elements enter the competition or as surface-mediated properties change.

This hypothesis accounts for the known phenomenology of synaptic rearrangement, though admittedly it has not been tested very systematically. There seem to be sharp differences between the recognitions of motor, sensory, autonomic, and various central axons for their respective target cells in a wide range of situations, including growth and regeneration of peripheral nerves (e.g., Weiss and Edds, 1945; Guth, 1958; Sidman and Singer, 1960), innervation of sympathetic ganglia *in vitro* (Olson and Bunge, 1973), innervation of host iris by various organs grafted into the anterior chamber of the eye (Olson and Seiger, 1972a), grafts of peripheral tissues into the brainstem, to be innervated by selected monoaminergic axons of the host (Björklund and Stenevi, 1971), or grafts of various combinations of neural and peripheral tissues into a neutral setting in the tadpole tail (Weiss, 1950). It is premature to speculate whether this range of specificities is derived, like immunological diversity, from a single class of cell surface molecules (the nervous system's equivalent of the immunoglobulins) or from a different class for each major neural recognition system.

A test of the duration of time that specificity may be retained was devised by Guth and Bernstein (1961). They confirmed earlier evidence that nerve fibers in thoracic sympathetic rami T1-3 innervate postganglionic neurons of the superior cervical ganglion, which in turn innervate the pupillodilator muscle of the iris. Fibers of rami T4-7 innervate predominantly neurons causing vasoconstriction of the ear. After section of the mixed cervical sympathetic trunk, fibers regenerated and reinnervated the ganglion, with retention of the original specificities as tested by stimulation of individual thoracic rami. Next, when rami T1-3 were crushed and rami T4-7 stimulated 1 month later, the pupil dilated. This was thought to reflect collateral sprouting of T4-7 axon terminals so as to innervate postganglionic neurons projecting to the iris. However, as 6 months elapsed, the original functional specificities were again established. This intriguing result was interpreted to indicate that the crushed fibers had regenerated and had again taken over their original synaptic sites on the postganglionic neurons. This specificity implies that the receptors on the target neurons persist essentially unaltered qualitatively for 6 months, and that the original axons could still compete effectively for their proper sites.

The active competition hypothesis is compatible with the view expressed earlier that both the presynaptic and the postsynaptic elements may be modified continuously by functional activity at synapses. This takes most extreme form when the total synaptic innervation is prevented or is removed, in which case either the postsynaptic or the presynaptic cell may fail to develop fully or may degenerate (reviewed in Cowan, 1970; Jacobson, 1970). At a more subtle and positive level, the total functional activity of the postsynaptic cell may influence the form and metabolic status of the presynaptic ones. Retrograde effects have been widely recognized during relatively early stages of development of the nervous system (e.g., Harrison, 1935; Detwiler, 1936; Hamburger and Levi-Montalcini, 1949; Weiss, 1955; Levi-Montalcini, 1964; Giacobini et al., 1973), and a theoretical model applicable to mature, functioning synaptic systems has been presented by Changeux et al. (1973). A possible cytological and molecular basis for such retrograde influences at the synapse is provided by the evidence for endocytosis by axon terminals of the peripheral (Ceccarelli et al., 1973; Heuser and Reese, (1973) and central nervous systems (Turner and Harris, 1973) and for retrograde intra-axonal transport of protein molecules all the way from axon terminal to parent cell body (Kristensson, 1970; Kristensson et al., 1971; Kristensson and Olsson, 1971; LaVail and LaVail, 1972; LaVail et al., 1973).

EXPERIMENTAL ANALYSIS WITH MUTANT MICE

Mutations serve as experiments of nature that give us the prospect of dissecting the development of a complex organ such as the nervous system into a series of causally related events. No other means of experimental intervention is as likely, in general, to perturb development in such precise and potentially definable ways.

Among vertebrates, the mouse is the animal of choice. One third to one fourth of all known mutations in this species, about 120 of approximately 400 identified loci, produce major behavioral disturbances. These genetic loci are distributed on almost every chromosome and show relatively little tendency to cluster. Even if one allows for strong biases in sampling, since certain classes of behavioral mutations are relatively easy to recognize and maintain, it appears that a major fraction of the genome affects behavior and, directly or indirectly, the nervous system. However, large as it may be, the number is finite, and it is logically necessary that genes affect general processes rather than influencing directly the form and precise connections of each class of cells.

Study of a small number of mutations affecting one region, the cerebellar cortex, has given some insights into the importance of cell interactions, presumably surface mediated, in the development of the nervous system. Four autosomal recessive mutations (Table 2) will be described individually and then will be considered collectively in relation to some general aspects of cerebellar development.

Four Cerebellar Mutants: The Problem of Primary Gene Action

Neurological mutants are recognized by their abnormal behavior, and this phenotypic expression is usually far removed from primary gene action. Thus far, no neurological mutation in mammals has been traced back through earlier phenotypic

TABLE 2

Gene	Symbol	Chromosome	Genetic Background[a]
Weaver	*wv*	–	C57BL/6J*
			B6CBAF1[+b]
Staggerer	*sg*	9	C57BL
Reeler	*rl*	5	C57BL/6J
			C3H/HeJ
			B6C3F1[+b]
Nervous	*nr*	8	BALB/cGr*
			C3H/HeJ

[a] Asterisk (*) indicates that the mutation occurred intially in this strain and is therefore coisogenic with it.

[b] B6C3F1 is the F1 hybrid from a cross between the C57BL/6J and C3H/HeJ strains, and B6CBAF1 is the F1 hybrid from a cross between the C57BL/6J and CBA/J strains.

expressions to a primary chemical abnormality, but some interesting cell interactions have been uncovered along the way.

Weaver, wv

This mutation occurred in the inbred C57BL/6J strain (Lane, 1964). It is maintained in the original strain in our laboratory and as a hardier F_1 hybrid between C57BL/6J and CBA/J at The Jackson Laboratory. The *wv/wv* homozygote is small, hypotonic, and ataxic. Its cerebellum is small because most granule cells die while they are still residents of the external granular layer (Sidman, 1968). Other parts of the nervous system and other organs appear normal. The +/*wv* heterozygote appears normal behaviorally, but Rezai and Yoon (1972) showed that its granule cells migrate at a slower than normal rate across the molecular layer. These authors suggested that granule cell death in *wv/wv* mice is secondary to a failure of migration. This migration failure in turn was related to an earlier abnormality of Bergmann glial fibers, as demonstrated in Golgi preparations and electron micrographs (Rakic and Sidman, 1973a, b).

This Bergmann glial abnormality is of crucial interest with reference to the earlier-suggested role of radially oriented fibers in the surface guidance of migrating neurons (see p. 226). In +/*wv* or *wv/wv* mice, no abnormalities have been recognized before the 3rd postnatal day. At this time the first abnormalities of Bergmann glial cells are seen in heterozygotes, though the changes do not become frequent and obvious until a few days later (Rakic and Sidman, 1973b). The profiles of Bergmann fibers are irregular in contour and enlarged 2–4 times in diameter. Their cytoplasm is abnormally electron lucent and partially vacuolated, microtubules appear broken and irregular, and mitochondria are enlarged. During the migration period, many granule cell somas in the molecular layer are round, whereas those in the process of translocation to the granular layer are elongated in parallel with Bergmann fibers. The migration failure is proportional to the glial abnormality at different levels of the cerebellum, as well as in heterozygotes compared to weaver homozygotes.

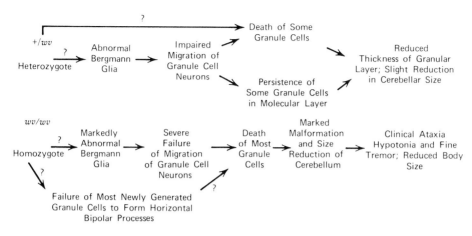

Figure 3. Pedigree of causes in the weaver mutant phenotype.

From these and related observations, a chain of causal events can be constructed (Fig. 3). The question marks on the arrows in Fig. 3 emphasize that the primary event, and even the primary target cell of the *wv* genetic locus, remain unknown. As reviewed in the section entitled "Cerebellar Granule Cell: Surface Mosaicism at the Translocation Stage," newly generated granule cell neurons normally form bipolar processes oriented parallel to the external surface of the cerebellum in the longitudinal axis of the folium, but most cells in *wv/wv* mice fail to do so. It is not clear whether this effect is independent of the Bergmann glial defect, is secondary to it, or is its cause. Another point of some importance has been made by Bignami and Dahl (1974) on the basis of selective immunofluorescent staining of the glial cells. They found the number of Bergmann fibers in *wv/wv* mice to be normal or close to normal, while confirming their abnormal configuration. The reduced average rate of migration measured by H³-thymidine autoradiography (Rezai and Yoon, 1972; Rakic and Sidman, 1973a), then, may relate not so much to a diminished number of Bergmann cells as to some abnormality of their surfaces or to the failure of granule cells to respond appropriately to Bergmann fiber surfaces.

Staggerer, sg

This mutation has been placed, along with marker genes (dilute and short-ear) onto the C57BL background (Sidman et al., 1965), for sharper comparison with weaver and reeler. Like these other mutations, it produces a syndrome featuring hypotonia and ataxia and displays a small cerebellum deficient in granule cell neurons (Sidman et al., 1962). The rate of cell proliferation in the external granular layer is reduced (Yoon, 1972). Many granule cells, however, are generated and migrate effectively to the granular layer. During and especially after the migration phase, these granule cells die, and almost all have disappeared by the middle of the 5th week after birth. To complicate the picture, it has been recognized that even before the stage of granule cell death the Purkinje cells appear to have small somas and stunted dendritic arbors, with virtually no dendritic spines of the type normally contacted synaptically by granule cell axons (Sidman, 1968, 1972). Granule cell axons do form some synapses on Purkinje dendritic shafts, but these are transient and neither cell retains any morphological sign of the former special contact (Landis and Sidman, unpublished observations).

These data, like those for weaver, suggest that here also the cell which dies may not be the direct target of the mutant genetic locus. The Purkinje cell disorder appears to antedate the granule cell degeneration, and the possibility has been posed elsewhere (Sidman, 1972) that basically normal young granule cells may degenerate if their axons fail to establish effective contact with Purkinje cell dendrites.

Reeler, rl

The reeler behavioral phenotype is similar to the weaver and staggerer phenotypes. The organization of the rl/rl brain is virtually identical whether the disorder is expressed on the C57BL/6J, C3H/HeJ, or B6C3F$_1$ hybrid background though on the hybrid background the mouse is generally larger, lives much longer, and shows a milder behavioral disorder (Caviness et al., 1972).

A difference between reeler and the two mutations just considered is that the reeler locus affects neurons in most parts of the cerebral cortex, as well as in the cerebellar cortex (Meier and Hoag, 1962; Hamburgh, 1963). The idea that the widespread defects might all result from a common developmental disturbance in a basic cell-cell recognition mechanism was expressed by Sidman (1968) and has been refined further since then. The initial impression that cortical neurons were random in position in mature reeler mice gave way during an extensive analysis of cortical architectonics (Caviness and Sidman, 1972, 1973a) to the view that, as neurons approach the developing cortical plate, they come to rest at systematically and reproducibly abnormal positions (Caviness and Sidman, 1973b; Caviness, 1973b). The mutant allele is responsible for the formation of an orderly cortex, but one with a markedly different order from normal. Cortical regions are undisturbed in their topological relations to one another; each contains the normal classes of cells, and cells of each class appear to arise and migrate on normal schedules along the normal radial vectors. In relatively simple cortical regions (except for the olfactory bulb, which is normal), the polymorphic and pyramidal layers are more or less inverted (Caviness and Sidman, 1972). In more complex granular isocortical areas the polymorphic layer lies externally, in the position normally occupied by the molecular layer. In addition, a broad pyramidal zone extends inward to the cerebral white matter, and granule cells are segregated within the outer part of the pyramidal field, close to their normal depth from the external surface (Caviness and Sidman, 1973). To express the abnormality in the most general terms, it is a laminar malposition of neuron somas relative to one another and to the molecular layer in all cerebral and cerebellar cortical regions which feature at least one class of uniformly polarized neurons receiving heterogeneous inputs.

The aberrant neuronal positions are already evident at embryonic day 15 and earlier (Sidman, 1968; Sidman, Caviness, and So, unpublished observations), but the particular cell interactions influenced by the reeler locus remain to be elucidated, and the molecular mechanism is even more hidden. It now seems likely that the abnormalities expressed in aggregation tissue cultures of 17-day reeler embryo isocortex (DeLong and Sidman, 1970) represent some secondary consequence of action of the reeler allele, though it remains possible that an earlier-expressed abnormal mechanism is still present in isocortex at late embryonic stages.

Nervous, nr

This is the most recently recognized of the four mutations under consideration (Sidman and Green, 1970). Affected mice on the original Balb/cGr background are hyperactive at 2–3 weeks of age and then show a moderate ataxia, which stabilizes before they reach 2 months. The cell type bearing the brunt of the affection is the Purkinje neuron of the cerebellar cortex. These cells die progressively during the period from about 18 to 50 days after birth. Many granule cell neurons and interneurons also degenerate in the lateral parts of the cerebellar cortex, but appear to be relatively normal more medially. Degenerating neurons have not been recognized elsewhere in the central nervous system.

The earliest abnormality detected to date is a striking transformation of Purkinje cell mitochondria from an elongate serpentine form to a round shape, beginning at about postnatal day 9 (Landis, 1973a). By day 15 all Purkinje cells display this abnormality; subsequently most of them die, but the others gradually recover. No derangement of synaptic inputs or other evidence of abnormality exogenous to the cell type destined to degenerate has been recognized, in contrast to weaver, staggerer, and, probably, reeler. The changes in position of some granule cells and the death of some others are thought to be secondary to the Purkinje cell disorder (Landis, 1973b). In addition to Purkinje cells, neurons of some other classes develop the abnormal rounded shape of mitochondria, but these cells do not die (Landis, 1973c). It remains to be established how closely the abnormal mitochondrial phenotype reflects the primary action of the *nr* genetic locus.

Cerebellar Abnormalities Based on Disorders of Cell Interaction

Although we do not know the primary chemical expressions of these mutations, we can use the abnormal phenotypes nonetheless to test hypotheses about cell interactions. It is particularly advantageous to be able to make comparisons among several mutants. A set of developmental features is summarized in Table 3 and will be analyzed in this section.

Cell Proliferation in the External Granular Layer

One of the most remarkable general phenomena of ontogeny is the proportionate development of bodily parts, a feature so successfully handled by the organism almost every time that it is usually taken for granted. A modest example that is relatively accessible for study is the proliferation and maintenance of external granule cells in numbers appropriate to the volume or surface area of a given cerebellum. Even when a large percentage of cells in the external granular layer is decimated by X-irradiation (Altman et al., 1969) or by the mitolytic drug methazoxymethanol (Shimada and Langman, 1970), the residual cells proliferate excessively and may restore the total complement.

Purkinje cells, Bergmann glial fibers, and afferent axons constitute most of the mass of the developing molecular layer, and from the data summarized in Table 3, line 1, it appears likely that, of these, the Purkinje cell may be the dominant exogenous influence on cell proliferation in the external granular layer. In weaver and nervous mutant mice, the Purkinje cells appear to be basically normal during the period of external granule cell proliferation. By contrast, they are abnormal through-

TABLE 3. SELECTED DEVELOPMENTAL FEATURES[a] IN FOUR CEREBELLAR MUTANTS

Feature	wv	sg	rl	nr	Comment
1. Cell proliferation in external granular layer	OK	↓	↓	OK	Probably influenced by Purkinje cells
2. Fissure formation	OK	↓	↓↓	OK	Influenced by ratio of volume of outermost and deeper cortical zones
3. Synaptogenesis	OK	(↓)	OK	OK	Except for absence of parallel fiber: Purkinje dendritic spine synapse in sg, these genetic loci do not affect synaptogenesis
4. Purkinje cell basic shape and polarity	OK	OK	OK	OK	Properties are intrinsically controlled by other genetic loci or depend on prenatal exogenous factors
5. Purkinje cell dendritic branchlets	↓	↓↓	↓↓	↓	Dependent on parallel fibers and other factors
6. Purkinje cell dendritic branchlet spines	OK	↓	OK	OK	Independent of parallel fibers
7. Granule cell survival	↓↓↓	↓↓↓	↓↓	↓	Influenced by early contact with Purkinje cell dendritic spines
8. Interneuron dendrites	↓↓	OK	↓	OK	Dependent on local milieu of parallel fibers and other inputs

[a] The arrows indicate that the feature in question is diminished. Comparison of numbers of arrows should be made horizontally only; that is, the four mutants are being compared for a given feature, but no comparison is intended between features.

out the postnatal period and probably earlier in staggerer, and the external granular layer in that mutant is already reduced soon after birth (Yoon, 1972; Landis and Sidman, unpublished observations). In reeler, Purkinje cells are basically normal, but a majority of them lie so deep in the cerebellum that it is doubtful that they effectively influence the external granular layer. The reeler cerebellar cortex strengthens the argument further by presenting internal controls, for in areas where Purkinje cells lie in relatively normal positions the thickness of the external granular layer is relatively normal (see Fig. 3 in Sidman, 1972).

It is doubtful that the Bergmann glial fibers exert a unique influence on cell proliferation, though in principle they would seem to be better candidates than Purkinje cells because they actually penetrate the external granular layer, whereas Purkinje cell processes normally do not and would have to act at more of a distance. The reasons for tentatively excluding Bergmann fibers are that in sg they appear normal even though external granule cell proliferation is reduced, and in wv many Bergmann fibers are qualitatively abnormal but external granule cell proliferation is normal.

A possible role of afferents is not tested with the present mutants, but surgical undercutting of the neonatal cerebellar cortex was reported not to influence the external granular layer (Hámori, 1969).

Fissure Formation

The cellular basis for the gross morphogenetic events of fissure and gyrus formation has not been discovered. The common idea that a cortical surface folds when the volume of neural tissue increases out of proportion to the volume of the cranial vault cannot be sustained when one considers that the weaver homozygote develops essentially normal, though miniaturized, fissures and folia despite the fact that its small cerebellum lies in a relatively normal cranium. Another possibility is that a variable proliferation rate along the cerebellar surface might lead to local buckling (Mares and Lodin, 1970), but even if such a nonuniform cell generation pattern were found to be more pronounced in *wv* and *nr* than in *rl* or even *sg* (Table 3, line 2), it would be difficult to distinguish cause from effect.

On the basis of studies on the developing human cerebral cortex, Richman et al. (1973) have suggested that differential volumes of superficial and deep cellular laminae may influence the formation of convolutions, a high ratio of superficial to deep volumes correlating with the formation of excessive small gyri and a low ratio with too few gyri. Their idea is supported by data on the mutant cerebella. The low ratio of superficial to deep volumes caused by the abnormally deep position of Purkinje cells in reeler is associated with a near absence of fissures (Fig. 16 in Sidman, 1968). Similarly, a moderately low ratio, reflecting the reduced external granular layer in staggerer (Yoon, 1972), correlates with the abnormally shallow fissures and their diminished number in staggerer. In weaver the total volume is much reduced, but the proportions are normal (Rakic and Sidman, 1973b), while in nervous both the total volume and the ratio of superficial to deep zones are normal during the period of external granule cell proliferation.

Synaptogenesis

One of the most surprising findings in the study of these mutants by electron microscopy has been the relative rarity of nonspecific synaptic contacts. Neither the selective loss of a cell type nor the malpositioning of a cell leads to novel types of synaptic relationships. In heterozygous weaver mice, for example, mossy fiber afferents were identified in the early postnatal period in their expected position below or at the level of Purkinje cell somas. Subsequently, many mossy terminals were found in the molecular layer, where they formed synaptic contacts with dendrites of heterotopic granule cells (Rakic and Sidman, 1973c). Similar observations were made in *nr/nr* mice (Landis, 1973b). In the reeler cerebellum, all classes of normal synaptic inputs were recognized on abnormally positioned Purkinje cells (Rakic and Sidman, 1972). These findings make it unlikely that the specificity of mossy afferents for granule cell dendritic targets is attained either by a predetermined addressing mechanism or by a strict temporal mechanism. Furthermore, it implies that granule cells in the molecular layer somehow can attract mossy fibers a distance, either directly or through the mediation of other cells.

In homozygous weaver mice on the inbred C57BL/6J background, most granule cell neurons degenerate while still located on the outer side of the molecular layer (see "Weaver, *wv*"), and mossy fibers, in the absence of their usual targets, commonly make synaptic contacts with dendrites of Golgi II neurons (Rakic and Sidman, 1973c). Rather than representing an exception to the specificity rule, this relationship fits the active competition hypothesis set forth in the section entitled "Transynaptic Interactions during Development," for mossy fibers normally do synapse with Golgi II dendrites, and when no granule cell dendrites are available, this apparently represents the best fit. Similarly, when the dendritic arbor of the Purkinje cell is reduced in surface area, as in staggerer and nervous, climbing fiber inputs retain their developmental distribution on persistent somatic thorns for weeks longer into the postnatal period than is normal (Landis, 1973a; D. Landis and Sidman, unpublished observations). Finally, it should be emphasized that in none of the mutants deficient in granule cells have we seen mossy fibers forming synaptic contacts with Purkinje cells. Such contacts have been reported, however, in the granule cell-depleted cerebellum of the virus-infected ferret (Llinas et al., 1973), where perhaps a wider range of cell damage has altered the synaptic hierarchies more drastically.

Purkinje Cell Properties

In all four mutants, Purkinje cells assume shapes that are recognizably representative of that unique class of neurons. Even the most abnormally positioned and misoriented Purkinje cells in reeler develop a smooth-surfaced ovoid cell body with dendrites commonly originating in a single main trunk from a point on the soma more or less opposite the axon hillock (see Fig. 20 in Sidman, 1968, and Fig. 3 in Sidman, 1972). The shape and polarity, then, must be controlled either by other genes acting intrinsically or by mediation of other cells at a stage of development preceding the emergence of Purkinje cell dendrites. No mutations have been recognized that bear directly on this issue. By contrast, the available mutants show that the normal orientation of Purkinje dendrites toward the external surface of the cerebellum in the sagittal plane is strongly influenced by the local milieu, particularly by the orientation of granule cell axons (see also Altman, 1973).

In addition, the volume and the branching pattern of Purkinje cell dendrites are markedly influenced by the local milieu. The smaller branches (referred to as "branchlets" in Table 3, lines 5 and 6) are virtually absent in staggerer and on the heterotopic cells in reeler, and are moderately reduced in weaver and nervous. The secondary branches often "droop" inwards toward the level of the Purkinje cell somas (e.g., Fig. 1C in Rakic and Sidman, 1973c). Similar effects have been described when external granule cells are reduced in number by nongenetic means (e.g., Altman and Anderson, 1972). But while a decreased number of parallel fibers in the local milieu may be a major determinant, it is not the only one. In staggerer, the Purkinje cells may be abnormal even before genesis of granule cells, and the stunted dendritic arbor may be one expression of this. Likewise, in nervous all the Purkinje cells express an abnormal phenotype by postnatal day 15 (Landis, 1973a), and the deficit in late-forming dendritic branchlets may, in part, be another of its manifestations.

The dendritic spines of the Purkinje cells, the specialized surface sites destined normally to be contacted synaptically by parallel fibers, are under different controls. It would seem consonant with general views of cell interaction that the so-called postsynaptic thickening, a series of filamentous densities on the cytoplasmic aspect of the surface membrane of the dendritic spine, and even the shape or the very existence of the spine itself, might depend on some influence of the presynaptic element, in this case, the parallel fiber. Such a relationship has been described for dendritic spines on pyramidal cells of the visual cortex (Szentágothai, 1965; Globus and Scheibel, 1967; Valverde, 1967). For the Purkinje cell dendritic spines, however, the opposite is true. Persistence of abundant spines, including the postsynaptic thickenings in the absence of parallel fibers, has been found not only in reeler (Rakic and Sidman, 1972) and weaver (Rakic and Sidman, 1973a; Hirano and Dembitzer, 1973) but also in experimentally induced deficits of granule cells (Herndon et al., 1971; Hirano et al., 1972; Altman and Anderson, 1972; Llinas et al., 1973). The weaver mutation has been of unique value in establishment of the point that the spines are not only maintained, but actually are formed, in the absence of parallel fibers. The critical evidence was obtained in $+/wv$ heterozygotes, where Purkinje dendrites extend abnormally along Bergmann fibers right through the level of the external granular layer to reach the external surface of the cerebellum. There they form numerous spines, at a distance of more than 20 μm from the nearest parallel fiber and with an apparent barrier of Bergmann glial cytoplasm interposed between them (See Fig. 11 in Rakic and Sidman, 1973c). Once formed, the specialized membrane thickenings on Purkinje cell dendritic spines may persist without innervation, enveloped by glial cytoplasm, for more than a year (see Fig. 14 in Rakic and Sidman, 1973c).

Electron microscopic observations on staggerer (Sotelo, 1973; Landis and Sidman, unpublished observations) suggest that potential presynaptic, as well as postsynaptic, elements may express specialized surface properties in the absence of innervation in the usual sense. During the 2nd and 3rd postnatal weeks, before granule cell degeneration, parallel fiber axons were found making synapse-like contacts with astrocytic processes in the most external parts of the molecular layer. The axons displayed focal swellings containing a few vesicles and specialized segments of membrane with an electron-dense fuzz on the cytoplasmic side and a dense material in the widened cleft between axon and glial cell surface. These specialized axonal features were not seen in the normal cerebellum, and were not observed in the staggerer cerebellum in the second month after birth, when the granule cells had degenerated. It should be emphasized that neither the axonal specializations described in this paragraph nor the dendritic ones described in the preceding paragraph are formed by processes that are literally isolated from contact with other cells, for in each case there is a closely apposed glial process.

Granule Cell Survival

Although all four mutants feature granule cell death, this important phenotypic expression appears in each case to be an indirect consequence of some cell interaction. A unifying hypothesis is that the young granule cell may depend for its survival on the establishment of effective contact with Purkinje cells. In weaver, few granule cells either make horizontal bipolar processes or migrate inward along Bergmann glial fibers (see "Weaver, wv"), nor do they form functional contacts with Purkinje cell dendrites in the 5 days representing their average survival time. In staggerer, the granule cells seem basically normal; before they degenerate they receive the normal types of synaptic inputs and axons in turn making normal-appearing synaptic contacts with dendrites of Golgi II cells and with interneurons of the molecular layer, as well as curious specialized contacts with astrocytes. However, parallel fibers make only transient synapses on Purkinje cell dendritic shafts, the Purkinje cells fail to form dendritic spines, and no special relationship with parallel fibers is maintained. It appears that, as a consequence, the granule cells die. In reeler, the basic problem seems to be an abnormality of cell position in cortical structures, based on a derangement of cell recognition early in development. This in itself would seem to be no cause for cell death, and no loss of cells has been recognized in the abnormal cerebral cortex. However, the granule cell of the cerebellar cortex is a special case, for its initial genesis and its subsequent maintenance seem to be impaired in areas of the reeler cerebellar cortex where most of the Purkinje cells lie in abnormally deep positions. Even in nervous, loss of granule cells clearly follows the alteration in Purkinje cells and is probably dependent on it. The mildness of the granule cell loss in nervous may reflect either the persistence of a near-threshold number of Purkinje cells or a partial escape of granule cells with time from their dependence on contact with Purkinje cells. That the latter is the more important factor may be inferred from cell behavior in a new mutant named Purkinje cell degeneration, *pcd* (Mullen et al., 1974), in which virtually all Purkinje cells degenerate at even later stages, but most granule cells are maintained.

Dendrites of Interneurons of the Cerebellar Molecular Layer

Anaylsis of the developing cerebellar cortex in fetal and neonatal rhesus monkeys led to the concept that granule cells and interneurons of the molecular layer are derived from different clones, and that the volume, orientation, and branching pattern of basket and stellate cell dendrites are highly determined by the number and orientation of parallel fibers and other inputs in their local milieu (see "Interneurons of the Cerebellar Molecular Layer," p. 228). The data obtained in the mutants are consistent with these ideas. Interneurons are formed and are maintained in all four mutants. Their dendritic patterns, however, are much reduced in weaver and reeler, where the parallel-fiber bed is scanty, but not in staggerer or nervous, where numerous parallel fibers form (Table 3, line 8). It would seem that granule cells and interneurons have differentiated along separate paths, in that they respond to different sets of intercellular signals. Each in its individual way is highly dependent on contact relations with its neighbors.

NEW EXPERIMENTAL APPROACHES

Chimerism

The dissociation of an organ rudiment into a suspension of single cells, followed by their reassembly *in vitro* into aggregates and the subsequent sorting of cells into histotypic patterns, has been used for many purposes, including the preparation of interspecies chimeras (Moscona, 1965). As reviewed elsewhere in this volume, the dissociation-aggregation procedure is of great value in the analysis of surface-mediated interactions of developing cells in retina and other parts of the nervous system. The grafting of tissues *in vivo* serves as another way to challenge developing cells by exposing them to new neighbors. Regenerating axons of neurons with different neurohumoral specificities, when grafted into the eye or brain, were shown to be quite discriminating in their behavior toward different classes of target tissues (see the section entitled "Transynaptic Interactions during Development"). It is feasible also to graft cubes of immature cerebellar cortex between young individuals of the same species and obtain incorporation of donor granule cell neurons into host cerebellum (Altman and Das, 1972).

A more general method for producing mammalian chimeras, particularly interesting because it involves no direct manipulation of the organ to be studied, is to fuse 8- or 16-cell mouse embryos and allow the product to develop into a single mouse with cells of different genotypes in its various organs (Tarkowski, 1961; Mintz, 1962). This type of experimental chimerism has been used effectively to study a wide range of biological problems (e.g., Mintz, 1971), but its potential for analysis of neural development has not yet been exploited.

The difficulty lies with the limited number of cellular phenotypic markers available for the nervous system. The ideal marker would be a stable chemical or morphological property controlled by two or more alleles at a single genetic locus and demonstrable on all cells throughout life. The histocompatibility or other surface antigens would seem to be suitable candidates, but thus far it has not been feasible to demonstrate them on individual cells, except in suspension. It is hoped that this technical limitation will prove to be merely temporary. Meanwhile, certain of the neurological mutations are being used, though each is expressed on limited classes of cells.

The retinal degeneration mutation, rd, causes loss of retinal photoreceptor cells postnatally, and chimeras have been prepared between rd/rd and wild-type mice (Mintz and Sanyal, 1970; Wegmann et al., 1971). A mosaic distribution of normal and photoreceptor-free retinal patches is obtained, with an orderliness possibly suggestive of derivation of photoreceptor cells from a limited number of clones (Mintz and Sanyal, 1970). However, at the borders between patches are zones of varying width with reduced numbers of photoreceptor cell nuclei and apparently short rod outer segments (Wegmann et al., 1971). An independent marker is needed to determine if these intermediate zones contain genetically normal cells that are partially compromised by their proximity to mutant cells, or the reverse. It is possible that photoreceptor cells may be interacting laterally within a homogeneous lamina, a phenomenon not likely to be recognized except in the special case of a chimera.

Cell interactions appear to be of such deep importance in the development and function of the vertebrate nervous system, if the arguments presented throughout this paper are valid, that one might predict greater difficulty in obtaining clear-cut mosaic patterns in the mouse nervous system than in certain other mouse organs or in, say, the fruit-fly nervous system. This prediction may be proving valid, on the basis of preliminary data obtained from chimeric combinations of normal mice with mice bearing the mutations listed in Table 2 (Mullen, personal communication).

To obtain the independent check on cellular genotypes essential to these studies, we are turning to histochemical visualization of glucuronidase activity, as already applied to mosaic mouse livers by Condamine et al. (1971). Allelic differences in enzyme activity have not been detectable histochemically except in liver, but a much more sensitive new procedure has been developed that retains most of the enzyme activity through fixation, dehydration, embedding, and sectioning of tissues (Feder, personal communication). It has been applied to chimeric tissues and to normal brain (Fig. 4) and may allow recognition of mosaicism in the nervous system of chimeric mice (Mullen, Feder, and Sidman, unpublished observations). It is likely that there will still be formidable problems to overcome; for example, there may be competition between cells of different genotypes for position and even for survival, and the cellular ratios may change with time, as implied by the change from mutant to normal behavior at several weeks of age in the case of a chimeric mouse with a

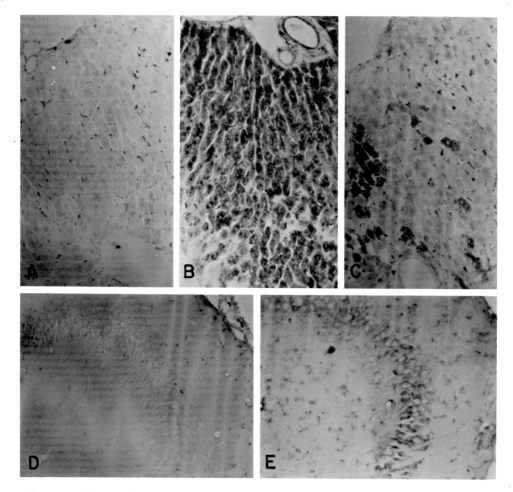

Figure 4.   Glucuronidase stains; the red color of the original slides is rendered in black. Figure 4A: liver of C3H/HeJ mouse, with almost no demonstrable glucuronidase activity in parenchymal cells and high activity in Kupfer cells. Figure 4B: liver of C57BL/6J mouse, with high enzyme activity in parenchymal cells. Figure 4C: liver of chimeric mouse made by fusing eight-cell embryos of the above two strains; there are cellular patches of high activity, some as small as one cell, in a predominantly low-activity organ. Figure 4D: a sector of the hippocampus from the brain of a C3H/HeJ mouse, with almost no demonstrable enzyme activity. Figure 4E: the same region from the brain of a C57BL/6J mouse, with low to moderate activity in all neurons and glial cells.

mixture of normal cells and cells bearing the sex-linked mutation jimpy, *jp* (Eicher and Hoppe, 1973). Or there may be transfer of glucuronidase molecules, a lysosomal enzyme, from cells with high activity to cells genetically endowed to express low activity.

Immunocytological Characterization of Surface Components

If one's index of suspicion is sufficiently raised from all the preceding material suggestive of developmental control through surface-mediated cell interactions, it may be accepted that the time has come to deal directly with localization and characterization of surface components. Serological methods would seem pertinent for this task (Bennett et al., 1972), though other probes than antibodies, such as toxins with high binding affinity for neurotransmitter receptors, also are valid (e.g., Giacobini et al., 1973; Greene, 1974).

A logical place to begin is with genetically defined alloantigens known to be shared by the nervous system and some other organ. The other organ is the source of cells for the cytotoxicity test that measures the antibody. The H-2 antigen, the major histocompatibility system in the mouse, was our first object of study. A sensitive assay for H-2 antigen activity in mouse brain proved to be a quantitative absorption technique in which a predetermined dilution of antiserum was incubated with varying volumes of packed brain homogenate, and the absorbed antiserum was then titrated in serial dilutions for residual complement-mediated toxicity toward mouse lymphocytes (Schachner and Sidman, 1973). The representation of H-2 was found to reach the adult level within 4 weeks after birth and to be about equal in various parts of the brain. In the same way, Thy-1, a cell surface antigen shared by nervous system and thymus, was measured quantitatively in normal mice and in 15 mutants in the hope of finding a marker for specific brain cell types (Schachner, 1973a). Virtually no differences were found, even in brains of myelin-deficient mutants and in cerebella selectively lacking particular cell classes. It was concluded that Thy-1 is not differentially distributed among classes of brain cells.

To test the representation of a range of cell surface markers shared by brain and some other cells on a single neural cell type, the C1300 mouse neuroblastoma was analyzed for H-2, Thy-1, NZB (a naturally occurring autoantibody to thymocytes and brain), Pc (shared with plasma cells), Sk (shared with skin cells), several lymphocyte-specific antigens, and GSCA (the Gross cell surface antigen, found in cells carrying leukemia virus) (Schachner, 1973b). The list of antigens demonstrable on this peripheral nervous system neuronal tumor cell grown by serial transplantation *in vivo* is similar to that of normal brain tissue, except for the presence of GCSA in the tumor. A tissue culture subline, NB41A, was deficient in Thy-1 and NZB specificities, but, disappointingly, no antigenic differences were found between attached "differentiated" and nonattached proliferating neuroblastoma cells.

To demonstrate and measure antigenic specificities unique to the nervous system, it is necessary to use a cytotoxicity test with individual nervous system cells in suspension, rather than the usual lymphocytes or thymocytes. A cell suspension obtained by trypsinization and trituration of 30–50 mg fragments of 3- to 10-day-old mouse brain was used for this purpose (Schachner and Sidman, 1973). This system serves, of course, only for assay of antibody against antigens represented on the surfaces of cells that enter the test suspension. A similar assay can be made with cell suspensions prepared from brain tumors.

The first nervous system-specific surface antigen has now been recognized and given the symbol NS-1 (Schachner, 1974). A methylcholanthrene-induced glioblastoma originating in the C57BL/6J mouse strain was used to raise alloantibodies. It is found that NS-1 occurs on cells from brains of all mouse strains studied, and in higher concentration in central or peripheral tissue rich in white matter than in gray matter. Concentration is reduced in brains of two genetically independent myelin-deficient mutant mice, quaking (*qk*) and myelin synthesis deficiency (*msd*). In the normal brain, concentration is just measurable on the day of birth and reaches the adult level at about 4 weeks. It is represented on three of four mouse glial tumors but not on C1300 neuroblastoma cells. It was also found in brains of rats, cats, and human beings, but not chickens. Not all the reactivity of the antiserum against mouse cells was eliminated by repeated absorption with human cerebral gray and white matter, but the residual activity was then removed by absorption with mouse brain. This indicated that the human analog of NS-1 does not contain all the antigenic specificities carried by the mouse antigen.

This approach to the analysis of surface components of nervous system cells is likely to yield dividends in addition to the definition of new surface constituents. The antisera may serve for perturbing cell interactions *in vitro* or even *in vivo*. They already are serving for visualization and selection of antigen-bearing cells in suspension; Schachner and Hämmerling (1974) prepared a hybrid antibody with antimouse immunoglobulin and antisheep red blood cell specificities, and developed a mixed adsorption hybrid antibody (MAHA) test. A suspension of mouse brain cells is reacted with appropriately diluted antiserum followed by hybrid antibody, by centrifuging the cells through a discontinuous density gradient containing these reagents and washes in alternate sequence. This method avoids repeated washing and resuspending of fragile cells and yields populations of free brain cells and cells enveloped in a rosette of sheep red blood cells. These are readily separated from each other by resuspending them in medium 199, layering them on medium 199 containing 25% fetal bovine serum, and allowing them to sediment at 1g for 45 minutes. This method, like the velocity sedimentation method of Barkley et al. (1973), yields enriched populations of cells that are viable for tissue culture and other uses.

These new approaches increase the analytical and experimental prospects for unraveling the critical events in the genesis of the mammalian nervous system.

REFERENCES

Akert, K. (1973). Dynamic aspects of synaptic ultrastructure. *Brain Res.* **49**, 511–518.

Akert, K., K. Pfenninger, C. Sandri, and H. Moor (1972). Freeze-etching and cytochemistry of vesicles and membrane complexes of the CNS. In *Structure and Function of Synapses* (G. Pappas and D. P. Purpura, eds.). Raven Press, New York, pp. 67–86.

Altman, J. (1973). Experimental reorganization of the cerebellar cortex. IV. Parallel fiber reorientation following regeneration of the external granular layer. *J. Comp. Neurol.* **149**, 181–192.

Altman, J., and W. J. Anderson (1972). Experimental reorganization of the cerebellar cortex. I. Morphological effects of elimination of all microneurons with prolonged x-irradiation started at birth. *J. Comp. Neurol.* **146**, 355–406.

Altman, J., W. J. Anderson, and K. A. Wright (1969). Early effects of x-irradiation of the cerebellum in infant rats: decimation and reconstitution of the granular layer. *Exp. Neurol.* **24**, 196–216.

Angevine, J. B., Jr. (1970). Time of neuron origin in the diencephalon of the mouse: an autoradiographic study. *J. Comp. Neurol.* **139**, 129–188.

Angevine, J. B., Jr., and R. L. Sidman (1961). Autoradiographic study of cell migration during histogenesis of cerebral cortex in the mouse. *Nature* **192**, 766–768.

Barkley, D. S., L. L. Rakic, J. K. Chaffee, and D. L. Wong (1973). Cell separation by velocity sedimentation of postnatal mouse cerebellum. *J. Cell Physiol.* **81**, 271–280.

Bennett, D., E. A. Boyse, and L. S. Old (1972). Cell surface immunogenetics in the study of morphogenesis. In *Cell Interactions: Proceedings of the Third Lepetit Colloquium* (L. G. Silvestri, ed.). North-Holland Publishing Co., Amsterdam, pp. 247– 263.

Bennett, M. V. L.(1970). Comparative physiology: electric organs. Ann. Rev. Physiol. **32**, 471–528.

Bernstein, M. E., and J. J. Bernstein (1973). Regeneration of axons and synaptic complex formation rostral to the site of hemisection in the spinal cord of the monkey. *Int. J. Neurosci.* **5**, 15–26.

Berry, M., and W. A. Rogers (1965). The migration of neuroblasts in the developing cerebral cortex. *J. Anat.* **99**, 691–709.

Bignami, A., and D. Dahl (1974). The development of Bergmann glia in mutant mice with cerebellar malformation: reeler, staggerer, and weaver: Immunofluorescence study with antibodies to the glial fibrillary acidic protein. *J. Comp. Neurol.* (in press).

Björklund, A., and U. Stenevi (1971). Growth of central catecholamine neurons into smooth muscle grafts in the rat mesencephalon. *Brain Res.* **31**, 1–20.

Black, I. B., I. A. Hendry, and L. L. Iversen (1971). Trans-synaptic regulation of growth and development of adrenergic neurons in a mouse sympathetic ganglion. *Brain Res.* **34**, 229–240.

Blinkov, S. M., and I. I. Glezer (1968). *The Human Brain in Figures and Tables.* Plenum Press, New York.

Bloom, F. E. (1972). The formation of synaptic junctions in developing rat brain. In *Structure and Function of Synapses* (G. D. Pappas and D. P. Purpura, eds.). Raven Press, New York, pp. 101–120.

Boulder Committee ((1970). Embryonic vertebrate central nervous system: revised terminology. *Anat. Rec.* **166**, 257–262.

Bunge, M. G. (1973). Fine structure of nerve fibers and growth cones of isolated sympathetic neurons in culture. *J. Cell Biol.* **56**, 713–735.

Caviness, V. S., Jr. (1973). Time of neuron origin in the hippocampus and dentate gyrus of normal and reeler mutant mice: an autoradiographic analysis. *J. Comp. Neurol.* **151**, 113–120.

Caviness, V. S., Jr., and R. L. Sidman (1972). Olfactory structures of the forebrain in the reeler mutant mouse. *J. Comp. Neurol.* **145**, 85–104.

Caviness, V. S., Jr., and R. L. Sidman (1973a). Retrohippocampal, hippocampal and related structures of the forebrain in the reeler mutant mouse. *J. Comp. Neurol.* **147**, 235–253.

Caviness, V. S., Jr., and R. L. Sidman (1973b). Time of origin of corresponding cell classes in the cerebral cortex of normal and reeler mutant mice: an autoradiographic analysis. *J. Comp. Neurol.* **148**, 141–152.

Caviness, V. S., Jr., D. K. So, and R. L. Sidman (1972). The hybrid reeler mouse. *J. Hered.* **63**, 241–246.

Ceccarelli, B., W. P. Hurlbut, and A. Mauro (1973). Turnover of transmitter and synaptic vesicles at the frog neuromuscular junction. *J. Cell Biol.* **57**, 499–524.

Changeux, J-P., P. Courrège, and A. Danchin (1973). A theory of the epigenesis of neuronal networks by selective stabilization of synapses. *Proc. Nat. Acad. Sci. U.S.A.* **70**, 2974–2978.

Condamine, H., R. P. Custer, and B. Mintz (1971). Pure-strain and genetically mosaic liver tumors histochemically identified with the B-glucuronidase marker in allophenic mice. *Proc. Nat. Acad. Sci. U.S.A.* **68**, 2032–2036.

Cowan, W. M. (1970). Anterograde and retrograde transneuronal degeneration in the central and peripheral nervous system. In *Contemporary Research Methods in Neuroanatomy* (W. J. H. Nauta and S. O. E. Ebbesson, ed.). Springer, New York, pp. 217–251.

Cragg, B. G. (1967). Changes in visual cortex on first exposure of rats to light: effect on synaptic dimensions. *Nature* **215**, 251–253.

Cragg, B. G. (1969). The effect of vision and dark-rearing on the size and density of synapses in the lateral geniculate nucleus measured by electron microscopy. *Brain Res.* **13**, 53–67.

Cragg, B. G. (1972). The development of synapses in cat visual cortex. *Invest. Ophthal.* **11**, 377–385.

Das, G. D., and J. Altman (1972). Studies on the transplantation of developing neural tissue in the mammalian brain. I. Transplantation of cerebellar slabs into the cerebellum of neonatal rats. *Brain Res.* **38**, 233–249.

DeLong, G. R., and R. L. Sidman (1962). Effects of eye removal at birth on histogenesis of the. mouse superior colliculus: an autoradiographic analysis with tritiated thymidine. *J. Comp. Neurol.* **118**, 205–223.

DeLong, G. R., and R. L. Sidman (1970). Alignment defect of reaggregating cells in cultures of developing brains of reeler mutant mice. *Develop. Biol.* **22**, 584–600.

Detwiler, S. R. (1936). *Neuroembryology: An Experimental Study.* Macmillan, New York.

Edds, M. V., Jr. (1953). Collateral nerve regeneration. *Quart. Rev. Biol.* **28**, 260–276.

Eicher, E. M., and P. C. Hoppe (1973). Use of chimeras to transmit lethal genes in the mouse and to demonstrate allelism of two x-linked male lethal genes, *jp* and *msd. J. Exp. Zool.* **183**, 181–184.

Fischbach, G. D., and S. A. Cohen (1973). The distribution of acetylcholine sensitivity over uninnervated and innervated muscle fibers grown in cell culture. *Develop. Biol.* **31**, 147–162.

Fox, C. A., and J. W. Barnard (1957). A quantitative study of the Purkinje cell dendritic branchlets and their relation to afferent fibers. *J. Anat.* **91**, 299–313.

Fuxe, K., T. Hökfelt, and U. Ungerstedt (1970). Morphological and functional aspects of central monoamine neurons. *Int. Rev. Neurobiol.* **13**, 159–180.

Giacobini, G., G. Filogamo, M. Weber, P. Boquet, and J-P. Changeux (1973). Effects of a snake α-neurotoxin on the development of innervated skeletal muscles in chick embryos. *Proc. Nat. Acad. Sci. U.S.A.* **70**, 1708–1712.

Globus, A., and A. B. Scheibel (1967). The effect of visual deprivation on cortical neurons: a Golgi study. *Exp. Neurol.* **19**, 331–345.

Greene, L. A. (1974). The use of α-bungarotoxin to probe acetylcholine receptors in sympathetic neurons in cell culture. In *The Neurosciences: Third Study Program* (F. O. Schmitt and F. G. Worden, eds.-in-chief). MIT Press, Cambridge, pp. 765–771.

Guth, L. (1958). Taste buds on the cat's circumvallate papilla after reinnervation by glossopharyngeal, vagus, and hypoglossal nerves. *Anat. Rec.* **130**, 25–38.

Guth, L., and J. J. Bernstein (1961). Selectivity in the re-establishment of synapses in the superior cervical sympathetic ganglion of the cat. *Exp. Neurol.* **4**, 59–69.

Hamburger, V., and R. Levi-Montalcini (1949). Proliferation, differentiation, and degeneration in the spinal ganglia of the chick embryo under normal and experimental conditions. *J. Exp. Zool.* **111**, 457–501.

Hamburgh, M. (1963). Analysis of the postnatal developmental effects of "reeler," a neurological mutation in mice: a study in developmental genetics. *Develop. Biol.* **8**, 165–185.

Hámori, J. (1969). Development of synaptic organization in the partially agranular and in transneuronally atrophied cerebellar cortex. In *Neurobiology of Cerebellar Evolution and Development* (R. Llinas, ed.), American Medical Association Educational Research Foundation, Chicago, pp. 845–858.

Hámori, J. (1973). The inductive role of presynaptic axons in the development of postsynaptic spines. *Brain Res.* **62**, 337–344.

Harrison, R. G. (1935). The origin and development of the nervous system studied by the methods of experimental embryology. *Proc. Roy. Soc.* **B118**, 155–196.

Hartzell, H. C., and D. M. Fambrough (1973). Acetylcholine receptor production and incorporation into membranes of developing muscle fibers. *Develop. Biol.* **30**, 153–165.

Hendry, I. A. (1973). Trans-synaptic regulation of tyrosine hydroxylase activity in a developing mouse sympathetic ganglion: effects of nerve growth factor (NGF), NGF-antiserum, and pempidine. *Brain Res.* **56**, 313–320.

Herndon, R. M., G. Margolis, and L. Kilham (1971). The synaptic organization of the malformed cerebellum induced by prenatal infection with the feline leukopenia virus. *J. Neuropathol. Exp. Neurol.* **30**, 557–570.

Heuser, J. E., and T. S. Reese (1973). Evidence for recycling of synaptic vesicle membrane during transmitter release at the frog neuromuscular junction. *J. Cell Biol.* **57**, 315–344.

Hinds, J. W. (1972). Early neuron differentiation in the mouse olfactory bulb. II. Electron microscopy. *J. Comp. Neurol.* **146**, 253–276.

Hinds, J. W., and P. L. Hinds (1972). Reconstruction of dendritic growth cones in neonatal mouse olfactory bulb. *J. Neurocytol.* **1**, 169–187.

Hinds, J. W., and T. L. Ruffett (1971). Cell proliferation in the neural tube: an electron microscopic and Golgi analysis in the mouse cerebral vesicle. *Z. Zellforsch.* **115**, 226–264.

Hirano, A., and H. Dembitzer (1973). Cerebellar alteration in the weaver mouse. *J. Cell. Biol.* **56**, 478–486.

Hirano, A., H. M. Dembitzer, and M. Jones (1972). An electron microscopic study of cycasin-induced cerebellar alteration. *J. Neuropathol. Exp. Neurol.* **31**, 113–125.

Iversen, L. L. (1971). Role of transmitter uptake mechanisms in synaptic neurotransmission. *Brit. J. Pharmacol.* **41**, 571–591.

Jacobson, M. (1970). *Developmental Neurobiology*. Holt, Rinehart, and Winston, New York.

Katz, B., and R. Miledi (1965). The effect of calcium on acetylcholine release from motor nerve terminals. *Proc. Roy. Soc.* **B161**, 495–503.

Katz, B., and R. Miledi (1967). Tetrodotoxin and neuromuscular transmission. *Proc. Roy. Soc.* **B167**, 8–22.

Katzman, R., A. Björklund, Ch. Owman, U. Stenevi, and K. A. West (1971). Evidence for regenerative axon sprouting of central catecholamine neurons in the rat mesencephalon following electrolytic lesions. *Brain Res.* **25**, 579–596.

Kornguth, S. E., and G. Scott (1972). The role of climbing fibers in the formation of Purkinje cell dendrites. *J. Comp. Neurol.* **146**, 61–82.

Kristensson, K. (1970). Transport of fluorescent protein tracer in peripheral nerves. *Acta Neuropathol.* **16**, 293–300.

Kristensson, K., and Y. Olsson (1971). Retrograde axonal transport of protein. *Brain Res.* **29**, 363–365.

Kristensson, K., Y. Olsson, and J. Sjostrand (1971). Axonal uptake and retrograde transport of exogenous proteins in the hypoglossal nerve. *Brain Res.* **32**, 399–406.

Landis, S. (1973a). Ultrastructural changes in the mitochondria of cerebellar Purkinje cells of nervous mutant mice. *J. Cell Biol.* **57**, 782–797.

Landis, S. (1973b). Granule cell heterotopia in normal and nervous mice of the BALB/c strain. *Brain Res.* **61**, 175–189.

Landis, S. (1973c). Changes in neuronal mitochondrial shape in brains of nervous mutant mice. *J. Hered.* **64**, 193–196.

Lane, P. (1964). Personal communication. In *Mouse News Letter* **30**, 32.

LaVail, J. H., and M. M. LaVail (1972). Retrograde axonal transport in the central nervous system. *Science* **176**, 1416–1417.

LaVail, J. H., K. R. Winston, and A. Tish (1973). A method based on retrograde intraaxonal transport of protein for identification of cell bodies of origin of axons terminating within the CNS. *Brain Res.* **58**, 470–477.

Levi-Montalcini, R. (1964). Events in the developing nervous system. *Prog. Brain Res.* **4**, 1–29.

Livingston, R. B., K. Pfenninger, H. Moore, and K. Akert (1973). Specialized paranodal and interparanodal glial-axonal junctions in the peripheral and central nervous system: a freeze-etching study. *Brain Res.* **58**, 1–24.

Llinas, R., D. E. Hillman, and W. Precht (1973). Neuronal circuit reorganization in mammalian agranular cerebellar cortex. *J. Neurobiol.* **4**, 69–64.

Lund, R. D., and J. S. Lund (1971). Synaptic adjustment after deafferentiation of the superior colliculus of the rat. *Science* **171**, 804–807.

Lynch, G., D. A. Matthews, and S. Mosho (1972). Induced acetylcholinesterase-rich layer in rat dentate gyrus following entorhinal lesions. *Brain Res.* **42**, 311–318.

Mares, V. and Z. Lodin (1970). The cellular kinetics of the developing mouse cerebellum. II. The function of the external granular layer in the process of gyrification. *Brain Res.* **23**, 343–352.

Marin-Padilla, M. (1971). Early prenatal ontogenesis of the cerebral cortex (neocortex) of the cat (*Felis domestica*): A Golgi study. I. The primordial neocortical organization. *Z. Anat. Entwicklungs-Gesch.* **134**, 117–145.

Meier, H., and W. G. Hoag (1962). The neuropathology of "reeler," a neuromuscular mutation in mice. *J. Neuropathol. Exp. Neurol.* **21**, 649–654.

Miledi, R., and L. T. Potter (1971). Acetylcholine receptors in muscle fibers. *Nature* **233**, 599–603.

Mintz, B. (1962). Formation of genotypically mosaic mouse embryos. *Am. Zool.* **2**, 432.

Mintz, B. (1971). Genetic mosaicism *in vivo*: development and disease in allophenic mice. *Fed. Proc.* **30**, 935–943.

Mintz, B., and S. Senyal (1970). Clonal origin of the mouse visual retina mapped from genetically mosaic eyes. *Genetics* **64**, Suppl., 43–44.

Molliver, M. E., and H. Van der Loos (1972). The ontogenesis of cortical circuitry: the spatial distribution of synapses in the somesthetic cortex of newborn dog. *Ergeb. Anat. Entwicklungsgesch.* **42**, 1–54.

Molliver, M. E., I. Kostovic, and H. Van der Loos (1973). The development of synapses in cerebral cortex of the human fetus. *Brain Res.* **50**, 403–407.

Morest, D. K. (1968). The growth of synaptic endings in the mammalian brain: a study of the calyces of the trapezoid body. *Z. Anat. Entwicklungsgesch.* **127**, 201–220.

Morest, D. K. (1969). The growth of dendrites in the mammalian brain. *Z. Anat. Entwicklungsgesch.* **128**, 290–317.

Morest, D. K. (1970). A study of neurogenesis in the forebrain opossum pouch young. *Z. Anat. Entwicklungsgesch.* **130**, 265–305.

Moscona, A. A. (1965). Recombination of dissociated cells and the development of cell aggregates. In *Cells and Tissues in Culture* (E. N. Willmer, ed.). Academic Press, New York, pp. 489–529.

Mugnaini, E., and P. F. Forgstrønen (1967). Ultrastructural studies on the cerebellar histogenesis. I. Differentiation of granule cells and development of glomeruli in the chick embryo. *Z. Zellforsch.* **77**, 115–143.

Mullen, R. J., E. M. Eicher, and R. L. Sidman (1974). Personal communication. In *Mouse News Letter,* **50**, 38.

Olney, J. W. (1968). An electron microscopic study of synapse formation, receptor outer segment development, and other aspects of developing mouse retina. *Invest. Ophthal.* **7**, 250–268.

Olson, L., and A. Seiger (1972a). Brain tissue transplanted into the anterior chamber of the eye. I. Fluorescence histochemistry of immature catecholamine and 5-hydroxytryptamine neurons reinnervating the rat iris. *Z. Zellforsch.* **137**, 175–194.

Olson, L., and A. Seiger (1972b). Early prenatal ontogeny of central monoamine neurons in the rat: fluorescence histochemical observations. *Z. Anat. Entwicklungsgesch.* **137**, 301–316.

Olson, M. I., and R. P. Bunge (1973). Anatomical observations on the specificity of synapse formation in tissue culture. *Brain Res.* **59**, 19–33.

Palay, S. L., C. Sotelo, A. Peters, and P. M. Orkand (1968). The axon hillock and the initial segment. *J. Cell Biol.* **38**, 193–201.

Peters, A., C. C. Proskauer, and I. R. Kaiserman-Abramof (1968). The small pyramidal neuron of the rat cerebral cortex: the axon hillock and initial segment. *J. Cell Biol.* **39**, 604–619.

Peters, A., S. L. Palay, and H. deF. Webster (1970). *The Fine Structure of the Nervous System: The Cells and Their Processes.* Harper & Row, New York.

Poliakov, G. I. (1961). Some results of research into the development of the neuronal structure of the cortical ends of the analyzers in man. *J. Comp. Neurol.* **117**, 197–212.

Rahaminoff, R. (1974). Modulation of transmitter release at the neuromuscular junction. In *The Neurosciences: Third Study Program* (F. O. Schmitt and F. G. Worden, eds.-in-chief). MIT Press, Cambridge, pp. 943–952.

Raisman, G. (1969). Neuronal plasticity in the septal nuclei of the adult rat. *Brain Res.* **14**, 25–48.

Raisman, G., and P. M. Field (1973). A quantitative investigation of the development of collateral reinnervation after partial deafferentiation of the septal nuclei. *Brain Res.* **50**, 241–264.

Rakic, P. (1971). Neuron-glia relationship during granule cell migration in developing cerebellar cortex: A Golgi and electronmicroscopic study in *Macacus rhesus. J. Comp. Neurol.* **141**, 283–312.

Rakic, P. (1972a). Mode of cell migration to the superficial layers of fetal monkey neocortex. *J. Comp. Neurol.* **145**, 61–84.

Rakic, P. (1972b). Extrinsic cytological determinants of basket and stellate cell dendritic pattern in the cerebellar molecular layer. *J. Comp. Neurol.* **146**, 335–354.

Rakic, P. (1973). Kinetics of proliferation and latency between final division and onset of differentiation of cerebellar stellate and basket neurons. *J. Comp. Neurol.* **147**, 523–546.

Rakic, P. (1974). Neurons in rhesus monkey visual cortex: Systematic relation between time of origin and eventual disposition. *Science* **183**, 425–427.

Rakic, P., and R. L. Sidman (1972). Synaptic organization of displaced and disoriented cerebellar cortical neurons in reeler mice. *J. Neuropathol. Exp. Neurol.* **31**, 192.

Rakic, P., and R. L. Sidman (1973a). Weaver mutant mouse cerebellum: defective neuronal migration secondary to specific abnormality of Bergmann glia. *Proc. Nat. Acad. Sci. U.S.A.* **70**, 240–244.

Pakic, P., and R. L. Sidman (1973b). Sequence of developmental abnormalities leading to granule cell deficit in cerebellar cortex of weaver mutant mice. *J. Comp. Neurol.* **152**, 103–132.

Rakic, P., and R. L. Sidman (1973c). Organization of cerebellar cortex secondary to deficit of granule cells in weaver mutant mice. *J. Comp. Neurol.* **152**, 133–162.

Rakic, P. L. J. Stensaas, E. P. Sayre, and R. L. Sidman (1974). Computer aided three-dimensional reconstruction and quantitative analysis of cells from serial electron microscopic montages of fetal monkey brain. *Nature*, in press.

Ramon y Cajal, S. (1937). *Recollections of My Life* (E. H. Craigie, transl.). M.I.T. Press, Cambridge.

Ramon y Cajal, S. (1960). *Studies on Vertebrate Neurogenesis* (L. Guth, transl.). Charles C Thomas, Springfield, Ill.

Redfern, P. A. (1970). Neuromuscular transmission in new-born rats. *J. Physiol.* **209**, 701–709.

Rezai Z., and C. H. Yoon (1972). Abnormal rate of granule cell migration in the cerebellum of "weaver" mutant mice. *Develop. Biol.* **29**, 17–26.

Richman, D. P., R. M. Stewart, and V. S. Caviness, Jr. (1973). Microgyria, lissencephaly, and neuron migration to the cerebral cortex: an architectonic approach. *Neurology* **23**, 413.

Rosenbluth, J. (1962). Subsurface cisterns and their relationship to the neuronal plasma membrane. *J. Cell Biol.* **13**, 405–421.

Sandri, C., K. Akert, R. B. Livingston, and H. Moor (1972). Particle aggregations at specialized sites in freeze-etched postsynaptic membranes. *Brain Res.* **41**, 1–16.

Sauer, M. E., and B. E. Walker (1959). Radiographic study of interkinetic nuclear migration in the neural tube. *Proc. Soc. Exp. Biol. Med.* **101**, 557–560.

Schachner, M. (1973a). Representation of cell surface alloantigen Thy-1 (Θ) in brains of neurological mutants of the mouse. *Brain Res.* **56**, 382–386.

Schachner, M. (1973b). Serologically demonstrable cell surface specificities on mouse neuroblastoma Cl300. *Nature New Biol.* **243**, 117–119.

Schachner, M. (1974). NS-1 (nervous system antigen-1) a glial cell-specific antigenic component of the surface membrane. *Proc. Nat. Acad. Sci. U.S.A.* **71** (in press).

Schacner, M., and U. Hämmerling (1974). The postnatal development of antigens on mouse brain cell surfaces. *Brain Res.,* in press.

Schachner, M., and R. L. Sidman (1973). Distribution of H-2 alloantigen in adult and developing mouse brain. *Brain Res.* **60**, 191–198.

Shimada, Y., and D. A. Fischman (1973). Morphological and physiological evidence for the development of functional neuromuscular junctions *in vitro. Develop. Biol.* **31**, 200–225.

Shimada, M., and J. Langman (1970). Repair of the external granular layer after postnatal treatment with 5-fluorodeoxyuridine. *Am. J. Anat.* **129**, 247–260.

Sidman, R. L. (1961). Histogenesis of mouse retina studied with thymidine-H³. In *The Structure of the Eye* (G. K. Smelser, ed). Academic Press, New York, pp. 487–506.

Sidman, R. L. (1968). Development of interneuronal connections in brains of mutant mice. In *Physiological and Biochemical Aspects of Nervous Integration* (F. D. Carlson, ed.). Prentice-Hall, Englewood Cliffs, N.J., pp. 163–193.

Sidman, R. L. (1970). Autoradiographic methods and principles for study of the nervous system with thymidine-H³. In *Contemporary Research Methods in Neuroanatomy* (W. J. H. Nauta and S. O. E. Ebbesson, eds.). Springer, New York, pp. 252–274.

Sidman, R. L. (1972). Cell interactions in developing mammalian central nervous system. In *Cell Interactions: Proceedings of the Third Lepetit Colloquium* (L. G. Silvestri, ed.). North-Holland, Amsterdam, pp. 1–13.

Sidman, R. L. (1974). Cell-cell recognition in the central nervous system. In *The Neurosciences: Third Study Program* (F. O. Schmitt and F. G. Worden, eds.-in-chief). MIT Press, Cambridge, pp. 743–758.

Sidman, R. L., and M. C. Green (1970). "Nervous," a new mutant mouse with cerebellar disease (Symposium of the Centre National de la Recherche Scientifique, Orleans-la-Source, France). In *Les Mutants Pathologiques chez l'Animal Leur Intérêt dans la Recherche Biomédicale* (M. Sabourdy, ed.). CNRS, Paris, pp. 69–79.

Sidman, R. L., and P. Rakic (1973). Neuronal migration, with special reference to developing human brain: a review. *Brain Res.* **62**, 1–35.

Sidman, R. L., and M. Singer (1960). Limb regeneration without innervation of the apical epidermis in the adult newt, *Triturus. J. Exp. Zool.* **144**, 105–110.

Sidman, R. L., I. L. Miale, and N. Feder (1959). Cell proliferation in the primitive ependymal zone: an autoradiographic study of histogenesis in the nervous system. *Exp. Neurol.* **1**, 322–333.

Sidman, R. L., P. Lane, and M. Dickie (1962). Staggerer, a new mutation in the mouse affecting the cerebellum. *Science* **137**, 610–612.

Sidman, R. L., M. C. Green, and S. H. Appel (1965). *Catalog of the Neurological Mutants of the Mouse.* Harvard University Press, Cambridge.

Skoff, R. P., and V. Hamburger (1974). Fine structure of dendritic and axonal growth cones in embryonic chick spinal cord. *J. Comp. Neurol.* **153**, 107–148.

Sotelo, C. (1973). Permanence and fate of paramembranous synaptic specializations in "mutants" and experimental animals. *Brain Res.* **62**, 345–351.

Sotelo, C., and R. Llinas (1972). Specialized membrane junctions between neurons in the vertebrate cerebellar cortex. *J. Cell Biol.* **53**, 271–289.

Spira, A. W., and M. J. Hollenberg (1973). Human retinal development: ultrastructure of the inner retinal layers. *Develop. Biol.* **31**, 1–21.

Szentágothai, J. (1965). The use of degeneration methods in the investigation of short neuronal connexions. *Prog. Brain Res.* **14**, 1–32.

Tarkowski, A. K. (1961). Mouse chimeras developed from fused eggs. *Nature* **190**, 857–860.

Tennyson, V. M., R. E. Barrett, G. Cohen, L. J. Cote, R. Heikkila, and C. Mytilineou (1972). The developing neostriatum of the rabbit: correlation of fluorescence histochemistry, electron microscopy, endogenous dopamine levels and [³H] dopamine uptake. *Brain Res.* **46**, 251–285.

Thoenen, H. (1972). Comparison between the effects of neuronal activity and Nerve Growth Factor in the enzymes involved in the synthesis of norepinephrine. *Pharmacol. Rev.* **24**, 255–267.

Thoenen, H. (1974). Trans-synaptic enzyme induction. *Life Sci.* **14**, 223–235.

Turner, P. T., and A. B. Harris (1973). Ultrastructure of synaptic vesicle formation in cerebral cortex. *Nature* **242**, 57–59.

Valverde, F. (1967). Apical dendritic spines of the visual cortex and light deprivation in the mouse. *Exp. Brain Res.* **3**, 337–352.

Waxman, S. G. (1974). Ultrastructural differentiation of the axon membrane at synaptic and non-synaptic central nodes of Ranvier. *Brain Res.* **65**, 338–342.

Waxman, S. G., G. D. Pappas, and M. V. L. Bennett (1972). Morphological correlates of functional differentiation of nodes of Ranvier along single nerve fibers in the neurogenic electric organ of the knife-fish *Sternarchus. J. Cell Biol.* **53**, 210–224.

Wegmann, T. G., M. M. LaVail and R. L. Sidman (1971). Patchy retinal degeneration in tetraparental mice. *Nature* **230**, 333–334.

Weiss, P. (1950). Deplantation of fragments of nervous system in amphibians. I. Central reorganization and the formation of nerves. *J. Exp. Zool.* **113**, 397–461.

Weiss, P. (1955). Nervous system. In *Analysis of Development* (H. Willier, P. Weiss, and V. Hamburger, eds.). Saunders, Philadelphia, pp. 346–401.

Weiss, P., and M. V. Edds, Jr. (1945). Sensory-motor nerve crosses in the rat. *J. Neurophysiol.* **8**, 173–193.

Woodward, D. J., B. J. Hoffer, G. R. Siggins, and F. E. Bloom (1971). The ontogenetic development of synaptic junctions, synaptic activation and responsiveness to neurotransmitter substances in rat cerebellar Purkinje cells. *Brain Res.* **34**, 73–97.

Yntema, C. L., and W. S. Hammond (1947). The development of the autonomic nervous system. *Biol. Rev.* **22**, 344–359.

Yoon, C. (1972). Developmental mechanism for changes in cerebellum of "staggerer" mouse, a neurological mutant of genetic origin. *Neurology* **22**, 743–754.

Zwaan, J., and R. W. Hendrix (1973). Changes in cell and organ shape during early development of the ocular lens. *Am. Zool.* **13**, 1039–1049.

CHAPTER THIRTEEN

Complex Carbohydrates and Intercellular Adhesion

SAUL ROSEMAN

It is apparent that the surfaces of eukaryotic cells play an extremely important role in diverse developmental and physiological phenomena. These surfaces are responsible for the phenomenon of intercellular adhesion, for example, and cell adhesion is thought to be a primary process in the development of embryonic organs, the normal growth of tissues, the advent of malignancy, and so on (Abercrombie and Ambrose, 1962; Weiss and Mayhew, 1967). This elucidation of cellular adhesion on a molecular basis would prove invaluable to our understanding of these processes of cell recognition and adhesion. In recent years several lines of evidence have suggested that the surface glycocalyx, containing a variety of complex carbohydrates, may be involved in cellular adhesion. The surfaces are rich in these complex compounds, and they show the kind of diversity that would be necessary to achieve the specificity of intercellular action that is consistently observed.

Studies in our laboratory have been directed toward the elucidation of pathways of synthesis of the complex carbohydrates and their potential role in intercellular adhesion. For this purpose the biosynthetic pathways are briefly described, as well as three methods for the determination of intercellular adhesion, and the application of these methods to specific problems.

BIOSYNTHESIS OF COMPLEX CARBOHYDRATES

The complex carbohydrates may be divided into different classes on the basis of their components and structures. Representative classes include polysaccharides,

polysaccharides linked to protein as in the glycosaminoglycans, glycoproteins, and glycolipids. Only the glycoproteins and the gangliosides, a subclass of sphingoglyco-lipids, will be considered here. The glycoproteins comprise two major groups of substances, the serum type of glycoprotein and the mucins. The serum glycoproteins contain from one to several large branched oligosaccharide units attached to the protein chain via N-glycosylamine bonds, where the N atom is the amide nitrogen of asparagine in the protein chain, and the glycose unit is a β-N-acetylglucosaminyl (pyranosyl) residue. (All sugars are of the D configuration except L-fucose). In the mucins, on the other hand, as many as 800 short-chained oligosaccharides (Fig. 1), are linked to the polypeptide core through O-glycosidic bonds to serine and threonine, and the sugar residue at the linkage point consists primarily of N-acetyl-galactosamine. The sphingoglycolipids comprise a group of compounds containing mono- or oligosaccharides attached to N-acylated sphingosine (ceramide). Varia-tions occur in the oligosaccharide chains, as well as in the fatty acid moieties of the sphingosines. The gangliosides contain one or more sialic acid residues in the oligo-saccharide moiety.

The biosynthesis of these complex and diverse molecules has been extensively studied and reviewed (Roseman, 1968, 1970; Shachter and Roden, 1973). Never-theless, it will be necessary to recapitulate the stages of synthesis briefly here, in order for some of the later results to be understood. A general mode of synthesis for the complex carbohydrates has emerged, which may be summarized in the fol-lowing way. The complex carbohydrates are formed by stepwise addition of the monosaccharides from their respective sugar-nucleotides to the ends of oligosacchar-ide chains, each step being catalyzed by a specific glycosyltransferase. The battery of glycosyltransferases required for the synthesis of each complete molecule is local-ized in a multiglycosyltransferase system (Roseman, 1968). These reactions are illustrated schematically in Figs. 2 through 4.

Several points emerge from these studies. The monosaccharides are added se-quentially to the ends of the incomplete carbohydrate chains or at branch points, and each step is catalyzed by a different glycosyltransferase; in general, when a

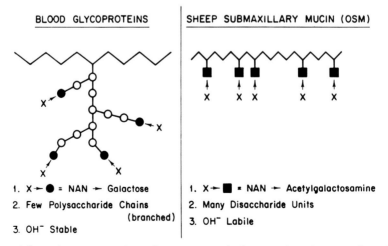

Figure 1. Schematic representation of two types of glycoproteins. See text for additional discussion.

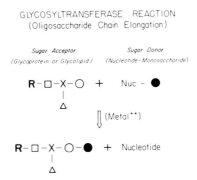

GLYCOSYLTRANSFERASE REACTION
(Oligosaccharide Chain Elongation)

Sugar Acceptor   *Sugar Donor*
(Glycoprotein or Glycolipid)   *(Nucleotide-Monosaccharide)*

R–□–X–O   +   Nuc – ●
      |
      △

⇓ (Metal⁺⁺)

R–□–X–O–●   +   Nucleotide
      |
      △

Figure 2. Schematic representation of glycosyltransferase reaction. R = protein or lipid. The incomplete oligosaccharide chain is represented by the branched tetrasaccharide containing the monosaccharides: □, X, △, 0.

$$R - Ⓐ - Ⓑ - Ⓓ - Ⓑ \xrightarrow[\text{Nucleotide-Ⓒ}]{\text{Transferase}} R - Ⓐ - Ⓑ - Ⓓ - Ⓑ$$
$$\quad\quad\quad | \quad\quad\quad\quad\quad\quad\quad\quad\quad\quad\quad\quad | \quad\quad\quad |$$
$$\quad\quad\quad Ⓒ \quad\quad\quad\quad\quad\quad\quad\quad\quad\quad\quad\quad Ⓒ \quad\quad Ⓒ$$

1. Sugar-nucleotides are glycose donors.

2. Sugars added as monosaccharides in a specific sequence.

3. Chain elongation at non-reducing ends or branch points.

R = protein or lipid

Figure 3. Elongation of oligosaccharide chain in a glycoprotein or glycolipid. Ⓐ, Ⓑ, Ⓒ, Ⓓ are monosaccharides. The reaction is catalyzed by a glycosyltransferase specific for the sugar-nucleotide, shown as nucleotide— C .

glycose moiety appears more than once in the same chain, a different glycosyltransferase is required for each transfer. This is a consequence of the fact that the glycosyltransferase shows high specificity for the acceptor molecule and may be influenced not only by the terminal monosaccharide residue, but by the penultimate one as well (Carlson et al., 1973a, 1973b; Bartholomew et al., 1973); indeed, different sialyltransferases may produce different structural isomers when the same acceptor is used. With this type of specificity, coupled with the fact that the product of one reaction provides the substrate for the next step, the sequential nature of the synthesis is ensured, eliminating the likelihood of "mistakes" in these complex structures. For completion of each of the molecules, however, each of the enzymes must be operating under optimal conditions. If any one is not, the sequence will not be completed, thus providing an explanation for the observed microheterogeneity of the mucins and oligosaccharide units in the serum type of glycoprotein, and for the fact that all of the possible intermediates from ceramide to tetrasialylganglioside are found in brain lipids.

Since the complex carbohydrates are synthesized by the action of glycosyltransferases, the distribution of these enzymes is of interest. They appear to be ubiquitous. We have studied the glycosyltransferases in a variety of tissues, including colostrum, liver, and embryonic brain tissue (Roseman, 1968, 1970), and they have

Each transferase specific for acceptor and its analogues:

1. *Different* transferase catalyzes each step.

2. Product of each step is substrate for next reaction.

3. *Different* MGT systems required for synthesis of glycoproteins, mucins, glycolipids.

Figure 4. Multiglycosyltransferase (MGT) systems. R = incomplete glycoprotein or glycolipid. The monosaccharide units are represented by (A), (B), (C), (D). The glycosyltransferases are represented by E_A, E_B, and so forth. Each enzyme requires the corresponding sugar-nucleotide as glycose donor. E_C and E_C^* (as well as E_B and E_B^*) represent *different* glycosyltransferases, although they utilize the same glycose donor.

also been found in cerebrospinal fluid and serum (Schachter and Roden, 1973; Den et al., 1970). A study of the subcellular location of the enzymes in embryonic chicken brain established that they are concentrated in the synaptosome-rich fraction; this fraction is rich in carbohydrates, and it has been postulated that such compounds may be involved in synaptic transmission or in the formation and attachment of synapses. In the liver, Schachter et al (1970) found that the glycosyltransferases are associated with the Golgi apparatus, suggesting that they may be involved in secretion of the plasma proteins. In view of the suggestion that the surface glycocalyx is involved in specific intercellular adhesion (discussed below), recent studies have attempted to determine whether the glycosyltransferases are in fact located on cell surfaces. Available evidence shows that galactosyltransferases appear to be present on the surfaces of embryonic neural retina cells (Roth et al., 1971b), fibroblasts grown in tissue culture (Roth and White, 1972), and intestinal cells (Weiser, 1973), as well as on human blood platelets which adhere to collagen (Jamieson et al., 1971).

CELL ADHESION

Despite the fundamental importance of cell adhesion and the considerable body of work that has been performed in this area of research, there is no generally accepted definition for the process. The manifold definitions are operational, that is, based on the method used by each investigator to study the phenomenon. Since cells in contact with other cells are capable of forming morphologically (or physiologically) detectable diverse structures such as desmosomes, gap junctions, tight junctions, and electrical connections, there is no a priori reason to suppose that each of the methods employed to study "adhesion" is measuring the same phenomenon. We have therefore elected to define cell adhesion as the first event that can be followed quantitatively, namely, the rate of formation of stable bonds between cells, or between cells and substrata (glass, tissue culture plastic, collagen, etc.). Specific or homologous intercellular adhesion is the rate at which homologous cells form stable intercellular bonds. Nonspecific or heterologous intercellular adhesion is the com-

TABLE 1. COMPARISON OF THREE METHODS FOR MEASURING INTERCELLULAR ADHESION

Method	Principle	Advantages	Disadvantages
Coulter counter (Orr and Roseman, 1969)	Measurement of rate of disappearance of single cells in a suspension in which aggregation occurs.	Simple, rapid, and accurate; can measure changes in rate of 10–15%; 50% or more of the cell population can be studied.	Cannot detect changes in a small percentage of the population or distinguish between specific and nonspecific adhesion; measures indirectly (i.e., by difference).
Collecting aggregate (Roth et al., 1971a)	Measurement of rate of attachment of labeled single cells to unlabeled cell aggregates.	Can distinguish between specific and nonspecific adhesion; permits independent treatment of cells and aggregates before measurement.	Considerable variation in reproducibility; may be measuring only a small percentage of total cell population.
Confluent layer (Walther et al., 1973)	Similar to collecting aggregate, but layer of cells on an artificial substratum replaces aggregates.	Reproducible, measuring up to 100% of cell population; can measure adhesion of cells to substratum or of cells to cells (specific and nonspecific); permits independent manipulation of single cells and cell layer.	Requires that collecting cells form a layer on substratum.

All methods determine the rate of adhesion of single cells to each other, or to aggregates or layers of cells of the same or a different type.

parable rate involving two different cell types. Cell-substrata cellular adhesion is defined similarly.

The methods used in this laboratory have recently been reviewed (Roseman et al., 1974), and only brief descriptions are given here. A summary of the advantages of each method is given in Table 1.

Our studies are based on the original observations by Moscona and Moscona (1952) and by Townes and Holtfreter (1955) that embryonic tissues can be dissociated into suspensions of single cells, and the aggregation of the single cells can be studied *in vitro*. We first measured the rate at which single cells in a suspension formed aggregates with each other by quantitatively determining the number of single cells remaining by means of the Coulter electronic particle counter (see Table 1). For certain purposes, this is the most convenient and accurate method available. However, because of the several disadvantages listed, and particularly because of the fact that this method was unable to distinguish between specific and nonspecific adhesion, we turned to the original method of Roth and Weston (1967), who had shown that intercellular adhesion could be studied using unlabeled cell aggregates as collecting fragments for single cells labeled with [³H]thymidine; the number of adhering cells was measured by radioautography. We extended and simplified this method by using ³²P-labeled single cells and measuring the amount of radioactivity by liquid scintillation techniques (Roth et al., 1971a).

This collecting aggregate assay could clearly distinguish between specific and nonspecific adhesion, but because of the variability of the results and other problems

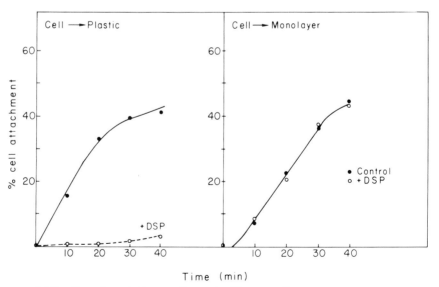

EFFECT OF DENATURED SERUM PROTEIN ON ADHESION
(BHK, 37°)

Time (min)
DSP = Denatured Serum Protein (40μg/ml)

Figure 5. Kinetics of cell adhesion to tissue culture plastic and to a homologous cell layer. The marked inhibition of denatured serum protein on adhesion to plastic is shown. No effect was observed as the adhesion of the BHK cells to a homologous cell layer.

SPECIFIC ADHESIVE PROPERTIES OF CHICK EMBRYONIC CELLS

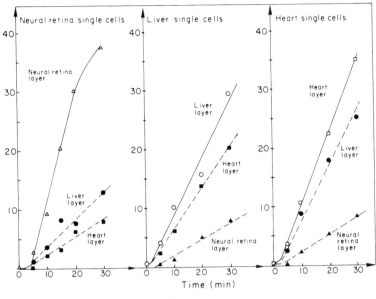

Open symbols = homologous adhesion
Closed symbols = heterologous adhesion

Figure 6. Specific and nonspecific intercellular adhesion of chick embryonic cells by the cell layer assay method.

encountered with the method, a further modification was introduced in which the aggregates were replaced by a confluent layer of cells. The suspension of labeled single cells was gently shaken above this layer, and after a specified time the remaining single cells were removed and the labeled single cells adhering to the cell layer were determined by counting. This method has clearly established (Walther et al., 1973) that the kinetics of adhesion of cells to the monolayer and to a glass or plastic substratum is different (Fig. 5), as well as distinguishing between specific and nonspecific adhesion (Fig. 6). The assay is particularly well adapted to studies of the cell surface, since the single cells and cell layer or substratum can be treated independently, for example, with enzymes or compounds such as lectins, known to affect adhesion. An investigation of the effect of concanavalin A on the adhesion process provided the basis for the development of a quantitative assay for concanavalin A-mediated cell agglutination (Rottmann et al., 1974.).

STUDIES ON THE INTERCELLULAR ADHESIVE PROCESS

Once methods for measuring intercellular adhesion had been developed, it became possible to test some of the hypotheses that had been put forward for the mechanism of the process. Considering the obvious importance of the phenomenon, few theories had, in fact, been offered. Any hypothesis would have to permit the type of reversible interaction that is found, for example, during the development of the embryo. Here, at certain stages of differentiation, cells that adhere to one another must separate, migrate to other locations in the embryo, and then once more adhere to each other or to other cells. The neural crest cells are one obvious example of this process. Perhaps the most imaginative working concept was proposed independently and simultaneously by Tyler (1947) and Weiss (1947), who suggested that apposing cell surfaces contained antigen-like and antibody-like molecules that interacted to form complexes, and this binding resulted in intercellular adhesion. The theory was attractive in that cell surfaces were known to be rich in substances capable of acting as antigens, and the only requirement was that they contained antibody-like or binding proteins as well. However, the main objection was that dissociation of the cells such as occurs in the differentiation of the embryo would require changes in, or even destruction of, the antigenic or antibody sites, and possibly rebuilding of these in order to permit further adhesion. We therefore proposed (Roseman, 1970) a modification of this theory in which enzymes and substrates (rather than antigens and antibodies) on opposing cell surfaces would interact with one another to result in intercellular adhesion. This hypothesis was developed further on the basis of the known fact that the surfaces of cells contain a wide variety of complex carbohydrates, and that these are in some way involved in intercellular adhesion, as described in the next section. We suggested that the substrates and enzymes involved in the adhesive process were the complex carbohydrates and their corresponding glycosyltransferases. As described in the section entitled "Biosynthesis of Complex Carbohydrates," the diversity of the complex carbohydrates and the specificity that is exhibited by the glycosyltransferases toward the carbohydrates would provide the kind of variation necessary to achieve the differences in adhesive specificity observed among cells.

An alternative hypothesis was also offered. Not only do hydrogen bonds exist between chains of carbohydrates, but also such bonds can be very firm. Cellulose and chitin are examples of this type. Conceivably, intercellular adhesion could result from hydrogen bonding between the oligosaccharide chains in the glycocalyx on opposing cell surfaces. Again, the variety of oligosaccharides is so great that it could readily account for selective adhesion. Since little is known of the three-dimensional conformation of these oligosaccharides, and essentially nothing of their relative capacities to form hydrogen bonds, the experimental approaches currently available to test this model are almost nonexistent. We can only hope that this is not the mechanism involved in intercellular adhesion.

The Tyler-Weiss and the two additional hypotheses described above are shown in Fig. 7. As indicated above, it is important that selective adhesion be modified during certain stages in embryonic development. One advantage of the glycosyltransferase-carbohydrate model is that such modification could occur, that is, by glycose transfer (Fig. 8). The appropriate sugar-nucleotide generated in the cytoplasm would act as the glycose donor to the transferase (shown as penetrating the membrane), and the latter would add this glycose unit to the incomplete chain. The enzyme, no longer capable of binding to the product, would dissociate from it, yielding either cell-cell dissociation or, in the case where the next transferase in the sequence was on the surface, leading to a more adhesive bond. The rest of this section deals with experiments suggesting that carbohydrates are indeed involved in intercellular adhesion.

By studying the interaction of single cells in specified media, it is possible to determine some of the requirements for aggregation. With the use of the Coulter

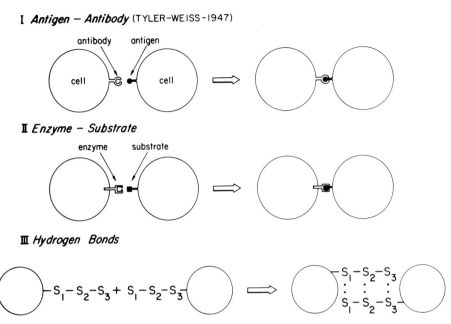

Figure 7. Three hypotheses for the mechanism of intercellular adhesion. S_1, S_2, S_3 represent monosaccharide units in the oligosaccharides of the complex carbohydrates in the glycocalyx.

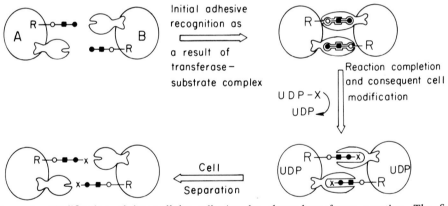

Figure 8. Modification of intercellular adhesion by glycosyltransferase reaction. The first step represents adhesion as a result of the binding of complex carbohydrate chain ($-\bigcirc-\square-\bullet$) to the corresponding glycosyltransferase. In the second step internally generated sugar-nucleotide (UDP-X) provides X to the enzyme on the internal face of the membrane. The transferase utilizes X to complete the reaction, and in the third step (cell separation) the product of the enzymatic reaction dissociates from the enzyme.

counter procedure (Orr and Roseman, 1969) it was found that single cells obtained by dissociating the "embryoid" bodies of teratoma would adhere to each other in a complete synthetic tissue culture medium, but were unable to aggregate in a glucose-salts medium (Oppenheimer et al., 1969). Of the 51 components present in the tissue culture medium, only L-glutamine promoted adhesion when added to the glucose-salts medium. L-Glutamine could, however, be replaced by D-glucosamine

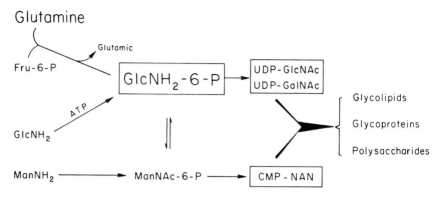

Figure 9. Abbreviated pathway for the synthesis of complex carbohydrates. GlcNH₂-6-P is D-glucosamine 6-phosphate, the key intermediate in the biosynthesis of the nitrogen-containing sugar-nucleotides: GlcNAc, N-acetylglucosamine; GalNAc, N-acetylgalactosamine; NAN, sialic acid. The other abbreviations are as follows: Fru-6-P, fructose 6-phosphate; GlcNH₂, glucosamine; ManNH₂, mannosamine; ManNAc-6P, N-acetylmannosamine 6-phosphate.

and D-mannosamine, but not by other substances such as D-galactosamine. These results can be explained by referring to the pathways of biosynthesis of the complex carbohydrates shown in Fig. 9. L-Glutamine is the amide donor in the synthesis of glucosamine 6-phosphate (Ghosh et al., 1960), the precursor of all the nitrogen-containing sugars, including the sialic acids. D-Glucosamine and D-mannosamine, being substrates for hexokinase, could replace L-glutamine, but no other components of the medium could perform this function. Thus, adhesion was prevented in the absence of synthesis of the complex carbohydrates, unequivocally linking these compounds in some way with intercellular adhesion.

The ability of D-glucosamine to increase the adhesiveness of teratoma cells has recently been extended to two tissue culture cell lines, mouse fibroblasts (3T3), and virally transformed fibroblasts (SV40/3T3). The results (Hellerqvist et al., 1974), shown in Fig. 10, indicate that about a twofold increase in the rate of adhesion by the cell layer assay. There was, however, a difference between the results with the tissue culture and those with the teratoma cells. In the latter case, the glucosamine effect was almost immediate (generally, a slight lag was observed). By contrast, the effect with the fibroblasts was evident only when the cells were grown for about one generation in the presence of low concentrations of the hexosamine. The fibroblast effect was concentration dependent, about 1mM giving optimal results; no effect was detected at 0.05 mM concentration. Furthermore, the effect was clearly on the monolayer, not on the cells used for preparation of the single cell suspension; in fact, glucosamine seemed to decrease the adhesiveness of the single cells (about 40%). It should be emphasized that all these experiments were conducted at the normally high glucose concentrations used for cell growth, that is, Dulbecco's modified Eagle's medium containing 10% calf serum. The effect was glucosamine specific. The following sugars did not influence the adhesive rates: galactosamine, galactose, N-acetyl-D-glucosamine, and N-acetyl-D-galactosamine. Finally, at 1mM concentration, glucosamine had no detectable effect on the morphology or growth rate of the cells used in these studies.

We have recently taken an entirely different approach to the problem. The idea

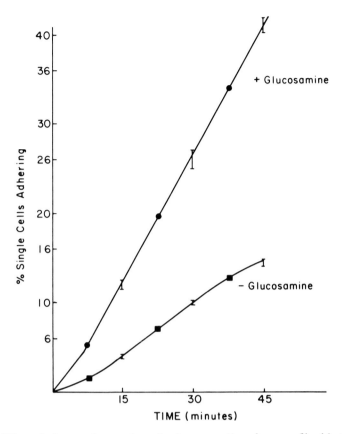

Figure 10. Effect of glucosamine on the adhesive properties of mouse fibroblast (3T3) cells grown in culture. See text for details.

is that, if cells respond by changes in growth rates, motility, adhesiveness, etc., when they come into contact with the surfaces of other cells, then perhaps they might show such responses to artificial synthetic analogs of cell surfaces. As a first approach to the problem (Chipowsky et al., 1973), monosaccharides were covalently linked to Sephadex beads 10–40 μ in diameter. The monosaccharides were β-pyranosides of N-acetylglucosamine, glucose, and galactose. Controls consisted of Sephadex beads per se, CNBr-activated but otherwise untreated beads, and beads containing the "linker" arm (6-aminohexanol) without the sugar moiety. When the beads were mixed with the suspensions of SV40/3T3 cells and observed microscopically, little or no effect was observed with all but one type, the galactose beads. With the latter, large aggregates rapidly formed, the aggregates containing both beads and cells. While the assay is visual and subjective, and requires modification to a reliable quantitative method, it appears that two important and discrete phenomena rapidly followed admixture of the galactose beads with the cells: (*a*) adhesion of the cells to the beads; (*b*) a marked increase in the adhesiveness of cells attached to the beads for other cells in the suspension. If the latter interpretation proves to be correct, it suggests that the adhesiveness of a given preparation of cells may not be a constant property and may be rapidly altered. In recent experiments we have found that this may indeed be true, and are being forced to the conclusion that intercellular adhesion is a much more complex process than hitherto suspected.

The experiments described above indicated carbohydrate specificity for intercellular adhesion. They also suggest cell specificity in that BHK fibroblasts showed no response with any of the beads tested, while 3T3 cells appeared much less reactive than did SV40/3T3.

As is obvious, the derivatized bead approach appears to be highly promising not only in attempts to define the molecular basis for intercellular adhesion, but also in regard to other equally interesting phenomena of cell interactions, such as motility and growth regulation.

One of the hypotheses described above suggests that intercellular adhesion occurs by binding of cell surface glycosyltransferases to their corresponding acceptors. This idea requires, therefore, that glycosyltransferases be present on cell surfaces, and, as noted above, this prediction has been supported by experimental evidence with several different cell lines. Our own view is that the results are indicative but do not unequivocally establish the existence of cell surface glycosyltransferases.

To sum up, some evidence has been provided implicating cell surface carbohydrates in the process of intercellular adhesion. The results also suggest that glycosyltransferases exist on cell surfaces. There is no available evidence to support the view that enzyme-carbohydrate or carbohydrate-carbohydrate complexes are responsible for adhesion.

MEMBRANE MESSENGER THEORY

The preceding section was concerned with the molecular mechanism by which eukaryotic cells adhere to one another and to substrata such as glass and plastic. Whereas no definite conclusions were offered, certain suggestions were made, for example, that glycosyltransferases and complex carbohydrates on apposing cell surfaces interacted to result in intercellular adhesion, and some evidence was provided to support them. This leads to a broader topic: How do eukaryotic cells produce the profound responses that are frequently observed once adhesive bonds have been established? For example, how are the far-reaching changes that occur in growth rate, metabolic patterns, differentiation, permeability, enzyme induction, motility, and so forth actually brought about? While some signals from the environment (e.g., steroid hormones) exert their effects directly on the cell after penetration of the cell membrane, many act on the membrane per se. How, in fact, does the plasma membrane transmit or translate extracellular signals?

A typical example of direct interaction between external stimuli and the cell membrane is contact inhibition of growth. It is a common experience in tissue culture that suspensions of dissociated diploid cells will not divide, although they will continue to metabolize at approximately normal rates in suitable media. When such cells are placed over a substratum to which they can adhere, such as glass, they attach to it, flatten out, and grow; growth continues in the monolayer until the cells come into contact on all sides with other cells (confluency), at which point growth stops. Although such cells usually will not divide in suspension culture, some cell lines will grow under these conditions if the viscosity of the medium is substantially increased. Thus one may ask: How are the physical environmental stimuli, contacts with glass or with a more viscous medium or with other cells, translated by the plasma membrane and transmitted to the cytoplasm as signals to grow or to stop growing?

A simple and general explanation for the diverse but specific effects exhibited by external stimuli on cells is shown in Fig. 11. It is suggested that the cell membrane acts as a transducer which translates the external stimulus to an internal signal (or signals) via a *membrane messenger*. This idea represents an extrapolation of the familiar Jacob-Monod nuclear messenger theory and a generalization of the "second messenger" hypothesis of Sutherland et al. (1965), in which the "second messenger" is cyclic AMP. The latter workers proposed that the profound responses observed when target cells interacted with various hormones resulted from changes in the internal levels of cyclic AMP (or other nucleotides). The hormones interacted with membrane receptors, which in turn stimulated or inhibited the appropriate enzymes (adenyl cyclase, cyclic AMP phosphodiesterase), thereby regulating the level of cyclic AMP in the cytoplasm. More recently, Rasmussen and Tenenhouse (1968) have invoked Ca^{2+} as a "second messenger." Previous suggestions are not sufficient to explain the diverse responses which can be elicited from a single cell type by different environmental stimuli. However, the generalization shown in Fig. 11 is that membranes contain a variety of receptors, specific for different stimuli, and each receptor then releases a different membrane messenger. Two questions then arise: What are the messengers, and how, in fact, do they work?

If membrane messengers exhibit a wide spectrum of physiological functions, it is

Membrane Messenger Hypothesis

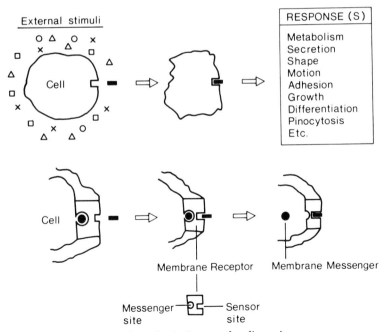

Figure 11. Membrane messenger hypothesis. See text for discussion.

reasonable to suppose that the messengers themselves are diverse, ranging from low molecular weight substances to macromolecules. One interesting possibility is RNA. Significant quantities of RNA have frequently been detected in even the most extensively purified animal cell plasma membranes (Glick and Warren (1969), and such preparations are capable of synthesizing protein. If the membrane receptor proteins respond to a specific external stimulus to release membrane-bound RNA, such RNA could have two important functions. First, it could serve as messenger RNA, resulting in a burst of *specific* protein synthesis as a consequence of the external signal. In this way, the cell could generate specific proteins without activating the corresponding genes. Second, membrane messenger RNA could regulate the transcription of selected genes. The latter idea is not unique, since at least one species of RNA, His tRNA is thought to be involved in regulating the histidine operon in *Salmonella* (Brenner and Ames, 1971), while "reverse transcriptase" utilizes RNA (viral RNA) as a template for DNA sythesis (Temin and Mizutani, 1970; Baltimore, 1970; Goodman and Spiegelman, 1972).

Some membrane messengers may be proteins. For example, adenyl cyclase is already fairly well established in this role. In terms of the present concept (Fig. 11), it is adenyl cyclase rather than cyclic AMP which is the membrane messenger, for it is the enzyme which is bound to the membrane receptor that is activated when the latter forms a complex with the appropriate hormone. Membrane messenger proteins may be enzymes, such as adenyl cyclase, but may also have entirely different functions, such as regulation of enzyme activity in the cytoplasm or, even more interestingly, of gene expression in the nucleus. Models for the latter exist, as, for example, the repressor proteins of the familiar Jacob-Monod hypothesis. The presence of membrane-bound activator or repressor proteins that can selectively regulate the transcription of certain genes, for example, provides an attractive mechanism for explaining the interesting observation that induction of synthesis of glucose-6-P permease in *Escherichia coli* is achieved from without; that is, external but not internal glucose-6-P (or 2-deoxyglucose-6-P) is required for induction (Dietz and Heppel, 1971).

Membrane messengers may also be substances of low molecular weight. As indicated above, calcium ion has already been postulated to serve such a function.

If membrane messengers are enzymes, activators, inhibitors, mRNA, and so forth, this would suggest that the messengers act by amplification. A few molecules of a membrane messenger released to the cytoplasm will show more than stoichiometric effects because they act on enzymes or enzyme systems. On the other hand, membrane messengers may act stoichiometrically, such as the repressor protein described above. Stoichiometry may be very important in some phenomena. For example, the mouse 3T3 cell line will grow on a glass plate until it reaches confluency, when growth stops (contact inhibition of growth). During the growth phase, the cells remain in contact with other cells in the culture, except on one side of the cell surface. When this surface comes into contact with another cell, growth stops. In this case, perhaps a minimum number or threshold level of membrane messengers must be released before the cell responds to the external stimulus.

One final question concerns the mechanism by which membrane messengers are generated and maintained. The simplest idea is based on the experimental observation (Warren, 1969) that the proteins of animal cell membranes show a surprisingly high rate of turnover, and this rate of turnover is maintained when *cell growth has stopped*. Turnover may be required to keep the normal complement of receptor proteins and messengers fully available to meet fluxes in the environment.

SUMMARY

The general mechanisms for the biosynthesis of the oligosaccharide chains of glycoproteins, mucins, and certain sphingoglycolipids are briefly reviewed, with emphasis on the glycosyltransferases. It is proposed that the complex carbohydrates, known to occur in the glycocalyx of eukaryotic cells, participate in intercellular adhesion. The adhesive process would occur by binding of the carbohydrates on apposing cell surfaces to each other via hydrogen bonds, or by the binding of the carbohydrates to the corresponding cell surface glycosyltransferase. Evidence is reviewed indicating that carbohydrates are indeed involved in intercellular adhesion and that glycosyltransferases are present on cell surfaces.

Cell recognition and adhesion are only the first of a series of responses that cells manifest towards each other. Phenomena which may be regulated by cell surfaces include motility, growth control, secretion, pinocytosis, exocytosis, and differentiation. Since a single cell can respond in diverse ways to different external stimuli, it is suggested that the membrane and glycocalyx act as transducers of such stimuli. Each stimulus is translated by a different membrane messenger to the cytoplasm or to cell organelles. Possible types of membrane messengers are considered.

ACKNOWLEDGMENT

This review was written in collaboration with Dr. Pamela Talalay. We express our deep appreciation to Mrs. Dorothy Regula for her expert assistance in preparing the manuscript. The studies reported here were supported by grants from the National Institutes of Health, the American Cancer Society, and the National Cystic Fibrosis Research Foundation, Publication 780 from the McCollum-Pratt Institute.

REFERENCES

Abercrombie, M., and E. J. Ambrose (1962). The surface properties of cancer cells. *Cancer Res.* **22**, 525–548.

Baltimore, D. (1970). RNA-dependent DNA polymerase in virions of tumour virus. *Nature* **226**, 1209–1211.

Bartholomew, B. A., G. W. Jourdian, and S. Roseman (1973). The sialic acids. XV. Transfer of sialic acid to glycoproteins by a sialyltransferase from colostrum. *J. Biol. Chem.* **248**, 5751–5762.

Brenner, M., and B. N. Ames (1971). The histidine operon and its regulation. In *Metabolic Pathways*, 3rd ed., Vol. V: *Metabolic Regulation* (H. J. Vogel, ed.). Academic Press, New York, pp. 349–387.

Carlson, D. M., G. W. Jourdian, and S. Roseman (1973a). The sialic acids. XIV. Synthesis of sialyl-lactose by a sialyltransferase from rat mammary gland. *J. Biol. Chem.* **248**, 5742–5750.

Carlson, D. M., E. J. McGuire, G. W. Jourdian, and S. Roseman (1973b). The sialic acids. XVI. Isolation of a mucin sialyltransferase from sheep submaxillary gland. *J. Biol. Chem.* **248**, 5763–5773.

Chipowsky, S., Y. C. Lee, and S. Roseman (1973). Adhesion of cultured fibroblasts to insoluble analogues of cell-surface carbohydrates. *Proc. Natl. Acad. Sci. U.S.A.* **70**, 2309–2312.

Den, H., B. Kaufman, and S. Roseman (1970). Properties of some glycosyltransferases in embryonic chicken brain. *J. Biol. Chem.* **245**, 6607–6615.

Dietz, G. W., and L. A. Heppel (1971). Studies on the uptake of hexose phosphates. II. The induction of the glucose 6-phosphate transport system by exogenous but not endogenously formed glucose 6-phosphate. *J. Biol. Chem.* **246**, 2885–2890.

Ghosh, S., H. Blumenthal, E. Davidson, and S. Roseman (1960). Glucosamine metabolism. V. Enzymatic synthesis of glucosamine 6-phosphate. *J. Biol. Chem.* **235**, 1265–1275.

Glick, M. C., and L. Warren (1969). Membranes of animal cells. III. Amino acid incorporation by isolated surface membranes. *Proc. Natl. Acad. Sci. U.S.A.* **63**, 563–570.

Goodman, N. C., and S. Spiegelman (1972). Distinguishing reverse transcriptase of an RNA tumor virus from the other known DNA polymerases. *Proc. Natl. Acad. Sci. U.S.A.* **68**, 2203–2206.

Hellerqvist, C., W. T. Rottmann, B. Walther, and S. Roseman (1974). Cell Adhesion. In *Proceedings of the Los Alamos (New Mexico) Symposium (1973): Mammalian Cells: Probes and Problems* (in press).

Jamieson, G. A., C. L. Urban, and A. J. Barber (1971). Enzymatic basis to platelet:collagen adhesion as the primary step in haemostatis. *Nature New Biol.* **234**, 5–7.

Moscona, A., and H. Moscona (1952). The dissociation and aggregation of cells from organ rudiments of the early chick embryo. *J. Anat.* **86**, 287–301.

Oppenheimer, S. B., M. Edidin, C. W. Orr, and S. Roseman (1969). An *L*-glutamine requirement for intercellular adhesion. *Proc. Natl. Acad. Sci. U.S.A.* **63**, 1395–1402.

Orr, C. W., and S. Roseman (1969). Intercellular adhesion. I. A quantitative assay for measuring the rate of adhesion. *J. Membrane Biol.* **1**, 109–124.

Rasmussen, H., and A. Tenenhouse (1968). Cyclic adenosine monophosphate, Ca++, and membranes. *Proc. Natl. Acad. Sci. U.S.A.* **59**, 1364–1370.

Roseman, S. (1968). Biochemistry of glycoproteins and related substances. In *Proceedings of the 4th International Conference of Cystic Fibrosis of the Pancreas (Mucoviscidosis)* Part II (E. Rossi and E. Stoll, eds.). S. Karger, New York, pp. 244–269.

Roseman, S. (1970). The synthesis of complex carbohydrates by multiglycosyltransferase systems and their potential function in intercellular adhesion. *Chem. Phys. Lipids* **5**, 270–297.

Roseman, S., W. Rottmann, B. Walther, R. Öhman, and J. Umbreit (1974). Cell Ádhesion. In *Methods in Enzymology* (in press).

Roth, S., and J. A. Weston (1967). The measurement of intercellular adhesion. *Proc. Natl. Acad. Sci. U.S.A.* **58**, 974–980.

Roth, S., and D. White (1972). Intercellular contact and cell surface transferase activity. *Proc. Natl. Acad. Sci. U.S.A.* **69**, 485–489.

Roth, S. E. J. McGuire, and S. Roseman (1971a). An assay for intercellular adhesive specificity, *J. Cell Biol.* **51**, 525–535.

Roth, S., E. J. McGuire, and S. Roseman (1971b). Evidence for cell-surface glycosyltransferases: their potential role in cellular recognition. *J. Cell Biol.* **51**, 536–547.

Rottmann, W. L., B. T. Walther, C. G. Hellerqvist, J. Umbreit, and S. Roseman (1974). A quantitative assay for concanavalin A-mediated cell agglutination, *J. Biol. Chem.* **249**, 373–380.

Schachter, H., and L. Roden (1973). The biosynthesis of animal glycoproteins. III. In Metabolic conjugation and metabolic hydrolysis **3** (W. H. Fishman, ed.) Academic Press, New York, 1–49.

Schachter, H., I. Jabbal, R. Hudgin, L. Pinteric, E. J. McGuire, and S. Roseman (1970). Intracellular localization of liver sugar nucleotide glycoprotein glycosyltransferases in a Golgi-rich fraction. *J. Biol. Chem.* **245**, 1090–1100.

Sutherland, E. W., E. Øye, and R. W. Butcher (1965). The action of epinephrine and the role of the adenyl cyclase system in hormone action. In *Recent Progress in Hormone Research*, Vol. XXI (G. Pincus, ed.). Academic Press, New York, pp. 623–642.

Temin, H. M., and S. Mizutani (1970). RNA-dependent DNA polymerase in virions of Rous sarcoma virus. *Nature* **226**, 1211–1213.

Townes, P. L., and J. Holtfreter (1955). Directed movements and selective adhesion of embryonic amphibian cells. *J. Exp. Zool.* **128**, 53–120.

Tyler, A. (1947) An auto-antibody concept of cell structure, growth and differentiation. *Growth* (Symposium) **10**, 6–7.

Walther, B. T., R. Öhman, and S. Roseman (1973). A quantitative assay for intercellular adhesion. *Proc. Natl. Acad. Sci. U.S.A.* **70**, 1569–1573.

Warren, L. (1969). The biological significance of turnover of the surface membrane of animal cells. *In Current Topics in Developmental Biology*, Vol. 4 (A. Moscona and A. Monroy, eds.). Academic Press, New York, pp. 197–222.

Weiser, M. M. (1973). Glycosyltransferases and endogenous acceptors of the undifferentiated cell surface membrane. *J. Biol. Chem.* **248**, 2542–2548.

Weiss, L., and E. Mayhew (1967). The cell periphery. *New Engl. J. Med.* **276**, 1354.

Weiss, P. (1947). The problem of specificity in growth and development. *Yale J. Biol. Med.* **19**, 235–278.

CHAPTER FOURTEEN

Cell Differentiation Analyzed by Somatic Cell Hybridization

NILS R. RINGERTZ

Somatic cell hybridization was introduced relatively recently but has already proved to be an extremely powerful experimental technique with applications in cell biology, genetics, developmental biology, tumor biology, and virology. Basically the technique involves spontaneous or induced fusion of two different cells into a hybrid cell. The first hybrid cells were discovered in 1960 by Barski, Sorieul, and Cornefert in Paris. These authors discovered a third cell type in a mixed culture of two different mouse cell lines. Analysis of the chromosomal content of the new cells revealed that they were hybrid cells containing chromosomes characteristic of both of the original cell types. The hybrid cells, which must have formed from spontaneous cell fusion, showed the ability to multiply *in vitro*.

Spontaneous cell fusion is, however, a very unusual phenomenon. On the basis of results obtained by Okada (1962), Harris and Watkins (1965) introduced Sendai virus-induced fusion as a technique for fusing different cells into multinucleated hybrid cells.

One of the most remarkable features of virus-induced cell fusion is the ease with which cells from completely different species or tissues can be fused. *Interspecific hybrids* can be formed by fusing cells from different species, while *intraspecific hybrids* result from the fusion of cells originating from the same species but from different tissues.

When Sendai virus, inactivated by ultraviolet irradiation, is added to a mixture of two types of mononucleated cells (A and B cells, respectively), some A cells will fuse with some B cells. The multinucleated hybrid cells which can be observed after virus-induced fusion are known as *heterokaryons* and contain both A-type and B-type nuclei within a common cytoplasm. In addition to these, there are some other multinucleated cells which contain only one type of nuclei (A or B). These are not hybrid cells and are now known as *homokaryons*. Most heterokaryons have limited viability, but some give rise to mononucleated hybrid cells which are capable of *in vitro* multiplication for long periods of time. This type of cell is commonly referred to as a "hybrid" or *synkaryon*. Intraspecific hybrids often contain a complete set of both A and B chromosomes. In man × mouse hybrids the human chromosomes are selectively eliminated (Weiss and Green, 1967).

A new modification of the cell fusion technique which promises to be of great interest is based on the reintroduction of nuclei into enucleated cytoplasms of cells representing another genotype or phenotype. We refer to these cells as *reconstituted cells* (Ege et al., 1973).

In the following, I shall indicate how hybridization of somatic cells has been used to analyze gene regulation and cell differentiation in eukaryotic cells. For more complete reviews of this field the reader is referred to monographs by Harris (1970), Ephrussi (1972), and Ringertz and Savage (1975).

REGULATION OF NUCLEIC ACID SYNTHESIS IN HETEROKARYONS

In heterokaryons formed by the fusion of actively growing cells (tumor cells or established cell lines) with highly differentiated but metabolically less active cells (lymphocytes, macrophages, nucleated erythrocytes), the inactive nuclei of the differentiated cells undergo a marked activation process (for references see Harris, 1970, and Ringertz and Bolund, 1974a). RNA synthesis is initiated or accelerated, and in some cases DNA replication begins. Fusion of human tumor cells (HeLa) with nucleated avian erythrocytes is an example of the activation, in heterokaryons, of dormant, inactive cell nuclei (Harris, 1967). The nuclei present in mature hen erythrocytes are inactive with respect to nucleic acid and protein synthesis. The erythrocyte nucleus in heterokaryons formed by fusing chick erythrocytes with tumor cells (Fig. 1) undergoes a reactivation process and begins to synthesize RNA and DNA. Later the synthesis of new chick proteins in the cytoplasm of the hybrid cells is detected. It appears, therefore, that the reactivation of the chick erythrocyte nucleus involves transcription as well as translation, that is, the RNA produced by the chick erythrocyte nucleus specifies the synthesis of chick-specific protein. The signals activating the chick genes appear to operate at several different levels. Early changes in the physicochemical properties of chick deoxyribonucleoprotein, which are referred to as "chromatin activation," appear to be triggered by changes in the ionic environment of the erythrocyte nucleus by its contact with HeLa cell cytoplasm (for review see Ringertz and Bolund, 1974a, b). The chicken erythrocyte nucleus also increases in size; its dry mass has been found to increase more than fivefold during the first 48 hours of reactivation process. The mass increase of the chick erythrocyte nucleus during the reactivation process appears to be due to the migration of special types of proteins from the cytoplasm into the chick erythrocyte nucleus. One remarkable characteristic of this process is that the proteins migrating into the chick nucleus appear to be mainly human proteins. These proteins do not seem to represent a cross section of the proteins found in the

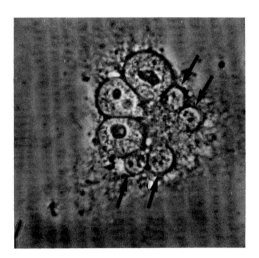

Figure 1. Chick erythrocyte × HeLa cell heterokaryon. Arrows indicate positions of four partially reactivated chick erythrocyte nuclei.

surrounding cytoplasm. Instead it appears that a special category of nuclear-specific human proteins is selectively concentrated in chick erythrocyte nuclei (Ringertz et al., 1971). By producing large numbers of heterokaryons and then reisolating the reactivated erythrocyte nuclei on sucrose gradients (Goto and Ringertz, 1974), it has been possible to examine this phenomenon biochemically (Appels et al., 1974; Bolund et al., 1974).

Reactivation of dormant cell nuclei has also been observed in other types of heterokaryons. Three major conclusions can be drawn from this work. The first is that the more active partner in the fusion always seems to stimulate the nucleus of the more inactive partner to synthesize more RNA and/or DNA. As yet, there are no indications that the inactive parental cell could inhibit or repress the more active nucleus. Second, the inactive nucleus seems to be activated to about the same level as the active nucleus but not beyond this level. Thus, in situations where the active cell is synthesizing RNA but not DNA (macrophage, myotubes), the inactive nucleus is stimulated to make only RNA. If, on the other hand, the active partner is synthesizing both types of nucleic acid, the inactive nucleus is also stimulated to synthesize *both* RNA and DNA. The third conclusion is that the signals which regulate the level of RNA and DNA synthesis do not seem to be species specific.

RECONSTITUTED CELLS

Another development of great interest for the future is the method of *enucleation,* developed by Prescott et al. (1972). This technique is based on the original observation of Carter (1967) that the drug *cytochalasin* causes nuclear extrusion and results in the formation of a few enucleated cells. Prescott et al. (1972) centrifuged cells adhering to a glass surface in the presence of cytochalasin and managed to increase the yield of enucleated cells by 90–100%. Such enucleated cells ("cytoplasms") remain viable for a couple of days and can be fused with cells with the aid of Sendal virus (Poste and Reeve, 1971; Ege et al., 1973). It is also possible

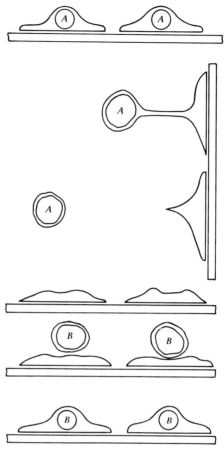

Figure 2.   Schematic illustration of enucleation and reconstitution of cells by fusing minicells (*B*) with enucleated cytoplasm (*A*). (From Ege et al., 1973.)

to put a new nucleus back into an enucleated cytoplasm without adding much of the cytoplasm from the nuclear donor (Fig. 2). The "nuclei" used in this type of experiment are surrounded by a narrow rim of cytoplasm and a plasma membrane and therefore should be referred to as *"minicells"* rather than nuclei. (Ege et al., 1974a). The minicells can be produced by centrifuging large numbers of cells on a bovine serum albumin gradient containing cytochalasin. By fusing minicells with enucleated cytoplasms on a glass slide, it is possible to *reconstitute* viable cells.

In our experiments (Ege, Peters, and Ringertz) we have used chick erythrocytes and lymphocytes as nuclear donors in reconstitutions with enucleated L-cell mutants. In both these systems the nuclei are activated to synthesize RNA and DNA. Minicell nuclei strongly prelabeled with ^{3}H-thymidine have also been used successfully as nuclear donors in reconstitutions with L-cell cytoplasms weakly prelabeled with ^{3}H-leucine (Fig. 3). By using this double-labeling technique it is possible to discriminate reconstituted cells from contaminating intact nuclear or cytoplasmic donors (Ege et al., 1974b). The phenotypes of reconstituted cells have not yet been analyzed, but it appears that the reconstitution technique will prove to be of great interest in the analysis of cell differentiation.

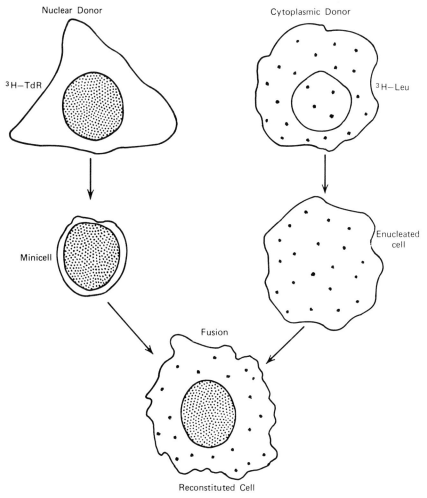

Figure 3. Labeling procedure to permit the recognition of reconstituted cells in a mixture of minicells and enucleated cells which could also be contaminated with a few intact nuclear or cytoplasmic donors. (Ege, Peters, and Ringertz, in preparation.)

GENE EXPRESSION IN HYBRID CELLS

Parental Cells and Markers

Four main types of cells have been used:

1. Totipotent or multipotent cells, for example, eggs, early cleavage stages, and teratoma cells (an unusual type of tumor cells arising in the gonads).
2. Normal diploid cell strains established from differentiated tissues.
3. Cell lines established from differentiated tumors.
4. Cell lines which have spent a very long time in culture and no longer exhibit any differentiated markers ("dedifferentiated" cells).

A great variety of properties have been analyzed in hybrid cells, including general characteristics such as morphology, growth rate, presence or absence of contact inhibition, electrophysiological and antigenic properties of the cell surface, and specific gene products (e.g., enzymes, hormones, immunoglobulins, pigments, and contractile muscle proteins). A distinction is made between properties common to all cells of a given organism (*constitutive markers*) and properties expressed by only certain determined cells as they enter the differentiated state (*facultative markers*) (Ruddle, 1972). Enzymes which are part of vital metabolic pathways common to all cells and are necessary for the survival of the single cell are termed "household proteins" (Ephrussi, 1972). Facultative markers, on the other hand, are more or less synonymous with "luxury proteins" (Ephrussi, 1972) since they represent properties which are expressed only by cells representing certain epigenotypes. Facultative markers are not essential to the survival of the single cell but are of vital importance to the organism as a whole. Enzymes of the citric acid cycle and oxidative phosphorylation and RNA and DNA polymerase are examples of constitutive markers. Hemoglobin, myosin, melanin, tissue-specific surface antigens, and the like are considered to be facultative markers.

Phenotypic Expression in Heterokaryons

The advantage with heterokaryons is that phenotypic expression can be studied from the time of fusion, before any selection or chromosomal losses have occurred. It is often possible to examine the role of gene dosage in phenotypic expression since virus-induced fusion results in heterokaryons containing varying numbers and proportions of the parental nuclei. Since single cells are analyzed, one can use culture conditions which favor cell differentiation even if cell multiplication is thereby inhibited.

Although only relatively few studies of phenotypic expression in heterokaryons have been made so far, different expression patterns have been recognizd. Most hybrid cells, whether heterokaryons or mononucleated hybrids, express properties characteristic of both parental cells (*coexpression* or *codominance*). Watkins and Grace (1967) found surface antigens characteristic of both parental cells in heterokaryons made by fusing human HeLa cells with Ehrlich cells. A similar case of coexpression was observed by Carlsson et al. (1973) in studies of hybrid myotubes. Myotubes made by fusing chick and rat myogenic cells synthesize both chick and rat myosin. In other types of heterokaryons, however, properties characteristic of one parental cell are *extinguished* and cannot be detected in the heterokaryons. Extinction of differentiated properties was observed by Gordon and Cohn (1970, 1971) in macrophage × melanoma heterokaryons, where the macrophage marker properties disappeared. In heterokaryons formed by fusing rat hepatoma cells with rat epithelial cells, Thompson and Gelehrter (1971) observed that the inducibility of tyrosine aminotransferase (TAT), a property characteristic of liver differentiation, disappeared in the heterokaryon. These examples illustrate the two most important patterns of phenotypic expression: *coexpression* and *extinction* of properties characteristic of the parental cells.

Phenotypic Expression in Mononucleated Hybrid Cells

Coexpression of constitutive markers has been demonstrated for a very large number of inter- and intraspecific hybrids. Enzymes of intermediary metabolism, surface antigens, and ribosomal RNA (for references see Ephrussi, 1972, and Ringertz and Savage, 1975) are among these markers. In addition to the parental enzyme forms, new hybrid (heteropolymeric) enzymes have been found. If the two parental cells produce dimeric enzymes which are termed AA and BB, respectively, the hybrids are found to contain an AB enzyme in addition to the parental forms. The degree to which the given constitutive markers is expressed may vary. Intermediate levels of expression have been reported for several constitutive markers, and in interspecific hybrids a relative dominance of the enzyme of one species has sometimes been observed. Constitutive markers located on X chromosomes are of particular interest since in normal cells all X chromosomes except one are irreversibly inactivated. Hybrid cells have several active X chromosomes, thus indicating that inactivation of supernumerary X chromosomes does not occur in hybrid cells (Siniscalo et al., 1969).

Extinction of differentiated functions has been observed in many different types of hybrid cells; in such cases, one of the parental cells expresses a facultative marker characteristic of histotypic differentiation and the other parental cell is undifferentiated and lacks this marker. Thus it has been found that melanoma hybrids lose their pigmentation and dopa oxidase (for review see Davidson, 1971), while hepatoma hybrids lose both the base-line and induced tyrosine aminotransferase activities, alcohol dehydrogenase, alanine aminotransferase, and so forth (for review see Ephrussi, 1972). From these studies it appears that two main factors determine whether or not a specific property of the parental cells will be expressed. The first factor is chromosomal loss, and the second one is a regulatory phenomenon. Depending on the types and species of cells involved, the chromosomal losses may selectively affect one of the parental cells. In fusions of differentiated and undifferentiated cells, one expects very different results, depending on whether the chromosomes of the undifferentiated parent or those of the differentiated parent are lost. Chromosomal segregation in hybrid cells has opened the possibility of chromosome mapping, and some time in the future this phenomenon will probably be of great help in the elucidation of regulatory mechanisms.

In the attempts which have been made to analyze gene regulation by somatic cell hybridization, chromosomal losses have so far been a complicating factor since it is very difficult to determine whether the loss of a certain phenotypic trait is due to a regulatory event, to a chromosomal loss, or to both. Only in cases where there is a *reappearance* of the extinguished property at a later stage is it possible to ascertain that the primary extinction is due to a regulatory event and not to the loss of a chromosome carrying a structural gene. Another problem in interpreting studies of phenotypic expression in hybrid and other cells concerns the degree to which a given marker is expressed and the sensitivity of the methods used for detection. The distinction between coexpression and extinction may in some cases be very clear-cut, but in other situations it is not. If coexpression or retention is taken to mean that the hybrid cell contains or produces the same amount of a marker molecule as do the parental cells and extinction means complete loss, there is also *intermediate expression* and *partial* or *incomplete extinction*. There may also exist a phenomenon which can be called *superexpression* since hybrid cells sometimes show a stronger expression of an individual marker than does either parental cell. The degree of expression, too, may be related to chromosomal losses since in nonhybrid cells of different ploidy there is a direct relationship between gene multiplicity and the expression of constitutive markers. Assume, for instance, that a hybrid cell is formed by the fusion of two diploid mouse cells. Theoretically the hybrid cell starts out with four chromosomes of each type. It seems likely that different phenotypes may result from the loss of one, two, three, or all of these chromosomes. In practice the situation is often more complex since the established cell lines used in this work are aneuploid and have rearranged and supernumerary chromosomes. That *gene dosage* is an important factor has also been shown in fusions where one parental cell type contributes two sets of chromosomes and other parental cell type just one set. This type of (2 + 1) or (1 + 2) hybrid may result from the fusion of three cells or from the fusion of a tetraploid cell with a diploid cell. Phenotypic studies on (2 + 1) hybrids have shown that these cells express markers which are not found in (1 + 1) hybrids, that is, hybrids formed from a fusion involving two different diploid or near-diploid cells.

Two other factors which must be considered when interpreting hybrid data are population dynamics and population heterogeneity. Within the parental cell population the tendency to fuse may vary, and the hybrid cells may arise from a small minority of variant cells. Thus, shortly after fusion, the hybrid cell population will consist of many small subpopulations, each arising from a single fusion event (one heterokaryon). With increasing time after fusion there will be a marked selection for the hybrids which show the greatest degree of chromosomal and metabolic fitness and grow most rapidly under *in vitro* conditions. The selection for rapid growth may well act as a selection against differentiating hybrids.

Retention of differentiated markers has been observed in both interspecific and intraspecific hybrids. Markers as different as serum albumin, hyaluronic acid and collagen synthesis, electrophysiological membrane properties, and hormone receptors have been retained in hybrids formed by the fusion of a differentiated cell with an undifferentiated cell from an established line.

Appearance of new properties has been observed in several cases. Some of these observations were made inadvertently, while in a few cases the experiments were designed to screen for a specific property. Hybrid cells have been found to synthesize complement factors, hyaluronic acid, and esterases in spite of the fact that neither of the parental cells exhibits these markers.

CELL ORGANELLES IN HYBRID CELLS

Nuclei in interspecific heterokaryons take up proteins and possibly other molecules from the cytoplasm. Chick nuclei in chick × human heterokaryons take up human nuclear antigens, while at the same time nuclear antigens migrate into the human nuclei. Thus it appears that at the heterokaryon stage both the chick and the human nuclei in the heterokaryon become molecular mosaics containing macromolecules typical of both species (Ringertz et al., 1971).

Mitochondria are equipped with their own genomes in the form of circular DNA molecules; however, many of the proteins found in mitochondria are specified by nuclear genes. At the moment very little is known about the method of protein migration from the cytoplasm into mitochondria. The technique of somatic cell hybridization has opened new ways of studying these relationships, as well as the multiplication of mitochondria. It has been particularly fortunate that the mitochondrial DNAs of different species, particularly human and rodent mitochodrial DNA, differ in buoyant density and thus can be separated by centrifugation. This technique has made possible the examination of the composition of mitochondrial populations in interspecific hybrids. The initial analysis of human × mouse hybrids demonstrated that human mitochondrial DNA was lost (Clayton et al., 1971; Attardi and Attardi, 1972). This result was all the more striking since the hybrid cells retained a considerable number of human chromosomes and synthesized several human isozymes. There may be several explanations for the rapid elimination of human mitochondrial DNA in human × rodent hybrids. It is possible that the retention of human mitochondria is dependent on a number of human nuclear genes. If some specific human chromosomes are lost, the human mitochondrial population can no longer reproduce. This means that mouse nuclear genes are unable to substitute for human nuclear genes in maintaining a state which permits propagation of human mitochondria in hybrids.

CONCLUDING REMARKS

The technique of somatic cell hybridization has opened up new possibilities of exploring the mechanisms which control gene activity, cell multiplication, and cell differentiation. Hybrids between differentiated and undifferentiated cells as a rule show extinction of the differentiated markers. After some time in culture, and probably as a result of the chromosome losses, individual markers of histotypic differentiation may be re-expressed. The fact that a single marker may be re-expressed at the same time that other markers remain extinguished indicates that the corresponding genes are independently regulated. Results obtained with some types of hybrids indicate, however, that groups of genes may also be coordinately re-expressed. The extinction and re-expression phenomena show that epigenotypes can be preserved for a long period of time in the absence of their expression. There are a few examples in which the fusion of a differentiated cell with an undifferentiated cell results in the activation of genes coding for differentiated products in the undifferentiated genome. Most of the studies of phenotypic expression in hybrid cells have been performed with established cell lines, some of which have been derived from differentiated tumors. Although these cells are very useful in experimental work, it will be valuable if similar studies can be performed on hybrids formed from normal, diploid parental cells.

REFERENCES

Appels, R., L. Bolund, and N. R. Ringertz (1974). *J. Mol. Biol.* (in press).

Attardi, B., and G. Attardi (1972). Fate of mitochondrial DNA in human-mouse somatic cell hybrids. *Proc. Nat. Acad. Sci. U.S.A.* **69**, 129–133.

Barski, G. S., S. Sorieul, and F. Cornefert (1960). Production dans des cultures *in vitro* de deux souches cellulaires en association, de cellules de caractére "hybride." *C. R. Acad. Sci. Paris* **251**, 1825–1827.

Carlsson, S.-A., O. Luger, N. R. Ringertz, and R. E. Savage (1973). Phenotypic expression in chick erythrocyte × rat myoblast hybrids and in chick myoblast × rat myoblast hybrids. *Exp. Cell Res.* **84**, 47–55.

Carter, S. B. (1967). Effects of cytochalasins on mammalian cells. *Nature* **213**, 261–266.

Clayton, D. A., R. L. Teplitz, M. Nabholz, H. Dovey, and W. Bodmer (1971). Mitochondrial DNA of human-mouse cell hybrids. *Nature* **234**, 560–562.

Davidson, R. L. (1971). Regulation of gene expression in somatic cell hybrids: A review. *In vitro* **6**, 411–426.

Ege, T., J. Zeuthen, and N. R. Ringertz (1973). *Nobel Symposium 23 on Chromosome Identification* (T. Caspersson and L. Zech, eds.). Academic Press, New York.

Ege, T., H. Hamberg, U. Krondahl, J. Ericsson, and N. R. Ringertz (1974a) Exp. Cell Res. in press.

Ege, T., U. Krondahl, and N. R. Ringertz. (1974b). Article in preparation.

Ephrussi, B. (1972). *Hybridization of Somatic Cells.* Princeton University Press, Princeton, N.J.

Gordon, S., and Z. Cohn (1970). Macrophage-melanocyte heterokaryons. I. Preparation and properties. *J. Exp. Med.* **131**, 981–1003.

Gordon, S., and Z. Cohn (1971). Macrophage-melanoma cell heterokaryons. IV. Unmasking the macrophage-specific membrane receptor. *J. Exp. Med.* **134**, 947–962.

Goto, S., and N. R. Ringertz (1974). *Exp. Cell Res.* **85**, 173–181.

Harris, H. (1967). The reactivation of the red cell nucleus. *J. Cell Sci.* **2**, 23–32.

Harris, H. (1970). *Cell Fusion.* Harvard University Press, Cambridge, Mass.

Harris, H., and J. F. Watkins (1965). Hybrid cells derived from mouse and man: Artificial heterokaryons of mammalian cells from different species. *Nature* **205**, 640–646.

Okada, L. (1962). Analysis of giant polynuclear cell formation caused by HVJ virus from Ehrlich's ascites tumor cells. I. Microscopic observation of giant polynuclear cell formation. *Exp. Cell Res.* **26**, 98–107.

Poste, G., and P. Reeve (1971). Formation of hybrid cells and heterokaryons by fusion of enucleated and nucleated cells. *Nature New Biol.* **229**, 123–125.

Prescott, D. M., D. Myerson, and J. Wallace (1972). Enucleation of mammalian cells with cytochalasin B[1]. *Exp. Cell Res.* **71**, 480–485.

Ringertz, N. R., and L. Bolund (1974a). Reactivation of chick erythrocyte nuclei by somatic cell hybridization, Vol. XIII. In *International Experimental Pathology* (G. W. Richter and M. A. Epstein, eds.). Academic Press, New York.

Ringertz, N. R., and L. Bolund (1974b). The nucleus during avian erythroid differentiation. In *The Cell Nucleus*, Vol. III (Busch, ed.) Academic Press, New York.

Ringertz, N. R., and R. E. Savage (1975). *Cell Hybrids.* Academic Press, New York (in press).

Ringertz, N. R., S.-A. Carlsson, T. Ege, and L. Bolund (1971). Detection of human and chick nuclear antigens in nuclei of chick erythrocytes during reactivation in heterokaryons with HeLa cells. *Proc. Nat. Acad. Sci. U.S.A.* **68**, 3228–3232.

Ruddle, F. H., (1972). Linkage analysis using somatic cell hybrids. In *Advances in Human Genetics, Vol. 3* (H. Harris and K. Hirschhorn, eds.), Plenum Publishing Co., New York pp. 173–235.

Siniscalco, M., B. B. Knowles, and Z. Steplewski (1969). Hybridization of human diploid strains carrying X-linked mutants and its potential for studies of somatic cell genetics [The Wistar Institute Symposium Monograph No. 9 (V. Defendi, ed.)]. In *Heterospecific Genome Interaction*, The Wistar Institute Press, Philadelphia, pp 117–132.

Thompson, E. B., and T. D. Gelehrter (1971). Expression of tyrosine aminotransferase activity in somatic cell heterokaryons: Evidence for negative control of enzyme expression. *Proc. Nat. Acad. Sci. U.S.A.* **68**, 2589–2593.

Watkins, J.F., and D. M. Grace (1967). Studies on the surface antigens of interspecific mammalian cell heterokaryons. *J. Cell Sci.* **2**, 193–204.

Weiss, M. C., and H. Green (1967). Human-mouse hybrid cell lines containing partial complements of human chromosomes and functioning human genes. *Proc. Nat. Acad. Sci. U.S.A.* **58**, 1104–1111.

CHAPTER FIFTEEN

Application of Somatic Cell Genetics to Surface Antigens

CAROL JONES, PAUL WUTHIER, AND THEODORE T. PUCK

Many studies indicate that cell surface components are of critical importance in normal and malignant differentiation. The classic experiments of Moscona illustrate the specificity of cell surface groups in directing recognition between cells of particular tissues. The present experiments were undertaken in an attempt to utilize methods of somatic cell genetics *in vitro* to explore cell surface antigens, to relate them to specific genetic loci, and to study molecular biological regulatory processes involving them.

The study of surface antigens in human cells is complicated by their great number and the difficulty in obtaining genetic information in man by classical genetic operations. The use of somatic cell methodologies *in vitro* makes possible rapid and accurate genetic analyses of several kinds. Single-cell plating can quantitate lethal action of specific antibodies on cells; mutant clones can be obtained with auxotrophic and other kinds of markers; and hybridization between Chinese hamster auxotrophic mutants and human cells in selective media can be used to retain the specific human chromosomes which complement each such deficiency of the Chinese hamster cells. This approach allows study of human cell surface antigens contributed to a human × Chinese hamster cell hybrid by one or a small number of human chromosomes. In most of these experiments, the CHO-K1 clone of the standard Chinese hamster ovary culture isolated in this laboratory was employed.

We previously demonstrated that under carefully controlled experimental conditions lethal antibodies produced by the rabbit in response to injection of human cells show no detectable cross-reaction with the Chinese hamster cell when tested by the single-cell plating method (Oda and Puck, 1961).

PRODUCTION AND ISOLATION OF AUXOTROPHIC MUTANTS

Auxotrophic mutants of the Chinese hamster ovary (CHO) cell have been produced by the use of mutangenic agents, followed by the bromodeoxyuridine-"visible" light procedure developed by Kao and Puck (1968; Puck and Kao, 1967), which isolates auxotrophs. As of this date 15 such mutants have been prepared (Kao and Puck, 1972a; in press) (Table 1). Four different mutants require addition only of glycine for growth. The mutant designated *gly A* has been shown to be deficient in the enzyme serine hydroxymethylase. Six other mutants are blocked in particular steps of the purine synthesis pathway. We have one mutant that requires glycine, adenine, and thymidine; one that requires adenine and thymidine; and one inositol-requiring, one serine-requiring, and one proline-requiring mutant. These mutants have all been shown to be recessive. Each mutant can complement any of the

TABLE 1. AVAILABLE CHINESE HAMSTER CELL AUXOTROPHIC, NONLEAKY MUTANTS

Mutant Designation	Growth Requirement	Mutagenic Agents Used to Produce Each Mutant Class	Number of Mutants Analyzed
gly A	Glycine	EMS, MNNG, X-ray	4
gly B	Glycine or folinic acid	EMS, MNNG, UV, NMUT	12
gly C	Glycine	EMS, UV	2
gly D	Glycine	EMS, UV	2
ade A	Hypoxanthine, adenine, inosinic acid, their riboside or ribotide, or AIC	ICR-191	3
ade B	Like *ade A*	ICR-191, NDMA, NMUT	20
ade C	Like *ade A*	Sydnone acetamide	1
ade D	Like *ade A*	EMS	1
ade E	Like *ade A*, but excluding AIC	ICR-191, MNNG	2
ade F	Like *ade E*	ICR-191	2
GAT⁻	Glycine, adenine, and thymidine	EMS, MNNG, UV, NMUT	2
AT⁻	Adenine and thymidine	EMS	1
ino⁻	Inositol	MNNG	1
ser⁻	Serine	EMS	1
pro⁻	Proline or △'-pyrroline-5-carboxylic acid	spontaneous	1

Abbreviations:		
	AIC:	5-aminoimidazole-4-carboxylic acid
	NMUT:	nitrosomethylurethane
	NDMA:	nitrosodimethylamine
	ICR-191:	an acridine mustard
	UV:	ultraviolet light
	EMS:	ethylmethane sulfonate
	MNNG:	*N*-methyl-*N*'-nitro-*N*-nitrosoguanidine

others. In addition, two drug-resistant CHO mutants have been prepared (Chasin, 1972). The availability of these mutants allows production of a variety of human × hamster hybrids by cell fusion (Puck et al., 1971; Kao and Puck, 1972b; Jones et al., 1972).

HYBRIDIZATION OF AUXOTROPHIC CHINESE HAMSTER MUTANTS AND HUMAN CELLS TO RETAIN SPECIFIC HUMAN CHROMOSOMES

Figure 1 shows the virus-mediated fusion of two cells to form a heterokaryon, which then develops into a hybrid cell. One of the cells can be any of the 15 auxotrophic mutants mentioned above, while the other can be a human cell, such as a human fibroblast cultivated from amniotic fluid, human lymphocyte, or cultured human lymphoblast. The hybridization is carried out in culture medium which lacks the nutrient required for the growth of the specific Chinese hamster auxotroph employed, and in which the parental human cell also grows poorly or not at all. The fused cell undergoes rapid and extensive loss of human chromosomes, while the Chinese hamster genome experiences minimal, if any, apparent change. The resulting hybrids, therefore, contain one or a limited number of human chromosomes, and the human component differs according to the nature of the Chinese hamster auxotroph employed. Thus an interesting series of cell hybrids can be created whose genomes have two parts: a constant part containing virtually all of the chromosomes of the CHO-K1 karyotype, and a variable part consisting of different human chromosomes.

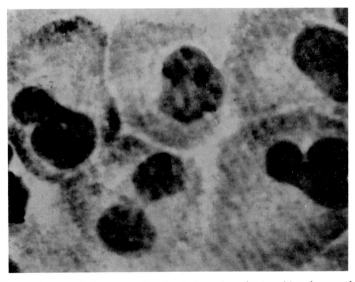

Figure 1. Photomicrograph demonstrating the fusion of nuclei in a binucleate or heterokaryon cell to form a single nucleus of the resulting hybrid cell. Such hybrids are readily obtained when tissue culture cells like the HeLa or Chinese hamster ovary are fused with the aid of Sendai virus inactivated by ultraviolet irradiation.

TEST OF HYBRIDS WITH ANTISERA AGAINST HUMAN CELLS

Figure 2 illustrates the method used to test cells for the presence of antigens which cause killing in the presence of specific antiserum and complement. The antiserum used in this experiment was produced in rabbits after injection of cultured human cells, but similiar antisera have also been produced in other animals such as the horse. When 200 human cells are placed in a petri dish containing growth medium alone and are incubated, the cells form large, discrete colonies. In the presence of an antiserum produced against Chinese hamster cells, no change is produced. However, in the presence of an antiserum produced against cultured human cells, all of the plated cells are destroyed. The Chinese hamster cell formed equal numbers of colonies in the presence and the absence of antihuman cell serum.

Starting with an antiserum to a cultured human fibroblast, we tested its killing effectiveness for the various human CHO hybrids available. It was found that hybrids from some of the CHO auxotrophs were killed by this antiserum, while others were unaffected. All the hybrids yielding a positive killing response were labeled A_L^+, while those which were negative were designated A_L^-. Antiserum prepared against A_L^+ hybrids cannot be tested directly against other hybrids because they both share the common Chinese hamster antigens. However, by adsorbing the antiserum with the parental CHO-K1 cell, antigens whose presence is due to human chromosomes can be tested for. Such antisera prepared against A_L^+ hybrids killed all other A_L^+ hybrids and none of the A_L^- hybrids. Thus the A_L behaved, at least in first approximation, like a stable immunogenetic marker. This formulation does not imply

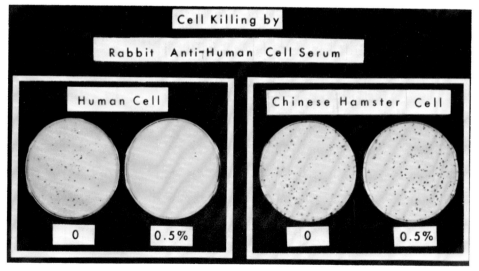

Figure 2. Demonstration of the species specificity exhibited in the killing of tissue culture cells by antiserum. An antiserum to human tissue culture cells was prepared by injecting these cells into a rabbit. The antiserum collected kills human cells rapidly and completely, but the same concentration of antiserum has no effect whatever on the plating efficiency of Chinese hamster cells.

that A_L is a single component. Indeed the experiments with human red blood cells described later suggest that this is not the situation.

In further development of this program, it was undertaken to prepare antisera to specific antigens or combinations of antigens whose corresponding genes are carried on specific human chromosomes. These antisera then form the selective media by which the presence of genes responsible for the appearance of such specific antigens can be detected. Such specific antisera have been prepared in the following ways: (a) use of specific human cells as antigens, (b) adsorption of antisera to various human cells by specific human × CHO hybrids so as to remove specific antibody activities from the antiserum, and (c) production of antisera by use of specific human × CHO hybrids as antigens injected directly into an animal. Antisera were prepared by injection into rabbits or other animals of human antigens consisting of cultured human fibroblasts obtained from biopsies of skin, amniotic fluid, or other tissues; cultured human lymphoblasts; cultured HeLa cells; and a variety of cells taken from various human tissues and used for immunization without previous cultivation *in vitro*. These sera were used directly or after partial or exhaustive adsorption with other cells, as described in the text.

There are also three ways by which specific antigenic activity can be demonstrated on specific cells: (a) the ability of the cells to be killed by a specific antibody preparation, indicating the presence of an antigen common to the given cell and the cell used to elicit the antiserum; (b) the ability of the cell when injected itself as an antigen to produce an antiserum with specific killing properties; (c) the ability of a given cell to adsorb out specific cell-killing antibodies from a given antiserum. In the discussion which follows, examples of all of these methodologies will be provided. Not all are equally applicable in all cases. However, the pattern of behavior which has been elicited so far in the testing of a variety of cells has been generally consistent with a relatively simple immunological and genetic picture for the observed phenomena.

The antigens here studied seemed to be primarily cell surface antigens, since the killing which is produced is accompanied by leakage of cytoplasmic constituents out of the enclosing membrane. Moreover, it is possible completely to absorb killing activity from an antiserum by use of a suspension of whole cells which presumably expose only the outer membrane to the antibody solution. Complement is required for this killing action. Control experiments have demonstrated that neither antiserum by itself nor complement by itself will kill the cells, nor will antiserum plus heated complement cause cell killing. In these experiments, normal rabbit serum has been used as a source of complement and each batch is carefully tested to ensure that only complement preparations are used which by themselves cause no killing of any of the cells tested.

In Fig. 3 are presented curves showing the killing action as a function of the antiserum concentration incorporated into the growth medium of antisera which have been designated as positive or negative. On the basis of data of this kind, we have arbitrarily labeled an antiserum as killing for a given cell type if 98% or more of the cells are killed by a concentration of 1% or less of the antiserum. Otherwise the reaction is designated as negative. It is interesting that we have had surprisingly few sera with activities intermediate between the two extremes shown in Fig. 3.

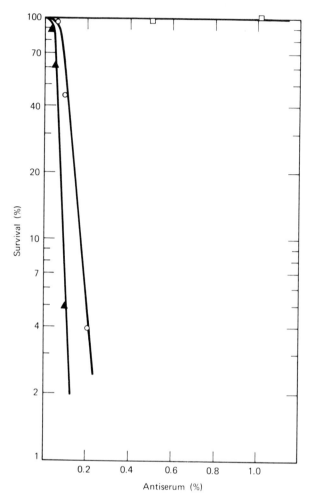

Figure 3. Survival curves demonstrating the basis for labeling an antiserum positive or negative with respect to the killing action for a given cell. The open squares represent typical results obtained when an A_L^- hybrid is treated with an antiserum containing antibodies to an A_L^+ cell. The solid triangles and the open circles are typical of the titration curves obtained when A_L^+ hybrids are treated wtih the indicated concentrations of the same antiserum. A cell is designated as being sensitive to killing by a given antiserum if more than 98% killing is obtained when the cell is treated with 1% or less of the antiserum in the presence of complement.

DEMONSTRATION OF ADDITIONAL ANTIGENIC MARKERS
BY SELECTIVE ADSORPTION

We assume that the antiserum which was prepared against human cells and was active against some of the human CHO hybrids contains an antibody against at least one antigen, which we arbitrarily name A_L. The next step, then, is to determine if additional antigenic sites can be identified by means of this antiserum. The procedure adopted is to adsorb this standard antiserum exhaustively with a variety of sensitive hybrids, one at a time. Each such adsorbed serum is then tested against all of the other hybrids to see whether these have or have not also lost the power to be killed by the adsorbed serum. A hybrid which loses its sensitivity to a given antiserum that has been adsorbed by another hybrid contains no unshared lethal antigens to antibodies present in that serum. However, if a hybrid retains the capacity to be killed, the antiserum must contain antibodies to an antigen not present on the first hybrid.

A typical experiment is shown in Table 2. When a given antiserum was adsorbed exhaustively with S3-HeLa cells, it lost its capacity to kill HeLa cells and two A_L^+ hybrids, called hybrids 1 and 2 in the table. However, when the antiserum was adsorbed with hybrid 1, the ability to kill hybrid 1 was lost, but lethal activity was retained for the HeLa cell and for hybrid 2. Therefore the latter hybrid contains an antigen absent from hybrid 1. The new antigen identified by such an experiment was arbitrarily named B_L. In this way it is possible to successively identify different antibody activities associated with different cell surface antigens, the basis of the definition in every case consisting in the availability of a specific hybrid capable of adsorbing lethal activities specific for some cells but not for others, from a multicomponent antiserum. In Fig. 4 is demonstrated an experiment in which a series of hybrids corresponding to all four possible antigenic combinations of A_L and B_L is treated with antisera specific for A_L and B_L, respectively.

Some hybrid cells were found to be $A_L^+B_L^+$, while others were $A_L^-B_L^+$, A_L^+, B_L^-, or $A_L^-B_L^-$. It was found possible to obtain clones negative for a given antigen supply by growing up a large population in the presence of the appropriate lethal antibody plus complement. Under these circumstances only the cells which had lost the specific antibody marker in question are able to grow and form colonies which can readily be isolated. Such subclones, like the one shown in the top left-hand plate of Fig. 4, when grown up and tested, were found to have lost the lethal antigen present in the original hybrid population. Thus, the colony arising from the top left-hand plate gave rise to the cell population demonstrated in the plates of the third row of Fig. 4. Similarly, when a surviving colony was isolated from a plate like that on the right-hand side of the third row, a population of $A_L^-B_L^-$ cells was obtained like that shown in the fourth row of the figure. Figure 4 demonstrates that a hybrid which is $A_L^+B_L^+$ is killed by either A_L or B_L antiserum; an $A_L^-B_L^+$ culture is killed only by A_L antiserum; an $A_L^+B_L^-$ hybrid is killed only by B_L antiserum; and, finally, an $A_L^-B_L^-$ cell is killed by neither antiserum.

It may be noted that, whereas A_L^- and B_L^- subclones can be obtained fairly regularly by the use of appropriate selective antisera, we have never yet obtained either

TABLE 2. SPECIFIC RETENTION OF KILLING ANTIBODIES AFTER EXHAUSTIVE ADSORPTION OF A SPECIFIC ANTISERUM BY CELLS

Absorbing Cells	Cell Killing by Specifically Adsorbed Antiserum		
	S3-HeLa A_L^+	Hybrid 1 A_L^+	Hybrid 2 A_L^+
S3-HeLa	0	0	0
Hybrid 1	+	0	+

The data show that, while all three cells are A_L^+, both S3-HeLa cells and hybrid 2 cells are sensitive to additional lethal antibodies present in the original antiserum, which do not kill hybrid 1.

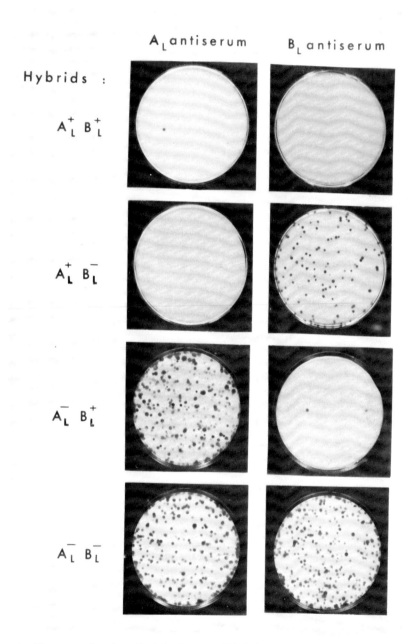

Figure 4. Demonstration that all of the four possible hybrids containing various combinations of A_L and B_L activity can be prepared, as well as specific antiserum which reacts individually to A_L or B_L antigenic activity.

an A_L^+ or a B_L^+ subclone from a clone that was originally lacking the respective antigen, as would be expected if simple genetic relationships are obeyed. The specific antisera for A_L and B_L antigens used in the experiments of Fig. 4 were obtained by adsorption of antisera against $A_L^+B_L^+$ cells with the appropriate cell culture. It should be possible also to prepare such specific antisera against A_L and B_L separately by directing immunization of animals with the $A_L^+B_L^-$ and $A_L^-B_L^+$ clones. Such immunization has been carried out with an $A_L^+B_L^-$ clone. The resulting antiserum had a high lethal titer for human fibroblast cells and, of course, also for all hybrids and for Chinese hamster cells alone because the cell used as the antigen still contained all or most of the Chinese hamster genome. However, by exhaustive adsorption with an A_L^- hybrid clone, the lethality for the Chinese hamster cell was removed while the killing titer for the human cells was unchanged. The resulting antiserum now kills only A_L^+, not A_L^-, human × CHO × hybrids and presumably contains antibodies only to substances that are due to genes on the human chromosomal complement of the hybrid. This procedure offers promise of providing reasonably pure antibody reagents to gene products of particular human chromosomes.

By extension of these procedures, we have prepared hybrids containing five other human antigenic markers. Their characterization by means of the three principal methodologies described earlier is currently in progress.

ANTIGENS ON CELLS TAKEN DIRECTLY FROM SPECIFIC TISSUES

Cells from various human tissue biopsies can be examined for their ability to adsorb antibodies to specific antigens here discussed. By exposing such specific antisera to measured cell numbers for constant periods of time, and then titrating the amount of killing activity for a specific hybrid remaining in the antiserum, it becomes possible to obtain a measure of the relative capacities of different cells to absorb specific antibodies, which in turn can be interpreted as a measure of the availability of a particular antigen on the cell surface.

Typical experimental results are shown in Table 3. It is apparent that the HeLa cell is the most effective, the human fibroblast less so, and the A_L^+ hybrid used still less effective in adsorbing A_L from an appropriate antiserum. The lymphocyte was far less active, being only about 1/300 as effective as the HeLa cell. These differences cannot be due to the smaller size of the lymphocyte alone since it is only 6 times smaller than the HeLa cell. The human red blood cell did not remove any A_L

TABLE 3. RELATIVE ADSORBING ABILITIES FOR SPECIFIC ANTIBODIES OF VARIOUS HUMAN CELLS AND CELL HYBRIDS

Cell Type	Number of Cells Needed to Adsorb 80% of the Anti-A_L Activity from Rabbit Antihuman Cell Serum
HeLa	1×10^6/ml
Human fibroblast	3×10^6
A_L^+ hybrid	3×10^7
Human lymphocyte	3×10^8
Human RBC	No adsorption at 5×10^9/ml
A_L^- hybrid	No adsorption at 1×10^8/ml

killing activity at any cell number tested. These experiments reveal how quantitative differences in the amounts of cell surface antigens on different tissues can be determined. In the case of the human red cell, however, further studies were carried out which revealed a set of interesting phenomena. While it is impossible to remove A_L killing activity from appropriate antisera by adsorption with human red blood cells, an antiserum produced after injection of human red cells into rabbits was capable of killing A_L^+ hybrids but not A_L^- hybrids. Moreover, A_L^+ hybrids were able to remove by adsorption all cell-killing activity from this antiserum. Finally, red cells can remove their own antibodies from an antiserum prepared against themselves.

It appears, then, that the red cell has available on its surface an antigen capable of eliciting antibodies which kill human cells containing the A_L^+ antigen, but yet is unable by itself to adsorb all of the killing antibodies elicited by the A_L antigen. Presumably the A_L marker is a complex of at least two different antigens, only one of which occurs on the red blood cell. Work is being directed toward the preparation of antisera containing these separate activities, and of hybrid cell strains with the individual antigens. In the meantime, we have named the red cell antigen which cross-reacts with A_L by antigenic action, but not by adsorption, A'_L.

DETERMINATION OF CHROMOSOMAL LOCATION OF SURFACE ANTIGENS

One can examine the chromosomes of the series of hybrid cells used in these studies and attempt to identify by cytologic means the human chromosomes which have been incorporated into their karyotypes. This procedure is laborious, however, and is open to some uncertainties because banding techniques which are satisfactory for the chromosomes of the pure human cell present difficulties when applied to the hybrid. In our studies so far, therefore, we have relied more on tests for linkage in order to determine which other markers invariably accompany the presence of a given antigen in a given human × CHO hybrid.

The human × CHO hybrids are examined for the presence of specific human enzymes. Cultures of hybrids are grown and harvested, and cell lysates prepared for gel electrophoresis. The resulting gels are then individually treated with substrates for specific enzymes, together with an indicator which changes color in the presence of the specific reaction. Therefore, discrete bands are obtained, indicating the presence of isozymes which have migrated with different velocities under the influence of the field. These migration patterns are often specific for isozymes of each of the parental cells and so can permit identification of the origins of different enzyme activities found in the hybrid.

Figure 5 presents the zymograms of the lactic dehydrogenase (LDH) enzyme derived from several different cell strains. Lactate is converted to pyruvate in the presence of NAD, and this oxidation-reduction reaction is coupled to a tetrazolium dye to form an insoluble, colored product. Lactic dehydrogenase is a tetramer composed of A and B subunits, and each of these tetramers migrates differently in an electrophoretic field so that five separate bands are possible. The identities of the different bands are explained in the legend of Fig. 5.

Hybrids containing various human antigens were subjected to a battery of different isozyme tests to determine whether any of these antigens are contained on the

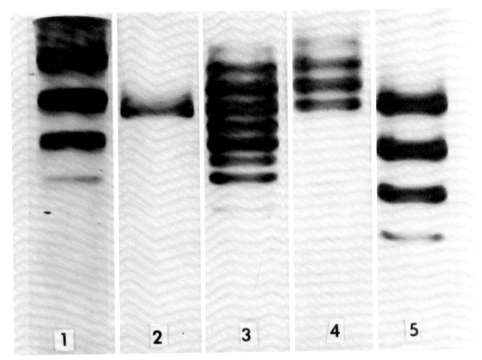

Figure 5. Typical zymograms for LDH patterns from the clonal cells here studied. (*1*) Normal human fibroblast. Four bands are visualized here, the fifth requiring a more intensive staining, which obscures resolution of the other bands. The pure A tetramer is the topmost band. (*2*) CHO-K1. Only one band, the A tetramer, occurs. (*3*) Human × Chinese hamster cell hybrid which possesses the human A^+B^+ phenotype. Ten of the 15 theoretically possible bands are readily visualized, and other nonoverlapping bands can be developed by more intensive staining. (*4*) Human × Chinese hamster cell hybrid which is A^+B^- for the human components. Four of the theoretical 5 bands are shown. (*5*) Human × Chinese hamster cell hybrid which is A^-B^+ for the human components. All 5 bands can sometimes be visualized on the same gel, although sometimes, as here, only 3 or 4 become prominent with the standard staining procedure.

same human chromosome as that furnishing genes for specific isozymes. In Table 4 is summarized the data obtained in examining the possibility that A_L may be contained on the same chromosome as either LDH-A or LDH-B. The data demonstrate clearly that the gene for A_L is syntenic with (i.e., carried on the same chromosome as) the gene for LDH-A, but that the LDH-B gene is contained on a different chromosome. Other workers (Boone et al., 1972) have demonstrated that the LDH-A subunit is coded for by a gene carried on human chromosome number 11.

TABLE 4. DEMONSTRATION THAT A_L IS CARRIED ON THE SAME CHROMOSOME (SYNTENIC WITH) AS LDH-A, BUT NOT LDH-B

Cell phenotype:	A_L	+	+	−	−	A_L	+	+	−	−
	LDH-A	+	−	+	−	LDH-B	+	−	+	−
Number of clones found:		67	0	0	101		46	18	49	55

We may then conclude that the gene for A_L is also carried on this chromosome. Similar tests for other human antigens described here are continuing.

Attempts to determine the chromosome which has the genetic information specifying B_L antigen are in progress by means of experiments similar to those described for A_L (Wuthier et al., 1973).

APPLICATIONS TO THE STUDY OF MUTAGENESIS

These markers lend themselves particularly well to the study of mutagenesis in mammalian cells. In order to assess the mutagenic effectiveness of a given agent, it is highly advantageous to measure the induction of a forward mutation since, in this case, a change almost anywhere along the genome can cause loss of the given gene activity. In measuring the reversion of a previous mutation, it is possible to score only changes in the DNA that take place at particular points so that the original gene activity can be restored. Therefore, the former method can scan a much more comprehensive range of gene changes.

The human antigens which are present on specific human×CHO hybrids constitute excellent genetic markers for measuring mutation. When the appropriate antiserum is incorporated into the growth medium, a selective medium is available in which only the cells that have lost the genetic basis for the antigen form colonies. Therefore, the procedure is simple, rapid, and quantitative. Moreover, this procedure will detect agents which cause chromosome loss and gross deletions, as well as single gene mutations. Since presumably all such agents could be important in human disease, this system could make possible detection of agents with a variety of deleterious actions on the genome. It is possible also to differentiate between agents that cause the loss of a whole chromosome and those producing a localized genetic change, by testing the colonies which grow up in the presence of the selective antiserum for retention of a syntenic marker. For example, cells which acquire, as a result of mutagenic action, the ability to form colonies even in the presence of A_L antiserum can then be tested to see whether or not they also have lost the LDH-A marker. If they have retained this marker, one can conclude that the genetic damage has been limited to only a portion of the chromosome. As more marker genes become identified, these determinations can be made with higher resolving power and the true linkage of the human genes should be readily ascertainable.

The series of hybrid cells described here constitutes a population with single gene markers that can be used for a variety of genetic experiments. As discussed earlier, the CHO system now affords 15 auxotrophic markers. In addition, 2 stable drug-resistant markers have been prepared. The availability of 7 additional antigenic markers raises to 24 the total number of genetic markers which are available in this cell family.

DISCUSSION

The cell surface antigens described here complement the system which has been elucidated by Moscona and furnish an important addition to the study of cell surface phenomena. The methods here employed might also be applicable for study of the Moscona antigens.

The number of antigens to be studied must be extended, and these studies should be carried out on adult, embryonic, and malignant cells. The aim is to study the cells after direct biopsy, as well as after prolonged growth in tissue culture. The biochemical natures of these antigens must be identified, and their biological roles in the cell and tissue economy elucidated. Their localization on the cell surface would appear to be possible by currently available methods, and the effects of various agents in changing the antigenic expression of cells must be carefully studied. The relationship of these markers to the human histocompatibility antigens also appears to furnish a fertile field for study.

SUMMARY

Hybrids between human and Chinese hamster cells have been produced and permit incorporation of one or a small number of human chromosomes into stable clones whose genomes otherwise contain all or most of the Chinese hamster cell chromosomes. This technique produces great simplification in the range of human antigens exhibited by each clone.

In these experiments, by virtue of the demonstrated lack of cross-reactivity between human and Chinese hamster antigens under the controlled conditions here described, single-cell plating has been used as a technique for demonstrating the presence of lethal antigens. It has provided a clear distinction between antibody-sensitive and antibody-resistant cells, and has permitted selective isolation of resistant clones from an initially sensitive parental population. Genetic analysis applied to human antigens of the human Chinese hamster cell hybrids reveals that these structures act as excellent genetic markers.

ACKNOWLEDGMENTS

This investigation is a contribution from the Rosenhaus Laboratory of the Eleanor Roosevelt Institute for Cancer Research and the Department of Biophysics and Genetics (Number 576), University of Colorado Medical Center, Denver, Colorado, and was aided by National Science Foundation Grant GB-38664.

REFERENCES

Boone, C., T. R. Chen, and F. H. Ruddle (1972). Assignment of three human genes to chromosomes (*LDH-A* to 11, *TK* to 17, and *IDH* to 20) and Evidence for Translocation Between Human and Mouse Chromosomes in Somatic Cell Hybrids. *Proc. Nat. Acad. Sci. U.S.A.* **69**, 510–514.

Chasin, L. A. (1972). Non-linkage of induced mutations in Chinese hamster cells. *Nature New Biol.* **240**, 50–52.

Jones, C., P. Wuthier, F. T. Kao, and T. T. Puck (1972). Genetics of somatic mammalian cells. XV. Evidence for linkage between human genes for lactic dehydrogenase B and serine hydroxymethylase. *J. Cell Physiol.* **80**, 291–298.

Kao, F. T., and T. T. Puck (1968). Genetics of somatic mammalian cells. VII. Induction and isolation of nutritional mutants in Chinese hamster cells. *Proc. Nat. Acad. Sci. U.S.A.* **60**, 1275–1281.

Kao, F. T., and T. T. Puck (1972a). Genetics of somatic mammalian cells. XIV. Genetic analysis *in vitro* of auxotrophic mutants. *J. Cell Physiol.* **80**, 41–50.

Kao, F. T., and T. T. Puck (1972b). Genetics of somatic mammalian cells: Demonstration of a human esterase activator gene linked to the *adeB* gene. *Proc. Nat. Acad. Sci.* **69**, 3273–3277.

Kao, F. T. and T. T. Puck (in press). Induction and isolation of auxotrophic mutants in mammalian cells. Chapter in *Methods in Cell Physiology*, Vol. VI: *Methods in Mammalian Cells* (David Prescott, ed.).

Oda, M., and T. T. Puck (1961). The interaction of mammalian Cells with antibodies. *J. Exp. Med.* **113**, 599–610.

Puck, T. T., and F. T. Kao (1967). Genetics of somatic mammalian cells. V. Treatment with 5-bromodeoxyuridine and visible light for isolation of nutritionally deficient mutants. *Proc. Nat. Acad. Sci. U.S.A.* **58**, 1227–1234.

Puck, T. T., P. Wuthier, C. Jones, and F. T. Kao (1971). Genetics of somatic mammalian cells. XIII. Lethal antigens as genetic markers for study of human linkage groups. *Proc. Nat. Acad. Sci. U.S.A.* **68**, 3102–3106.

Wuthier, P., C. Jones, and T. T. Puck (1973). Surface antigens of mammalian cells as genetic markers, II. *J. Exp. Med.* **138**, 229–244.

CHAPTER SIXTEEN

The Interrelations Between Development, Retrogenesis, Viral Transformation, and Human Cancer

NORMAN G. ANDERSON AND JOSEPH H. COGGIN

Cancer is a puzzle made up of many pieces; in fact, it has not been at all certain that cancer did not include several different puzzles. Quite recently, however, the addition of a few pieces to the many already in place has revealed a pattern or picture which has given new directions and new urgency to cancer research. Thus we are describing not a single, easily comprehended breakthrough, but rather the knitting together of a complex of ideas into a pattern not susceptible of facile summary or capsulation.

Cancer and differentiation have been closely linked through much of the history of cancer research, and the degree to which patterns and evidences of differentiation are lost in malignant cells (i.e., the cells appear dedifferentiated) is part of the basis for description and identification by pathologists. It is not difficult to see that explaining one unknown (cancer) in terms of another (differentiation) is not the final answer. In addition, it is quite evident that cancer cells are vastly different from embryonic cells, lacking in all but a very few instances the developmental potentialities of the latter. Superimposed on these ideas is an additional puzzle: The study of viral oncology has contributed brilliantly to the study of cancer in highly inbred domesticated animals such as chickens and cats and in laboratory animals, but has made few clinically useful contributions to solving the problem of human cancer. This suggests that some of our underlying concepts regarding embryogenesis, differentiation control mechanisms, virology, and oncology may require revision. In this paper we shall discuss such revisions, point out where they may require further examination, and offer one present view of the nature of the cancer problem.

DEVELOPMENT

The concept that embryonic cells respond to a complex series of external signals which specify specific cellular responses (Holtfreter, 1955) has given way to the view that cells develop specific competences (so-called developmental imminence) and respond in a specific manner to stimuli which may be specific but often are not (Bonner, 1960). The informational reservoir is therefore largely *within* the cell. There has been, however, a preformationistic reluctance to accept the possibility that the amount of information contained in an embryo or fetus is comparable to the amount of information required to construct it. Whether information theory has much to contribute to embryology (Apter and Wolpert, 1965) is an open question, and the problem may be quickly put into perspective for the biologist by simple extrapolation. The zygote yields an embryo which yields a man who *produces* information, the latter being the sum total of all he knows, writes, speaks, or composes. Not all of this is specified in the egg and sperm. Thus man creates and stores more information than he innately possesses. Embryonic cells are structure-forming machines and hence are capable of making structures which are not entirely prespecified in the architectural drawing sense. Hence the interest in explaining complex morphogenetic effects in terms of simple fields, or by the interaction of a few variables.

Yet the entire experience of modern molecular biology has been that each mechanism studied in detail (DNA replication, transcription, translation, mitochondrial respiration, etc.) is much more complex than initially imagined and is encrusted with control mechanisms. Therefore the number of different specific gene products which may be involved in early development might be expected to be very large, a view supported by the fact that the details of the structures formed, for example, the fine structure of facial patterns, are under such precise genetic control. Hence arises a paradox. Embryonic cells are described as being undifferentiated (or less differentiated than adult cells) but having more developmental potentialities. As these potentialities become restricted, cellular complexity in the form of the appearance of structures and substances characteristic of differentiated cells increases. Since the biochemistry of development has been studied largely in terms of the appearance of substances known to be present in differentiated cells, it is quite natural that differentiation has been viewed as the gradual turning on of more and more genes of the adult type, and restriction of developmental potential has been regarded as being due to or parallel with the appearance of adult gene products. However, data accumulated over the past few years begin to support a somewhat different picture: that a large library exists of genes which are concerned solely with early development and are inactive in adult tissues (Anderson and Coggin, 1971, 1972a, b). Does such a library really exist

It is natural to attempt to describe both cancer and differentiation in terms of substances already discovered. If the many different immunoglobulins are excluded, about 1000 types of proteins have been described in human and experimental animal species. If all nonrepeating DNA in the mouse were transcribed and translated, approximately 1 million different proteins would be produced. This suggests that ~0.1% of all different proteins has been described, and that, for each one known, 1000 remain to be described. For the purposes of this discussion it does not matter if these numbers are off by a factor of 5. The point remains that a large body of information resides in the nucleus, about which we know very little. Could a large fraction of it be concerned with early development?

When mRNA is isolated from embryos at different stages in development and its binding to DNA studied, it is found that a large fraction of RNA from one stage does not compete for binding sites with RNA from another stage (Schultz and Church, 1973), indicating, as pointed out by Davidson (1968), massive gene phasing. These phased genes appear to give rise to transient gene products which produce little or no cytologically visible alterations but do control or affect cell behavior. If such substances were largely confined to the cell membrane, and if many different molecular species were involved, the concentration of any single species would be expected to be very low indeed.

Cell proteins may, in this view, be classified into three major groups by function. The first group would include proteins which serve a function necessary and common to all nucleated cells. This group might be called the base set; it could include instances in which a function is served by only one protein, and instances in which several different isoenzymes or isoproteins are found, one of which must always be present.

The second group would include proteins having functions necessary to one or more differentiated cells, but never to all. These proteins would often tend to be in high concentration in the cells exhibiting them (hemoglobin, myosin, crystallin, for example) and would therefore be likely to be noticed. As previously noted, differentiation would tend to be equated with the appearance of such substances.

The third class of substances would be transient, phase-specific gene products concerned largely with morphogenesis and present in small amounts for a short period of time. Their presence is inferred from the massive gene phasing during development, from the dogma that the genome controls structure in great detail and does so only by controlling synthesis of macromolecules by conventional mechanisms, and from the informational requirements for morphogenesis. Why have these substances not received attention before if indeed they exist? In fact, is there any evidence for phasing at all during early embryogenesis from studies on protein composition, as contrasted to hybridization studies?

Phasing of isozymes, hemoglobins, and a variety of antigens has been reported during embryonic development and has recently been reviewed by Holleman and Palmer (1972). A surprising number of adult proteins have different fetal counterparts. At the very outset, sperm contain unique antigens, of which four appear to be autoantigenic in several species, including man (Shulman, 1971). Early embryos contain a variety of antigens which are present for limited time periods. The view that the program of development is mediated only by the selective switching on (but not the switching on and off) of genes is therefore clearly not justified.

Mechanisms for sequential gene activation based largely on the model of Jacob and Monod have been proposed by many authors (Anderson and Coggin, 1971; Sugita, 1963; Simon, 1965; Sherbet, 1966; Grigorev et al., 1967; Britten and Davidson, 1968; Venkatesan et al., 1971; Bablyoantz and Nicolis, 1972; Herstein and Frenster, 1972; Markowitz, 1972; Edelstein, 1972; Kauffmann, 1973; Glass and Kauffmann, 1973). In the Jacob-Monod model, the repeating element is the operon, consisting of an operator, structural genes, and regulator genes. Communication between these units occurs by diffusion, that is, random access. Britten and Davidson (1969) proposed a slightly different model in which an additional unit is described, consisting of a sensor gene and an activator gene. A detailed examination of the programming of such systems will be described elsewhere. However, the salient feature of interest for this discussion is that mechanisms now appear to exist for programming large arrays of genes in response to internally and externally generated signals and for turning them on and off in sequence.

EVOLUTION OF PROGRAMMING

The evolution of programming appears to occur by accretion and may be considered at two levels. First, the basic elements of the first reproductive mechanism, cell division, have common aspects over the entire animal and plant kingdoms. As multicellular organisms evolved, new processes were superimposed and retained (morula, gastrula, blastula stages). In higher animals remnants of the early kidney, notochord, and earlier heart structures appear transiently. This would appear to be done by retaining, little changed, many archaic genes and gene control subroutines. Thus (secondly) developmental modification would tend to be mediated by changes acting late in development, and few selective pressures would appear to act to modify phase-specific genes functioning only in early embryogenesis. Add-on programming therefore suggests that not only the morphological results, but also the molecular products of these ancient genes may be similar, if not identical, in many different species, and embryos therefore should have more antigenic similarities than adults. This suggests that the recapitulation of phylogeny during ontogeny has a molecular basis.

PROPERTIES MEDIATED BY TRACE PRODUCTS

We are left with the question of what specific properties may be conferred on cells by the numerous gene products postulated to be formed during development, and the mediation of morphogenesis is not a very enlightening answer.

Embryonic development involves many cell-cell interactions, including autorecognition leading to the reassembly of disaggregated tissues, and termination of interactions between cells, followed by movement of cells from one site in an embryo to another. Normal invasiveness is also illustrated by the trophoblast invasion of the uterine mucosa, which has all the appearance of a malignant process. Its cessation may be due to tissue maturation, an autoimmune response, or a specific inductive effect of the decidua.

Numerous other development-specific factors appear to exist, some of which have been described for the placenta and include placental hormones such as chorionic somatomammotropin, chorionic gonadotropin, placental alkaline phosphatase, and a placental factor stimulating capillary growth.

The technical problems associated with embryology should be stressed, however. If the dorsal lip of the blastopore produced as many factors as the pituitary, it is doubtful that these would be discovered in the near future because (*a*) only very tiny amounts of material are available, and (*b*) few assays are available.

RE-EXPRESSION OF ARCHAIC GENES

There is now little reason to assume that human malignant cells express anything more than the normal products of normal structural genes, although these genes may be acting at the wrong time and place. If malignancy can result from such re-expression, it would be useful to have a surveillance mechanism to identify the cells exhibiting such behavior. Burnet and others have proposed that cell-mediated immunity originated to serve this purpose. This works only if the malignant cells possess autoantigens. Why would the products of reactivated archaic genes of early development be autoantigenic? Or, in other words, how could matters be arranged so they would be?

The minimum requirement would appear to be that the adult not contain the autoantigens in question. This would require that the early embryo be physically, or at least immunologically, isolated from adult tissues, a requirement which may explain the origin and persistence of yolked eggs over most of the animal kingdom. These allow transient or phase-specific autoantigens to appear and disappear free of contact with the surveillance systems of the adult.

These considerations again suggest a rather large library of genes concerned solely with early development, which are repressed in adult tissues, and many of whose products would be autoantigenic in adults of the same species (Anderson and Coggin, 1972a; Castro et al., 1973). Note that these are obligatory autoantigens and are not to be confused with true transplantation antigens. The latter are detected by the mother in nonisogeneic matings. The former are obligatory autoantigens, occur in isogeneic matings, but would also be detected by the mother.

CANCER

As previously noted (Anderson and Coggin, 1972b), the salient properties of cancer cells to be explained are the following:

1. Possession of autoantigens different from classical transplantation antigens.
2. Ability to escape the host immune response.
3. Invasiveness and the ability to metastasize.
4. Alterations in cytoplasmic composition, including changes in isozyme ratio and appearance of isozymes and other proteins and hormones not found in the tissue of origin or, in some instances, in any adult tissue.
5. Stimulation of host responses, including an adequate blood supply and growth of supporting tissue.
6. An increased rate of growth and cell division in most instances.

In addition, we may here add another property stressed by Foulds (1969):

7. Progression, or the ability to change gradually in a variable and individualistic manner with time from preneoplastic through many stages of neoplastic growth, suggesting that malignancy may not be a single character.

Tumor Autoantigens

The problem with tumor autoantigens in experimental animals has been that onco-genic viruses appeared to produce virus-specific autoantigens, while chemical car-cinogens appeared to produce a wide variety of different, non-cross-protective antigens (and chemical carcinogens have been thought to produce their effect by activating latent viruses). Our concern is with human tumors. Since human onco-genic viruses which have been rigorously demonstrated to be causal in human can-cer have not been found, it might be expected that antigens cross-reactive between different human tumors would also *not* be found. Hence little effort was expended initially to find them, and when they were found, little attention was paid to them. Recently, however, a variety of antigens common to certain human tumor types have been described, and the question must be asked how this could occur since they appear to go contrary to predictions based on animal models.

Immune Escape

The problem of tumor immunity in animal models was long beclouded by the pres-ence of true transplantation antigens (Snell, 1953). When isogeneic mouse strains became available, it was found that weak antigens persisted in tumors which origi-nated and were transplanted within a strain. Further proof was provided by the study of autochthonous tumors, in which it was demonstrated that the host responds by either concomitant or sinecomitant immunity. An effective cell-mediated response appears to be mounted, since washed tumor-bearer lymphocytes will kill the tumor bearer's tumor cells. This response appears to be abrogated initially by the presence of circulating antibodies which may be cytostatic (Ambrose et al., 1971b), but appear to coat the tumor cells and protect them from lymphocyte attack. As the tumors grow (and during this phase growth appears to be very slow), the produc-tion and shedding of antigen appear to increase, finally reaching a point at which antigen excess occurs, first locally and then generally throughout the circulatory sys-tem (Ambrose et al., 1971b). The conditions of either circulating antigen-antibody complexes or of circulating antigen would also be blocking.

Invasiveness

Tumor invasiveness appears to correlate with the postulated periods of either local or general antigen excess and with the middle or later stages of tumor progression. The molecular basis of the surface changes is under very active investigation and includes, among other factors, changes in the binding of plant lectins (Moscona, 1971; Noonan and Burger, 1971; Pattillo et al., 1971) and in surface glycoproteins (Buck et al., 1970; Eveleigh, 1972).

Alterations in Intracellular Composition

The problem of compositional changes during human malignant transformation has centered around two questions. First, is there a single causal characteristic change? Since no single common-denominator molecular difference has been found to date (but may be in the future), a second question has occupied attention: Is there a characteristic pattern to the activity changes observed, or are the changes random, or at least completely variable from tumor to tumor? This is not the case, as shown by Weber (1972) and others, and is summarized in the molecular correlation concept. However, it must be noted that for the most part biochemical studies have centered around the two major classes of proteins—necessary enzymes, and substances characteristic of fully differentiated cells. As discussed in a subsequent section, when isozymes of individual enzymes (and not the sum of activity of several isozymes) are studied, a somewhat different picture emerges.

Stimulation of Host Response

Human tumors contain many nonmalignant cells. These are there, at least in part, as a result of host responses to the tumor. These responses appear to be mediated, again at least in part, by diffusible substances produced by tumor cells. The most studied of these is the tumor angiogenesis factor described by Folkman (1972) and his coworkers, which stimulates the ingrowth of capillaries into growing tumor masses. Other factors stimulating other types of cells and tissues are inferred from cytological studies (Foulds, 1969).

Increased Growth Rate

Factors which induce or suppress growth in specific tissues have long been sought, and the evidence for tissue-specific chalones has recently been reviewed (Bullough and Lawrence, 1968; Mathé, 1972). That such substances are specifically made (or are lacking) in tumors is suggested by the observation that some but not all tissues of tumor-bearing animals synthesize DNA more rapidly than do control tissues (Morgan and Cameron, 1973).

RETROGENESIS

The above-mentioned factors appear to include too many different substances for each to be synthesized by viral genes. The same may be said for mutations to structural genes. If we now ask how many factors observed in cancer cells result from the reactivation of embryonic or fetal phase-specific genes, the answer is that a surprising number of them do. Furthermore, the parallelism between the responses of the mother to the embryo and the cancer patient to his tumor is a close one which opens up many fruitful areas for investigation (Sinkovics et al., 1970).

Autoantigens

We have reviewed evidence, first, for transient phase-specific autoantigens occurring during normal development, and, second, for the occurrence of a variety of autoantigens in tumors. Could it be that some of these are the same? Can one immunize adult animals with antigens from isogeneic fetus? Immunization against tumor challenge has been done using either irradiated cells (so that the cells do not grow and form a tumor) or cell fractions (Ambrose et al., 1969; Coggin et al., 1969). The former are generally more successful since the cells appear to produce and release antigen over a period of many days, while the amount of antigen present at any instant (and therefore the amount present in a homogenate) is small. The key experiment therefore is to determine whether or not vaccination with irradiated fetal cells will protect adult animals against challenge with tumor cells. This work, begun in 1969 with hamsters (Coggin et al., 1970a, 1971), has now been extended to mice and guinea pigs, using some 40 different tumors (Coggin and Anderson, 1973). The level of immunity is always low and the tumor cell challenge dose carefully controlled; otherwise tumor immunity may be easily overridden.

Two points deserve emphasis. Little cross-protection is observed when animals immunized with one tumor are challenged with a different one. In contrast, irradiated fetal cell immunization protects against many different tumors, although the level of immunity is low. Are these results due to *one* fetal antigen, shared by the fetus and all tumors, which is different from the so-called tumor-specific transplantation antigens? This would mean that two different classes of antigens existed, one specific, and the other cross-reacting. The answer appears to lie in the experiments themselves. First, if there is one (or a very few) fetal antigen(s) also shared by tumors, and if immunity to this antigen gives protection against many tumors, then immunization with tumor cells (with any given tumor) should protect against challenge with any other tumor. Clearly this is not the case. One must conclude, therefore, that the fetal cells immunize against many *different* tumor antigens, while tumor cells may contain only one (or a few) of the assortment found on the fetus but may contain in addition antigens of viral origin. At this point it is not possible to state how many different autoantigens any given tumor may exhibit.

The possibility continues to be raised that weak true transplantation antigens may be present in challenge studies, either because the animals were not truly isogeneic and the tumor donor (or the donor of the cells transformed *in vitro* when that was the case) differed genetically from the recipient, or because the donor cells were of a different sex from the recipient and sex-associated antigens were involved. The answer to this problem is to see whether immunization with fetal cells prevents autochthonous tumor formation. To do this with spontaneous tumors requires the use of rather large numbers of animals and a long experimental time period. Hence we chose instead to use autochthonous tumors produced by injection of hamsters shortly after birth with SV40 or adenovirus 31 virus. After a relatively long latent period, tumors appeared. During the latent period a second infection with the same virus interrupted tumorigenesis. This then provided a suitable model system that included a defined latent period in which the immune system was clearly capable of responding, and in which the tumor cells obviously did not differ (except by being virus infected) from the host, but were in fact transformed host cells. When irradiated fetal cell vaccination was done during the latent period, the incidence of tumors was very markedly reduced (Ambrose et al., 1971b).

While weak true transplantation antigens are thus ruled out, this does not solve the problem of the relationship between fetal and tumor antigens completely. For example, one may ask whether immunization with tumors induces an immune response to fetal cells. This question has been approached in two ways. The first was to see whether immunization against different tumors interfered with pregnancy; indeed, smaller litters were seen in mice so immunized (Parmiani and Della Porta, 1973). The second approach was to study the effect of antitumor immunity on the colonization of the spleens of irradiated animals with fetal cells. Again antitumor immunity reduced the number of colonies (Salinas et al., 1972).

These results strongly suggest that a large number (probably hundreds) of fetal autoantigens exist. From this fetal autoantigen gene library, each tumor selects one or a very few genes for re-expression. This would explain why fetal cells are rather poor immunizing agents—too many antigens are present, and the animal responding to them cannot mount a very effective response to the one or two which may appear in a given tumor.

If our previous discussion suggesting that obligatory embryonic autoantigens are the products of ancient genes common to many species has some validity, it should be possible to demonstrate a cut-off point at which the antigen disappears or at least is no longer active. This occurs at approximately day 10.5 in the hamster system (Coggin et al., 1973). One might also expect that the same antigens might be common to many different species, and that from a serological viewpoint early embryonic antigens would be more similar than adult proteins. Experimentally, cross-species immunization is observed, and mouse or human fetal cells will protect hamsters against tumor challenge (Ambrose et al., 1971a).

CANCER AND PREGNANCY

The presence of true transplantation antigens on the fetus has long obscured the presence of obligatory autoantigens. The view that the placental barrier physically prevents immunizing doses of antigen from reaching the mother has proved untenable, especially in view of the finding that maternal cells are present in the fetus, and that circulating antibodies against transplantation antigens appear in the mother. Rather, some special mechanism has appeared to abrogate fetal rejection by the mother. From our present viewpoint there are two central questions: (1) Does the mother respond to fetal antigens with both circulating antibodies and sensitized lymphocytes and (2) do these also react with tumor cells?

Antifetal antibodies which coat fetal cells have been demonstrated in pregnant animals, and fetuses of multiparous animals appear to be coated with IgG. In addition, lymphocytes from pregnant animals will kill fetal cells. Postpartum hamster serum also reacts with SV40-transformed fetal cells (Duff and Rapp, 1970), and well-washed lymphocytes from pregnant animals will kill several types of tumor cells (Brawn, 1970).

This then brings into sharp focus the problem of the escape mechanism. Sinkovics et al. (1970) have proposed that circulating immunity was developed to protect the fetus of placental animals from destruction by a cell-mediated immune surveillance system. Blocking could actually occur in several ways. The antibodies could coat tumor or fetal cells and thus make them immunologically invisible to lymphocytes. If the amount of circulating antigen increased, antigen-antibody complexes could be formed which could blind lymphocytes, or a condition of antigen excess could exist in which lymphocytes would be coated with antigen and would hence not be able to recognize tumor cells. Serum from tumor-bearing animals will block cell-mediated killing of tumor cells by sensitized lymphocytes, as will pregnancy serum.

This summary of the experimental studies on the parallelism between immune escape in pregnancy and that in cancer shows the strong support for the view that in autochthonous tumors and in cancer both the same antigens and the same escape mechanisms are involved.

Invasiveness

Since invasiveness is characteristic of trophoblast cells, and since many embryonic cells invade other tissues, it has been logical to propose that the same mechanisms and genes are involved. While this is an interesting parallel, one would like more concrete molecular evidence. The finding that embryonic and tumor cells have certain lectin-binding properties in common is suggestive, as is the finding that antigens produced by the placenta (and possibly by trophoblasts) occur uniformly in tumor patients (Tal, 1971). Suggestive as these results are, much more must be known about the molecular basis of invasiveness before it will be possible to ascertain whether the mechanisms involved are identical in detail.

Isozymes, tRNA, and So Forth

Enzyme pattern changes and isozyme shifts when observed in cancer cells tend to be in the direction of the fetal state; a large body of literature has been built up supporting this view (Criss, 1971; Knox, 1972; Weinhouse, 1972). Transfer RNAs characteristic of fetal tissues have been found in tumors (Yang, 1971; Gonano et al., 1973), while three fetal serum proteins have been shown also to recur in certain cancers, of which the best known is α-fetoprotein (Abelev, 1971), associated with primary cancer of the liver. In addition, the list of tumor antigens shared with fetal tissues is long and is rapidly expanding (see the review by Coggin and Anderson, 1973).

Stimulation of Host Responses

The tumor angiogenesis factor of Folkman (1972) and his associates also occurs in the placenta.

Increased Growth Rate

The molecular mechanisms underlying rapid growth in embryonic and tumor cells are not known in detail. However, there is no reason to postulate that existing mechanisms already programmed into the genome are not involved. It is interesting that the characteristic growth hormone of the placenta, chorionic sommatomammotropin (or placental lactogen), recurs in about 8% of human cancer patients in a concentration sufficient to be detected in the circulation (Weintraub and Rosen, 1971).

Progression

Human cancer (e.g., cancer of the colon) appears to involve preneoplastic alterations which may occur over a relatively long period of time, a transition to a malignant state, followed by progression to new degrees of invasiveness, metastasis; and different morphologies, as has been described and discussed in detail by Foulds (1969). Biochemical differences are sometimes noted in surrounding tissues, which may exhibit metaplasia. Thus sucrase is not found in the gastric mucosa. In cancer of the stomach, however, the gastric mucosa may exhibit sucrase (a characteristic intestinal enzyme), while the tumor tissue does not (Sugimura et al., 1972). These changes suggest that cancer is not to be explained by a single molecular event, but rather appears to be due to abnormal functioning of a very complex program, much of which is still operative. The only source of such a program would be the differentiation program itself. Only a complex program would be able to generate the multiplicity of gradual changes actually observed, which are quite different from what might be expected with a one-step viral transformation.

CANCER AND EMBRYOLOGY

If development involves gene orchestration on a large scale, as hybridization experiments now suggest, and if thousands of specific substances appear and disappear in the early germ layers as they first form, in the primitive streak, in the dorsal lip of the blastopore, and in organ primordia consisting of only a few cells, there is little chance of isolating and characterizing them. Furthermore, their function may require that they be specifically positioned in specific cells at a specific stage of development, and their effects may be observable only in the behavior of other specific cells at a specific stage of development. There thus exist almost insurmountable barriers to the exploration of molecular events underlying vertebrate morphogenesis. However, if large segments of the program of differentiation are still functional in tumor cells, the possibility exists that many molecular factors important in early development may be produced by tumor cells in tissue culture. The effect of tumors and tumor-cell fractions on early development therefore deserves emphasis. Work already done in this area has been recently reviewed by Sherbet (1974).

PARANEOPLASTIC SYNDROMES

If a variety of phase-specific genes of early development are indeed derepressed in human cancer (and there is no present reason to believe all genes so derepressed have been described), one might expect to see a variety of effects on the cancer patient, many of them unexplained. As has been stressed by previous investigators, wasting and death often occur when the tumor mass is too small to explain the lethal course of the disease on the basis of tumor mass or anatomical effects. In many instances known hormones are secreted which often are not produced by the cell of origin (Table 1). In other instances a wide variety of unexplained symptoms are seen. These include selective pigmentation, unusual hair growth, and selective effects on taste receptors and mental function (see Hall, 1974). In addition, as previously noted, the rate of DNA synthesis in other tissues may be selectively affected.

TABLE 1. HORMONES FOUND IN HUMAN ECTOPIC HUMORAL SYNDROMES[a]

Adrenocorticotropic hormone	Gastrin
Melanocyte stimulating hormone	Insulin
Parathyroid hormone	Erythropoietin
Luteinizing hormone	Thyroid stimulating factor
Follicle stimulating hormone	Chorionic somatomammotropin
Antidiuretic hormone	

[a] From Liddle et al. (1969).

TUMOR CLASSIFICATION

If the characteristic feature of cancer cells is a re-expression of embryonic genes and other genes not active in the cell of origin, and if this does not occur in a completely random fashion, it may be possible to develop a tumor classification based on this re-expression. Conversely, if such a classification is possible, gene reactivation does not occur randomly! The pioneering work of Sherbet (1974) and of Metz (1972) has shown that in many cases of ectopic hormone secretion (i.e., where the tumor secretes a hormone not found in the cell of origin), the hormone is characteristic of another tissue produced by the same germ layer. In a lesser number of instances the hormone is the product of cells derived from a different germ layer. Although this approach to classification is in its infancy, the possibility exists that, were the entire program of differentiation known, the precise locus of error or misreading could be pinpointed. This raises the question, to be considered in detail elsewhere, of what the program of differentiation actually looks like, how much of it is reversible, and how many different sites could, if damaged, lead to misprogramming and cancer.

RETROGENESIS AND AGING

The "hold," latch, or clamping mechanisms which maintain the differentiated state are not inflexibly fixed in many tissues. In response to injury, many otherwise quiescent cells can round up, divide, and become mobile, returning to the differentiated state when repair is complete. There appear to be instances, therefore, in which the differentiation program is preprogrammed to be partially reversible. With aging one might expect to see a gradual breakdown in the differentiation "hold" mechanism, accompanied by instances of metaplasia, re-expression of some embryonic gene products, and, as is well known, an increased incidence of cancer. Thus fetal hemoglobin may be seen with advancing age (Popp, 1971), as well as an increased incidence of antibodies against tissue autoantigens, such as those against mitochondria and smooth muscle (Duheille et al., 1972). This has very serious implications for cancer immunodetection and immunotherapy. If human cancer antoantigens are really embryonic antigens, then *unique tumor-specific human tumor antigens do not exist.* Furthermore, the antigens of interest on tumor cells may occur in other pathological states, in hyperplasia and metaplasia, and in aging. This is indeed the case with α-fetoprotein and CEA. We question the animal model studies on which the concepts of absolute specificity are based. We also have questioned experiments purporting to show that a given antigen is *not* present in the embryo or fetus. In the hamster system the protective effect conferred by irradiated fetal cells disappears sharply at 10.5 days. Other antigens may be present for very short periods of time, in low concentration, in very tiny organ anlagen. Absorption of antisera with fetuses of one or unknown age is therefore insufficient evidence.

RETROGENESIS AND VIRAL ONCOLOGY

No major class of human tumors has thus far been unequivocally shown to be due to horizontal virus infection, although highly suggestive evidence is available for the Burkitt lymphoma and cervical cancer. For other tumors, including leukemia and breast cancer, vertical infection has been favored. Man has simply not followed animal models in many instances.

Two points stand out. First, viral carcinogenesis is largely a disease of inbred animals, and, second, virus infection always appears to cause extensive host cell reprogramming.

The present laboratory cancer models were developed by inbreeding, so that both copies of the differentiation program would be identical (except in males), and by breeding for low or high cancer incidence. The latter means that there are genes associated with cancer incidence; and since it turns out that high virus cancer incidence is what was being bred for, it follows that genes must exist which mediate viral transformation. This absolutely requires that either viral nucleic acid or a viral gene product *interact directly or indirectly with specific normal cellular genes involved in the control of other genes.* This obligatory parataxis is also a characteristic of viral infection generally. The mechanisms which could be operative involve either direct viral nucleic acid interactions with host cell DNA, interaction of derivative virus NA with host cell DNA, interaction of a viral gene product (repressor or derepressor) with host cell DNA, or interaction of a viral gene product with a normal host cell repressor or derepressor. This requires that very specific information be contained in the viral genome. The end result is the same as some discrete step in the differentiation process, that is, the turning on of some genes and the turning off of others. This viral reprogramming may be virus specific if the virus interacts with specific loci in the cell nucleus. The effect will be against the background of the phenotype of the cell infected, and the specific phenotype may strongly influence the course of oncogenesis. If the process of transformation is indirect and occurs through, for example, a virus-produced repressor, one should be able to transform cells transiently with the repressor (provided that it is gotten into the cell). What these considerations suggest is that the easiest way to make an oncogenic virus would be to incorporate within a viral genome a gene concerned with the control of some phase of differentiation. One may ask where else such information could come from. Viruses have no independent means for making genes, and it appears not implausible that they are picked up from cells somewhere.

This view—that oncogenesis requires the presence of a gene similar to or identical with a normal gene active in early development—offers one explanation of the finding that oncogenic virus antigens such as a gs antigen may occur universally in the embryos of a species. It will occur to the reader that the latter finding gives us two choices. Either all embryos are infected with oncogenic viruses and infection occurred eons ago, as proposed by Huebner and his coworkers (Huebner and Todaro, 1969), or all oncogenic viruses are infected by embryonic genes and such infection could occur anytime, especially in the presence of nucleotide analogs or their derivatives. Virus rescue and virus creation are difficult to distinguish. That such analogs can affect specific switching genes in development has been recently demonstrated (Weintraub et al., 1973).

The idea of primeval virus infection, the persistence of viral genomes over eons of time, has interesting implications which have been explored elsewhere. These center around the idea that viruses serve a significant role in evolution by transporting genes randomly over the entire plant and animal kingdoms for adoption and modification if useful, and for rejection if deleterious (Anderson, 1970). The arguments may be summarized as follows:

1. Viruses would have been eliminated from at least a few species if they served no useful function. (Not a strong argument, but an interesting one.)
2. Viruses cross species, phylum, and kingdom barriers; hence there probably exist infectivity routes, via intermediates, from any given species to any other.
3. Incorporation of host DNA into viruses is well known.
4. Whole viral genomes may be vertically transmitted.
5. Viral gene transmission could account for parallel evolution.
6. Use of the viral mail requires that the genetic code be universal, as it is.
7. Evolution by amino acid substitution appears to involve increments too small and too numerous to account for evolution.

All these arguments come to naught, however, if effective mechanisms to prevent permanent viral gene incorporation into host cells exist. It is noteworthy that these arguments suggest a saltatory molecular evolution. All available amino acid sequence data point to exactly the opposite—namely, the one-at-a-time substitution of amino acids over long periods of time.

Thus we have the outlines of a fascinating argument: deleterious cancer genes brought in from outer oncogene space which spread throughout a species uniformly, contrary to present concepts of evolution, versus point mutation evolutionary theory and data. The argument that the products of oncogenes are useful at some stage of early development and hence are incorporated, but become part of cancer later, is a different way of stating our major thesis. The only point of difference is that the genes were not originally viral, but only became so later.

CONCLUSIONS

The purpose of this exercise is to present a view of cancer which allows certain conclusions to be drawn. Before doing this, however, we should like to return to one point which interests us greatly; it is an example of Rule 1 of the administrative world, which reads: Exceptions Cause Problems.

We have noted the *end-add-on* nature of evolution with the maintenance of ancient mechanisms, patterns, and programs. An interesting subroutine inserted back in the early portion of the program of development is concerned with the placenta. Of all mammalian organs this is probably the most variable, indicating that experimentation is still in progress. The placental subprogram may therefore be inserted in such a way that it is quite easily executed as a package, thus explaining the ease with which genes for hormones, antigens, factors, and the like associated with the placenta are reactivated. Thus placentation—a recent and makeshift solution—may have been the origin of cancer in more ways than previously thought.

The major objective here has been to present a view of human cancer which integrates the maximum number of available facts. The inescapable conclusion is that reactivation of genes concerned with development is of central importance. How can this discussion be harmonized with the results and concepts which have arisen from animal model studies, where cross-reactive antigens have been considered as proof of virus infection and nonviral tumors have been thought to produce tumor-unique antigens?

The model of development adopted here assumes the existence of a large number of control genes important to early development, which could cause a variety of new and possibly malignant phenotypes if irreversibly switched on. Loss of control could occur as a result of mutation; it could result from the action of a chemical which, because of similarity to a natural coderepressor or derepressor, could activate a gene under conditions in which the mechanism to switch it off was no longer active; or the gene might be incorporated into a virus and introduced, free of normal controls, in the viral genome.

The salient feature of virus infection, as previously mentioned, is interaction with the host genome. This interaction could be mediated either by viral gene products, which may produce repressors or derepressors, or by nucleic acid hybridization. Either case requires that specific information be shared by the virus and the host genome. If the effect is produced via a depressor acting on a particular gene, the host makes a derepressor with an equivalent affinity and action. If the effect occurs via hybridization, the virus has nucleotide sequences in common with the host. Should such specific annealing to host DNA occur, strand breakage close to the point of initiation of annealing could result in incorporation of the viral genome into the host genome. Factors which produce single strand breaks should therefore strongly facilitate cell transformation by DNA tumor viruses, as is indeed the case (Coggin et al., 1970b), and should also assist in virus recovery.

If specific interaction between viral nucleic acids (or viral gene products) and host DNA is an obligatory requirement for oncogenesis, one would expect that alleles to the sensitive host genes could be found which would not participate in such an interaction and hence would appear as genes for resistance. This appears to be the case (Peters et al., 1973).

Oncogenic viruses therefore would be expected either to contain sequences also found in the host genome (identical in the case of DNA viruses, capable of producing an identical sequence via reverse transcription in the case of RNA viruses), or to produce a gene product either identical or very similar to a normal gene product of early development. Thus tumor viruses should produce or contain antigens found in the early embryo, as is the case with RNA mouse tumor viruses, or show sequence homologies, as is the case for SV40 virus (*untransformed* African green monkey cells were found to have 0.5 SV40 genome equivalent per cell, while

3T3 and BALB/3T3 mouse cell lines had 0.45 and 0.39 per cell: Gelb and Martin, 1971).

If viruses interact with specific genes in cells, either those specific genes or genes coding for substances (derepressors) interactive with those specific genes have been incorporated into the virus.

Thus, as suggested, oncogenic viruses appear to be infected with embryonic genes. This is in contrast to the view that all embryos are infected with oncogenic viruses. Since viruses have no visible means of generating new genes independent of cells, it requires no suspension of logic to assume that all viruses were originally of host cell origin. Given a variety of ubiquitous viruses and selective chemical methods for promoting transduction, one would expect to be able to produce a range of viruses containing host genes which would thus acquire oncogenic potential and exhibit some form of cross-reaction with the host. Requirements for tumor formation may be that (1) the gene transduced into the virus be present in the host, that is, the host must have the necessary genetic specificity; and (2) both copies of the differentiation program be identical or nearly so (i.e., highly inbred). It may thus turn out that one natural defense against viral cancer is heterozygosity per se, making viral cancer in inbred animals an interesting artifact.

Thus we suggest that a more detailed investigation of the transformation genes of oncogenic viruses will reveal that they originated as normal host genes concerned with the control of early development. Hopefully these concepts will serve to reintegrate the field of oncology and to direct it along lines of greater immediate benefit to the patient.

ACKNOWLEDGMENTS

The Molecular Anatomy (MAN) Program, with which one of us (N.G.A.) is associated, is sponsored by the National Cancer Institute, National Institute of General Medical Sciences, National Institute of Allergy and Infectious Diseases, National Aeronautics and Space Administration, and the U. S. Atomic Energy Commission. It is carried on at the Oak Ridge National Laboratory, operated for the U. S. Atomic Energy Commission by Union Carbide Corporation, Nuclear Division.

REFERENCES

Abelev, G. I. (1971). In *Advances in Cancer Research* Vol. 14 (G. Klein and S. Weinhouse, eds.). Academic Press, New York, p. 295.

Ambrose, K. R., E. L. Candler, and J. H. Coggin, Jr. (1969). *Proc. Soc. Exp. Biol. Med.* **132**, 1013.

Ambrose, K. R., N. G. Anderson, and J. H. Coggin, Jr. (1971a). *Nature* **233**, 194.

Ambrose, K. R., N. G. Anderson, and J. H. Coggin, Jr. (1971b). *Nature (London)* **233**, 321.

Anderson, N. G. (1970). *Nature* **227**, 1346.

Anderson, N. G., and J. H. Coggin, Jr. (1971). In *Proceedings of 1st Conference and Workshop on Fetal and Embryonic Antigens in Cancer.* USAEC Report CONF–710527, p. 7.

Anderson, N. G., and J. H. Coggin, Jr. (1972a). In *Embryonic and Fetal Antigens in Cancer,* Vol. 2. USAEC Report CONF–720208, p. 361.

Anderson, N. G., and J. H. Coggin, Jr. (1972b). In *Membranes and Viruses in Immunopathology* (S. B. Day and R. A. Good, eds.). Academic Press, New York, p. 217.

Apter, M. J., and L. Wolpert (1965). *J. Theor. Biol.* **8**, 244.

Babloyantz, A., and G. Nicolis (1972). *J. Theor. Biol.* **34**, 185.

Bonner, J. T. (1960). *Am. Sci.* **48**, 514.

Brawn, R. J. (1970). *Int. J. Cancer* **6**, 245.

Britten, R. J., and E. H. Davidson (1968). *Science* **165**, 349.

Britten, R. J., and E. H. Davidson (1969). *Science* **165**, 349.

Buck, C. A., M. C. Blick, and L. Warren (1970). *Biochemistry* **9**, 4567.

Bullough, W. S., and E. B. Lawrence (1968). *Eur. J. Cancer* **4**, 607.

Castro, J. E., E. M. Lance, P. B. Medawar, J. Zanelli, and R. Hunt (1973). *Nature* **243**, 225.

Coggin, J. H., Jr., and N. G. Anderson (1973). *Adv. Cancer Res.* (in press).

Coggin, J. H., Jr., L. H. Elrod, K. R. Ambrose, and N. G. Anderson (1969). *Proc. Soc. Exp. Biol. Med.* **132**, 329.

Coggin, J H., Jr., K. R. Ambrose, and N. G. Anderson (1970a). *J. Immunol.* **105**, 524.

Coggin, J. H., Jr., S. E. Harwood, and N. G. Anderson (1970b). *Proc. Soc. Exp. Biol. Med.* **134**, 1109.

Coggin, J. H., Jr., K. R. Ambrose, B. B. Bellomy, and N. G. Anderson (1971). *J. Immunol.* **107**, 526.

Coggin, J. H., Jr., K. R. Ambrose, and N. G. Anderson (1973). *Adv. Exp. Med. Biol.* **29**, 483.

Criss, W. E., (1971). *Cancer Res.* **31**, 1523.

Davidson, E. H., (1968). In *Gene Activity in Early Development*. Academic Press, New York.

Duff, R. J., and F. Rapp (1970). *J. Immunol.* **105**, 522.

Duheille, J., E. Cuny, and F. Penin (1972). In *Organisation des Laboratoires—Biologie Prospective*. L'Expansion Scientifique Francaise, Paris, p. 581.

Edelstein, B. B. (1972). *J. Theor. Biol.* **37**, 221.

Eveleigh, J. W. (1972). In *Embryonic and Fetal Antigens in Cancer*, Vol. 2. USAEC Report CONF–720208, p. 133.

Folkman, J. (1972). *Ann. Surg.* **175**, 409.

Foulds, L. (1969). In *Neoplastic Development*, Vol. 1. Academic Press, London and New York.

Gelb, L. D., and M. A. Martin (1971). In *Proceedings of 1st Conference and Workshop on Embryonic and Fetal Antigens in Cancer*. USAEC Report CONF–710527, p. 71.

Glass, L., and S. A. Kauffmann (1973). *J. Theor. Biol.* **39**, 103.

Gonano, F., G. Pirro, and S. Silvetti (1973). *Nature New Biol.* **242**, 236.

Grigorev, L. N., M. S. Polyakova, and D. S. Chernavskii (1967). *Mol. Biol. (USSR)* **1**, 349.

Hall, T. C., (1974). Paraneoplastic syndromes. *Ann. N.Y. Acad. Sci.* (in press).

Herstein, P. R., and J. H. Frenster (1972). In *Embryonic and Fetal Antigens in Cancer*, Vol. 2. USAEC Report CONF–720208, p. 5.

Holleman, J. W., and W. G. Palmer (1972). In *Embryonic and Fetal Antigens in Cancer*, Vol. 2. USAEC Report CONF–720208, p. 117.

Holtfreter, J. (1955). *Exp. Cell Res., Suppl.* **3**, 188.

Huebner, R. J., and G. J. Todaro (1969). *Proc. Nat. Acad. Sci. U.S.A.* **64**, 1087.

Kauffmann, S. A. (1973). *Science* **181**, 310.

Knox, W. E. (1972). *Enzyme Patterns in Fetal, Adult, and Neoplastic Rat Tissues*. S. Karger, Basel.

Liddle, G. W., W. E. Nicholson, D. P. Island, D. N. Orth, K. Abe, and S. C. Lowder (1969). *Recent Prog. Hormone Res.* **25**, 283.

Markowitz, D. (1972). *J. Theor. Biol.* **35**, 27.

Mathé, G. (1972). *Rev. Eur. Etud. Clin. Biol.* **17**, 548.

Metz, S. A. (1972). *The Ectopic Hormone Syndromes.* Thesis, Yale University.

Morgan, W. W., and I. L. Cameron (1973). *Cancer Res.* **33**, 441.

Moscona, A. A. (1971). *Science* **171**, 905.

Noonan, K. D., and M. M. Burger (1971). In *Proceedings of 1st Conference and Workshop on Embryonic and Fetal Antigens in Cancer.* USAEC Report CONF–710527, p. 59.

Parmiani, G., and G. Della Porta (1973). *Nature New Biol.* **241**, 26.

Pattillo, R. A., L. L. Hause, R. O. Hussa, and E. F. Walborg (1971). In *Proceedings of 1st Conference and Workshop on Embryonic and Fetal Antigens in Cancer.* USAEC Report CONF–710527, p. 313.

Peters, R. L., G. J. Spahn, L. S. Rabstein, G. J. Kellott, and R. J. Huebner (1973). *Science* **181**, 665.

Popp, R. A. (1972). In *Embryonic and Fetal Antigens in Cancer*, Vol. 2. USAEC Report CONF–720208, p. 67.

Salinas, F. A., J. A. Smith, and M. G. Hanna (1972). In *Embryonic and Fetal Antigens in Cancer*, Vol. 2. USAEC Report CONF–720208, p. 187.

Schultz, G. A., and R. B. Church (1973). In *Biochemistry of Animal Development*, Vol. 3 (R. Weber, ed.). Academic, New York (in press).

Sherbet, G. V. (1966). *Prog. Biophys. Med. Biol.* **16**, 89.

Sherbet, G. V. (1974). *Ann N.Y. Acad. Sci.* (in press).

Shulman, S. (1971). *Crit. Rev. Clin. Lab. Sci.* **2**, 393.

Sinkovics, J. G., P. J. DiSaia, and F. N. Rutledge (1970). *Lancet* **ii**, 1190.

Snell, G. D. (1953). *Physiopathology of Cancer* (F. Homburger, ed.). Hoeber-Harper, New York, pp. 338–391.

Sugimura, T., T. Matsushima, T. Kawachi, K. Kogure, N. Tanake, S. Miyake, M. Hozumi, S. Sato, and H. Sato (1972). *Gann Monograph on Cancer Research* **13**, 31.

Sugita, M. (1963). *J. Theor. Biol.* **4**, 179.

Tal, C. (1971). In *Embryonic and Fetal Antigens in Cancer*, Vol. 2. USAEC Report CONF–720208, p. 53.

Venkatesan, N., J. C. Arcos, and M. F. Argus (1971). *J. Theor. Biol.* **33**, 517.

Weber, G. (1972). *Gann Monograph on Cancer Research* **13**, 47.

Weinhouse, S. (1972). *Cancer Res.* **32**, 2007.

Weintraub, B. D., and S. W. Rosen (1971). *J. Clin. Endocrinol.* **32**, 94.

Weintraub, H., G. L. Campbell, and H. Holtzer (1973). *Nature New Biol.* **244**, 140.

Yang, W. K. (1971). *Cancer Res.* **31**, 639.

INDEX